Geographies of the Indian Subcontinent

Sukla Hazra • Anwesha Haldar
Rajarshi Dasgupta • Pinaki Sarkar
Subhasree Singha Ray • Panchali Majumdar
Editors

Geographies of the Indian Subcontinent

Cutting-Edge Methods in 21st Century Research

Springer

Editors
Sukla Hazra
East Calcutta Girls' College (Retired)
Kolkata, West Bengal, India

Anwesha Haldar
Department of Geography
East Calcutta Girls' College
Kolkata, West Bengal, India

Rajarshi Dasgupta
Department of Geography
East Calcutta Girls' College
Kolkata, West Bengal, India

Pinaki Sarkar
Department of Geography
East Calcutta Girls' College
Kolkata, West Bengal, India

Subhasree Singha Ray
Department of Geography
East Calcutta Girls' College
Kolkata, West Bengal, India

Panchali Majumdar
Department of Geography
East Calcutta Girls' College
Kolkata, West Bengal, India

ISBN 978-3-031-74812-7 ISBN 978-3-031-74813-4 (eBook)
https://doi.org/10.1007/978-3-031-74813-4

This Springer imprint is published by the registered company Springer Nature Switzerland AG
The registered company address is: Gewerbestrasse 11, 6330 Cham, Switzerland

Foreword

It gives me immense pleasure to write a foreword for the present book, entitled *Geographies of The Indian Subcontinent: Cutting-Edge Methods In 21st Century Research*, edited by Sukla Hazra, Anwesha Haldar, Rajarshi Dasgupta, Pinaki Sarkar, Subhasree Singha Ray, Panchali Majumdar, which is a worthy contribution to the geographical research arena.

India is a country with very diversified terrain conditions, land use land cover, and meteorological factors. This subcontinent represents a variety of ecosystems and rich biodiversity, and balancing our living style with ecological balance may only lead to a sustainable future.

The post-independence era recorded remarkable development and improvement in the quality of life, but environment degradation has been alarming. We need to understand human-environmental system dynamics. The advancement of geographical research and its methodologies will provide a more profound understanding of the sustainable development. The authors of the articles published in this book have explored the pathways for carrying natural and social sciences towards sustainability.

The book is well designed by categorizing into four subthemes comprising 16 chapters covering the most relevant approaches in present-day research of the geospatial, earth system, and social sciences. I congratulate all the authors who have contributed to this compilation.

Chairman, West Bengal Pollution
Control Board, Government of West
Bengal, Kolkata, India

Kalyan Rudra

Contents

Editors Biography

Sukla Hazra Ex-Principal, East Calcutta Girls' College, currently associated with the Techno India University as a Visiting Professor, did her M.A. and Ph.D. in Geography from the University of Calcutta. She undertook advanced training in GIS Mapping at Laval University, Quebec, Canada. She taught undergraduate and postgraduate students for the last 37 years at different colleges and universities. She was a Coordinator of Distance Education affiliated under Vidyasagar University on 'M.Sc. in Geography'. She was the Director of 'Centre of GIS and Remote Sensing' in the East Calcutta Girls' College under the University Grants Commission (UGC). She was a member of the NAAC Advisory Committee of the Kalinagar Mahavidyalaya, South 24 Parganas. She was a Visiting Professor of the West Bengal State University, Barasat, and undertook courses on Population Geography and Agricultural Geography. She is a member of the Executive Committee of the Indian National Cartographic Association. She undertook a few UGC-funded research projects and published several textbooks at school level and undergraduate level in Geography. Her areas of research interest are 'Applications of Remote Sensing', 'Urbanization', 'Environmental Studies', 'Sustainable Development', 'Agriculture', 'Food Security', and 'Wetland'. She has guided many doctoral students, and at present three researchers are working for their Ph.D. degree under her supervision. She published more than 50 research papers in different journals. She has also delivered lectures in seminars and symposia as invited speakers.

Anwesha Haldar is serving as an Assistant Professor at the postgraduate department of Geography, East Calcutta Girls' College, West Bengal State University since 2020. She was a university rank holder from Loreto College and went on to complete her masters at the University of Calcutta. She has done her Ph.D. on Sustainable Energy Management of Sundarban region, India, from the Department of Geography, University of Calcutta and has undertaken several specialized courses in Geospatial technologies and short-term courses under GIAN organized by IIT, Kharagpur and IIEST, Shibpur. She has taught Geography for the last 14 years at the undergraduate and postgraduate levels at the University of Calcutta, West Bengal State University, Rabindra Bharati University in the regular mode, as well as at the Directorate of Distance and Online Education and Vidyasagar University, Distance Education Course. She actively coordinates courses in remote sensing and GIS. Dr. Haldar is an enthusiastic field researcher and presented papers across India and abroad. She has also published in reputed national and international journals, in the fields of urban geomorphology, climatology, environmental geography, and hazard management studies. She is credited with four co-edited books on different aspects of geography. Presently, she is the managing editor of the *Indian Journal of Landscape Systems and Ecological Studies* and a significant member of the Indian Meteorological Society, Kolkata Chapter. She is currently working on gender issues and environment.

Rajarshi Dasgupta is currently a State Aided College Teacher-I in the Department of Geography, East Calcutta Girls' College. He has completed his undergraduate studies in Geography from the University of Calcutta, Kolkata, and postgraduate studies in Geography and Geology from Jawaharlal Nehru University, New Delhi, and University of Cincinnati, USA respectively. He is a physical geographer with research interests in Geomorphology and Environmental Management. His research utilizes techniques of geomorphology, low-temperature geochemistry, and social sciences to study earth surface processes and anthropogenic impacts on the natural environment, especially in the tropics. He has presented talks on these themes in conferences both within India and abroad. Several of his papers have been published in outlets brought out by leading publishing houses like Wiley, Elsevier, and Springer. He also teaches courses on the above-mentioned themes at both undergraduate and postgraduate levels and regularly supervises postgraduate students for their dissertations. A number of these student-led research have also been published in reputed journals published by the leading publishers mentioned above.

Pinaki Sarkar is an Assistant Professor at East Calcutta Girls' College, Kolkata, affiliated to West Bengal State University, teaching both undergraduate and post-graduate courses in Geography. He had graduated from Presidency College, Kolkata, in Geography (1997), was awarded National Scholarship for B.Sc. Geography (Honours) Examination in 1997, and was a research scholar in the Department of Geography, University of Calcutta, from where he had post-graduated (1999) and later received his Ph.D. degree in Science (2019) for his research on watershed management and socio-economic development of Gandheswari watershed area. He was awarded the degree of Bachelor of Education (B.Ed.) from All Bengal Teachers Training College, University of Calcutta (2002) and awarded UGC Junior Resource Fellowship, 2003. He is the life member of the Indian National Cartographic Association (LM-1777), The Geographical Society of India (L/S-148), and Bangiyo Bhugol Mancha (LM/011'07'156). He has published many research articles in reputed journals, and presented papers in the fields of rural development.

Subhasree Singha Ray graduated from the University of Calcutta and postgraduated in urban studies from Guru Ghasidas University. As a dedicated and passionate geography lecturer, she brings over 19 years of experience in higher education, specializing in social geography, computer applications, and GIS. She has developed and taught engaging courses that integrate fieldwork, technology, and real-world case studies. She is committed to fostering critical thinking skills and empowering students to apply geographical concepts to address global challenges.

Panchali Majumdar is a senior academician working in different fields in geography with special emphasis on micro-level planning. She was a student of Geography at the Presidency College, Kolkata and did her Postgraduate degree at the University of Calcutta with specialization in Regional Planning. She got her Ph.D. degree in 'Panchayet System of Organisation and Implementation of Rural Development Programme at Panchayet Level-A Geographical Analysis (Case study of Barrackpore Block II)'. Dr. Majumdar is currently an Associate Professor of Geography in East Calcutta Girls' College where she has groomed the department for near about three decades from a toddler to a self-sufficient one which is today expanded into a Postgraduate department with research facilities. Dr. Majumdar was a member of the Governing Body and has experienced three cycles of NAAC and also acted as a successful IQAC Coordinator. She is an invited guest faculty of West Bengal State University and a resource person at UGC-HRDC, Ranchi University. She is the executive council member of the Geographical Society of India and the Institute of Landscape, Ecology and Ekistics; an Institutional life member of the Indian Statistical Institute; Course Coordinator of IIRS Distance Learning (Nodal Centre East Calcutta Girls' College); and Treasurer of the Indian National Cartographic Association (INCA Kolkata Branch). Dr. Majumdar has organized and presented in a number of National and International Seminars and workshops. She also has published several national and international publications. Along with her academic profile, she is also a positive contributor to societal development and she has contributed to society as a National Service Scheme Program Officer.

Chapter 1
Geographies of the Indian Subcontinent: Cutting-Edge Methods in the Twenty-First-Century Research

Sukla Hazra, Anwesha Haldar, Rajarshi Dasgupta, Pinaki Sarkar, Subhasree Singha Ray, and Panchali Majumdar

Abstract This book explores the cutting-edge methods and approaches employed in contemporary research on the geographies of the Indian subcontinent, highlighting how interdisciplinary approaches and technological innovations have enriched research in geography and its allied subjects and have opened up new avenues in understanding this dynamic and multifaceted region. One of the most significant advancements in geographical research is the widespread adoption of geospatial technologies, yet both qualitative and quantitative field observations and experimentations remain an integral part of the information and data base that are fed into the machine learning algorithms. The analysis in quantitative research most often remains incomplete without statistical modelling where data alone cannot give reliable conclusions. The end objective of the research is to state inferences, facts with evidence, theories, proposals, analogies and plans for enhanced policymaking at every stage of governance.

1 Research in Geography and Its Allied Subjects in the Twenty-First Century

Geography as a discipline is spread across space, time and their interconnectedness that united the facts, realities, theories, approaches, experimentations, environmentalisms, regionalisms and their tools of study towards significant contributions in both fields of arts and sciences. It occupied a place in the minds of the generalist scholars of the classical period and beyond because it was an empirical reality (Matthews & Herbert, 2004). Geographical research involves both quantitative and qualitative methods, employing extensive as well as intensive research designs (Clifford et al., 2010). This introductory chapter of this book aims to understand

S. Hazra
East Calcutta Girls' College (Retired), Kolkata, West Bengal, India

A. Haldar (✉) · R. Dasgupta · P. Sarkar · S. S. Ray · P. Majumdar
Department of Geography, East Calcutta Girls' College, Kolkata, West Bengal, India

S. Hazra et al. (eds.), *Geographies of the Indian Subcontinent*,
https://doi.org/10.1007/978-3-031-74813-4_1

the journey of evolving research approaches in geographies of the Indian subcontinent since the beginning of the twenty-first century to the newest and most advanced area of research in an anthropocentric digital era: artificial intelligence technologies. As humans have emerged as perhaps the single most important driver of the Earth System processes, Nobel Laureate atmospheric physicist Paul Crutzen describes the changing times as the Anthropocene (Zalaciewicz et al., 2011; Steffen et al., 2015), which has not only moulded human approaches to environmental perceptions but gave rise to a variety of technological innovations. However, the advancements achieved in multiple facets also sprouted causes of concern and the need for their management. A good example is unbridled urbanisation leading to climatic changes through the heat island effects, destruction of forests, filling-up of water bodies, pressure on the resources and increased pollution levels, creating disruption in the balance of the environmental system and hence introducing a chaotic chain of events that have adverse effects on all its components. Throughout the year, thousands of people are affected by the furies of nature. Destruction of life and property is much more painful. Scientific understanding of the problems/disasters suggests that the prediction of extreme events is very important, for people without knowledge create more vulnerabilities. Depending on the magnitude and frequency of such events, they may be classified as normal or extreme, and they may be unique or repetitive, affecting an individual or communities as a whole. What is alarming is that the frequencies of such extreme environmental events are increasing, and in many cases, they are being catalysed by anthropogenic activities. Therefore, their mitigation would be better addressed because the socio-ecological systems of the Indian subcontinent with its large population, multiple ecosystems, developing economies and diverse cultural milieu have a cause-and-effect relationship with these extreme events. Thus, the aim of the proposed book is to highlight some aspects of this relationship by examining topics related to environmental hazards and socio-ecological systems using contemporary research techniques.

The Indian subcontinent has elicited interest from scientists and the general population for a long time, who have relentlessly tried to delve into the intricacies of the physical and social processes operative in this region. The rich history of the civilizations of this region makes the geographies of the lands even more stratified. However, in recent times, the control of human beings over nature has shown a great spurt where not only the socio-economic practices need to be critically understood for a sustainable future, but also the earth systems disruptions demand relevant management practices. The significance of compiling various themes in geography from the Indian subcontinent lies in the fact that processes have multidimensional and interdisciplinary implications on our lives. The methodologies used to study these aspects are even more thought-provoking and constantly being upgraded or remodelled to suit the requirements of the rapidly changing societies. Thus, this volume brings together the dynamicity in research practices and focus in a comprehensive way, where both theoretical and applied research papers across the full gamut of the subdiscipline highlights the major methodologies used in the research areas in geography.

From the last two decades of twentieth century, research on environmental pollution and sustainable development is knocking on the door of every corner of the earth regarding the human being and flora and fauna on one side and soil, water, river, glaciers, sea, etc. on the other side. At the same time, the ray of hope is that the technological changes leading to development are taking place at a very rapid pace. In the twenty-first century, various dimensions of geographical research mainly concerned with human-induced global changes gained predominance. Traditionally, there is a symbiotic link between geographers and their spatio-temporal aspects. Field experimentations and survey designs are as important as the region of interest; hence, with the changing approaches to perceive a particular phenomenon, results can widely change the interpretations. Thus, the research methods have been largely emphasised in this edited volume. Advancement in data collection, storage, processing, analysis and display has opened up new fields of research. The development of geospatial and statistical analysis engrave a landmark in cartography. During the last four decades, rapid evolution of digital mapping technology via personal computers has enormously changed the field of cartography. It restricts the inadvertent error of computing, mapping systems or techniques. Geographers are doing research on map production techniques, cartographic design and map reading by the public (Goodchild, 2019). The most interesting and potentially useful research is dynamic or animated cartography, which makes a paradigm shift in this field. Today, citizens and government are increasingly aware of problems and potential of a region. Understanding the ecosystem and the impact of human activities on the global environment are important components in government planning and policies. This region is the home of half of the world's population and diversity in the productive ecosystem, economy, culture, and habitat are noteworthy. With the advancement of satellite remote sensing technology, atmospheric conditions, ocean and land surface changes of large areas are computed with cloud computing technology. In this century, the central theme of research is primarily the pressure of growing population and depletion of natural resources. Also, water and air pollution have become a global concern. This prompted the researchers today to do big data analysis available from satellites to save green earth. The integrated approach in the mapping techniques sharpens the management of sustainable development (Dusan & Branislav, 2004).

Contemporary research on the geographies of the Indian subcontinent often adopts an interdisciplinary approach, drawing insights from fields such as anthropology, sociology, economics, and environmental science. This interdisciplinary perspective allows researchers to explore the complex interactions between human societies and the environment, taking into account social, cultural, economic, and political factors. In addition to traditional data sources, researchers are increasingly turning to social media platforms and big data analytics to study human geography in the Indian subcontinent. This necessity prompts the higher educational institutions to pursue interdisciplinary study and research. This type of work is directed to venture across the crosscutting methodologies in social sciences. Different types of hazards are studied in different academic disciplines. Interdisciplinary study is still lacking. The present book is an attempt to interlink all interested people in the realm of geospatial modelling and its application in various domains. The integration of geospatial technologies and remote sensing has opened up new frontiers for researchers to explore and comprehend the earth's dynamic processes. In this context, we delve into key concepts and techniques that form the foundation of advanced research methodology. Researchers from diverse fields are now collaborating to develop innovative solutions with the increasing availability of massive geospatial datasets and advanced computational capabilities. These technologies have reshaped the way we study our planet. Different attributes of environment have been discussed in this book through the next 17 chapters. It will be seen that no single method can be enough to analyse all environmental problems.

This book is the first collaborative research initiative by all the teachers of the department of geography at East Calcutta Girls' College dedicated to the students. This proposal was patronised and highly appreciated by none other than the late Professor Ram Babu Singh, popularly known as R.B. Singh (b. 3 February 1955 ~ d. 22 July 2021), who had been a distinguished stalwart not only in Indian geography but across the international geographical institutions, committees and organizations. He paved the way for future generations and opened up global possibilities for geography students coming from all sections of the society, and this philosophy is carried forward in the department, which encourages even the least resourceful students to strive to achieve their dreams. Through this book, the department of geography at East Calcutta Girls' College pays homage and tribute to Prof. R.B. Singh.

2 Major Sections and Issues Discussed in This Book

The book aims to address the various approaches and methodologies of geographical research that have been categorised into parts and chapters that have been chronologically arranged based on (a) *Field Observation and Experimental Designs*, (b) *Application of Geospatial Technologies*, (c) *Statistical Modelling*, and (d) *Policy Research*.

In the first part, the research papers are mainly based on field observation and analysis of the primary data. Field observation involves first-hand data collection through direct observation and interaction with the environment and its inhabitants while experimentation involves the systematic manipulation of variables to test hypotheses and elucidate causal relationships. In the context of the Indian subcontinent, fieldwork allows researchers to gain intimate knowledge of the region's diverse landscapes, cultures and societies, complementing and contextualizing data obtained through remote sensing and GIS technologies. Field data provide ground truth for remote sensing analysis, allowing researchers to validate and calibrate satellite-derived information. By immersing themselves in local communities and ecosystems, researchers can observe and document spatial patterns, cultural practices and environmental changes. Moreover, field-based experiments provide valuable context for interpreting spatial patterns and trends identified through remote sensing and GIS analysis. By deploying instruments and collecting empirical data, and combining multiple sources of data, researchers can validate theoretical models and refine predictions and develop holistic models of the region's landscapes, cultures and ecosystems, shedding light on complex dynamics and feedback loops operating in the Anthropocene world. In the first part, the works mainly underscore the importance of field observations and experimental design in the study of three critical issues: surface and ground water quality, urban fire hazards and socio-cultural interactions. These five Chaps. 2–6 are of significant concern as they directly impact human lives and environmental sustainability. By employing rigorous field observation, researchers can gain valuable insights and generate data that will result in effective mitigation and management strategies.

The chapter by Ghosal et al. tries to understand the extent of arsenic pollution and conducted a detailed field work on arsenic pollution on the eastern side of River Hugli in Bengal Basin where ground water is strongly controlled by subsurface deep and shallow paleochannels. This work highlights the proper methods of collection along with laboratory tests for identifying the sources of arsenic pollution.

Another paper on wetlands and oxbow lakes known as Baors in Bengal where Haldar et al. try to examine the conservation challenges as well as feasible methods of assessing the wetland ecosystem services (WES). Different methods of analysing WES relevant for Lower Ganga Delta region have been discussed. The fact that these water bodies are beneficial in preserving the regional ecological conditions including flora, fauna and human beings and are a critical component for sustainable management of the ecosystem has been studied through participant observation and matrix assessment methods.

Bhattacharjee studies the changes in surface coverage of water bodies due to unplanned urbanization especially in urban fringe areas. The study on water availability and its quality is focused on the changing environment. The analysis uses both the primary data collected from the field and data obtained by digital image processing. For the delineation of urban fringe, the author uses Otsu's threshold method. Also, the Shanon index is applied for selecting the optimal size of grid for this work. Finally, water qualities are determined through satellite image band ratio and accuracy assessment check by field data.

Study on environmental condition in the coalfield region is undertaken by Sarkar and Bhattacharya. The topical issues in contemporary problems are the transformation of different categories of land without considering its environmental impact. This work is exhaustively done on the change of forest and water bodies into fallow and excavated land by using Fragstats software. The result shows respirable particulate matter is much higher than the normal. Good governance only controls the situation.

As the world undergoes rapid physical and social changes, tribal communities face unique challenges in maintaining their socio-ecological equilibrium. Moyra and Hazra explore a holistic approach to examine the interspace between community liveability and socio-ecological environment in forest ecosystem. On the basis of 18 variables from four types of forest ecosystems, an ecosystem services index is calculated. In Indian context, this index explains the extent to which the tribal community at present is able to mitigate conflicts arising from external pressure such as land encroachment, resource extraction, infrastructural development, etc.

Multidimensional study of the environment in the fast-changing world needs set of tools, systems and data. The next part explores the application of different geospatial technologies to geographical research. Remote sensing and geographic information system (GIS) technology have revolutionised geographical research by providing new tools for data collection, analysis and visualisation. These technologies offer unparalleled insights into the spatial dynamics of the Indian subcontinent, allowing researchers to study land use changes, environmental processes and urban development patterns with unprecedented detail and accuracy. Application of geospatial technologies in the studies helps understand, analyse the spatial and temporal data for better decision making. The use of geospatial modelling and analysis can simulate the effect of various contamination, its ecological consequences and potential risk to humans. This part comprises four chapters.

Environmental degradation needs regular monitoring to reach nearly zero effects of climate change. The chapter by Chakraborty et al. uses remote sensing to retrieve aerosol optical depth (AOD) data of four different wavelengths. Sentinel images of 4 months of 3 consecutive years are analysed to understand the changing relation between aerosol concentration and land surface temperature. They identified presence of aerosols in the atmosphere, especially in urban metropolitan areas like Rajarhat in West Bengal. The chapter also discusses how AOD and landuse land cover along with land surface temperatures can be correlated, which is beneficial to planners as well as scholars who can work on such methodologies over varied spatio-temporal scales.

Seasonality, duration and intensity of floods are very important issues in monsoon countries like India. For assessment of flood situations, optical and radar satellite data are used today. Maity et al. study the flood situation in Assam in 2020. With the help of real-time SAR (Synthetic Aperture Radar) data, the footprint of flooded area was delineated and marked the hotspot. This type of analysis at regular intervals is useful for prediction and prevention of flood hazards.

Ghosh et al. consider the connectivity, accessibility, location layout, water availability, nature of thoroughfares and their proper utilisation as firefighting tools. A model on assessment of devastating consequences of urban fire on infrastructure, environment and communities and its prevention was developed. The model pertains to (1) right time alarm, (2) location of water body, (3) selection of shortest route and (4) firefighting extinguisher. Such experimental design can be employed by the government department and implement a better firefighting protocol in the densely populated country to save lives and properties.

Early warning and accurate forecasting in the disaster management research arena is the need of the hour. Sivakumar et al. have efficiently elaborated on the geoinformatics-based customised model of the Sikkim Earthquake Information System (SKIMEQIS) as an effective tool for seismic disaster resilience. This chapter is extremely informative as scholars get to understand the applications of programming language, GIS customizations and developing geospatial models.

The third part comprises chapters based on statistical modelling. Statistical modelling offers a powerful toolset for synthesising heterogeneous datasets and making informed predictions. It involves the development and application of mathematical models to analyse complex data and infer relationships between variables. In geographical research on the Indian subcontinent, statistical modelling techniques such as regression analysis, spatial autocorrelation, and machine learning play a crucial role in synthesising diverse datasets and identifying patterns. The combination of qualitative insights from fieldwork, empirical data from experiments, and quantitative analyses from statistical models enables researchers to develop robust frameworks for sustainable development and conservation efforts in the region.

Indians are deeply concerned about the monsoon and its erratic behaviour. Mukherjee studies the relationship between the onset and withdrawal of southwest monsoon and pre monsoon cyclonic disturbances. In this chapter, the authors explain significant transitions in the large-scale atmospheric and oceanic circulation in Indo Pacific region. To study its nature here Rough Set Theory is adopted and validated the result with consecutive 3 years atmospheric data of 2009, 2010, 2011 respectively. This model helps to predict the time of onset of monsoon, and mitigation is possible in case of disaster.

Another work on timely response for natural disasters was done by Satpati and Bhadra. A flood management technology has been conceived to give real-time spatial coverage. They used a novel approach by including local people in data gathering, and also with YOLO v 7, an object detection-pertained model from huge different types of data sources was applied. That work was tested with Google colab's virtual machine. Use of deep learning is crucial for recovery of natural disasters.

Construction of weir dam for irrigation in the Terai Himalaya of the Indian subcontinent has a major impact on agricultural development. Chakraborty and Hazra calculate evapotranspiration, effective rainfall, net irrigation requirement, crop water requirement and irrigation water supply for three different seasons: Kharif, Rabi and Zaid by Cropwat model, a decision support software. Total available moisture and readily available moisture data are calculated. Rainfall variability for last 40 years (1981–2021) is correlated with seasonal irrigational water requirements and climate change of the area, and the irrigational requirements are explained very nicely in this chapter.

Research methodology as a subject is taught at different levels. The work designed by Ghosh is specifically for newcomers and explains various methods of statistical analysis with examples and areas covered from simple to complex.

In Part IV, the papers are discussed on policy research. The twenty-first century, which is called Anthropocene era, is a period of paradigm shift in research methods and directed toward anthropocentric. Arguments on environment and development priorities and sustainable development targets have immense significance in geographical research. The synergistic combination of qualitative and quantitative approaches enables researchers to address complex challenges and promote sustainable development in the region. Policy documentation and analysis-based research involve examining existing policies, identifying gaps and challenges, and proposing solutions to address socio-environmental issues. In geographical research on the Indian subcontinent, policy analysis provides a critical lens through which to evaluate the effectiveness of governance structures and institutional mechanisms.

Mukhopadhyay and R. Pani analyse the carbon emission intensity of fossil fuel electricity in two populous developing countries, India and China, compared to the most developed country USA. The change of climate spread beyond the boundary without conforming to the country of origin. Authors calculate the yield variance of fossil fuels for electricity. The record of the last three decades shows the rapid growth of population along with increasing consumption/generation of electricity in China and India nearly seven and five times, respectively. This study reveals the thoughtfulness on the relation between population growth, income and technological development, with special consideration to the climate change all over the world.

As a concluding chapter, Haldar and Satpati studied the changing research methods in geography, i.e. the exploratory process marked by a gradual shift from a descriptive to an experimental and analytical approach. Interdisciplinary studies help identify the problem and methods of establishing the relationship between physical processes and social variables. Ethnographic work emphasises the point of inequality and concentrates on participatory management. Action research is started with collaborative work with stakeholders. Similarly, case studies, arts-based research, narrative work are getting importance in recent time. Technological changes, viz., big data analysis, machine learning and artificial intelligence help analyse large and complex data sets. It needs only some check for measuring reliability of the result.

It is our belief that various research works in this book will help the students to develop a conceptual framework of research and choose the appropriate methodologies. It aims to help to break the wall between qualitative and quantitative research and enlighten about the application of advanced technologies in research.

3 Relevance of the Book

Geography has undergone many paradigm shifts over the course of the subject's history. The advent of geospatial technologies and associated developments undoubtedly brought over one of the biggest shifts in trajectory that needs to be highlighted. Although this region was late to taste the fruits of these technological advances, over the past two decades, significant strides have been made in environmental research in this area. Consequently, it has resulted in many publications, both scientific as well as popular. The increasing human impacts during the twenty-first century, are manifested not only over social systems, ekistics and habitation patterns but also over landscapes, climate, hazards and environment at large, and perhaps nowhere on the earth are these changes more profoundly seen and felt than in the Indian Subcontinent. The book is aimed at highlighting the use of recent skill sets and innovative technologies in research methodologies on different sub-themes, as mentioned earlier. The novelty of this book lies in the fact that it tends to assimilate futuristic and unique ideas and methods of study in the subject based on advances in integrated and interdisciplinary research in the Indian Subcontinent. The idea of the compilation is to encourage students and young researchers in geography and its allied disciplines to undertake serious research-orientated studies for the enhancement of the subject as well as the society. Nevertheless, the proposed volume brings together the whole breadth of the subject of Geography through the lens of advanced research techniques, with a clear focus on the Indian subcontinent. Therefore, it would be a valuable addition to the repertoire of studies on the status of geographical research across the Indian subcontinent.

4 Conclusion

In conclusion, the book makes a good collection of holistic approaches that integrate diverse methodologies and perspectives. Through the integration of field observation, experimentation, statistical modelling, and policy research, contemporary geographical scholarship on the Indian subcontinent has made significant strides in understanding its socio-environmental dynamics and informing evidence-based decision-making. By integrating the different methodologies, researchers can generate comprehensive insights into the geographies of the Indian subcontinent, informing policy formulation, implementation, and evaluation processes. Through engagement with policymakers, stakeholders, and local communities, researchers can foster collaborative approaches to sustainable development and conservation, promoting resilience, equity, and sustainability in the face of environmental change. It is imperative that geographical research on the Indian subcontinent continues to embrace interdisciplinary collaboration, methodological innovation, and policy engagement. By working together across disciplines and sectors, researchers can contribute to a more inclusive, equitable, and sustainable future for the diverse peoples and ecosystems of the Indian subcontinent.

References

Clifford, N., Cope, M., Gillespie, T., French, S., & Valentine, G. (2010). Getting started in geographical. In *Key methods in geography*. Sage Publications Ltd.

Dusan, J., & Branislav, B. (2004). Elements of spatial data quality as information technology support for sustainable development planning. *Spatium, 11*, 77–83.

Goodchild, M. F. (2019). Geography and geographic information science: An evolving relationship. *The Canadian Geographer, 63*, 530–539.

Matthews, J. A., & Herbert, D. T. (2004). *Unifying geography. Common heritage, shared future* (pp. 377–381). Routledge. https://doi.org/10.4324/9780203611395

Steffen, W., et al. (2015). The trajectory of the Anthropocene: The great acceleration. *The Anthropocene Review, 2*, 81–98.

Zalaciewicz, J., et al. (2011). The Anthropocene: A new epoch of geological time? *Philosophical Transactions of the Royal Society A, 369*, 835–841.

Part I
Field Observations and Experimental Designs

Chapter 2
Impacts of Last Glacial Maximum Palaeosol (LGMP) on Arsenic Pollution in Groundwater of the Southern Bengal Basin Along a 12 km Traverse on the Western Bank of the River Hugli, West Bengal, India

Utsab Ghosal, Pradip K. Sikdar, Soumyajit Basu, and Utsha Dasgupta

Abstract Detail studies on arsenic (As) pollution on the eastern side of the River Hugli in the Bengal Basin suggest that the concentration of As in groundwater is strongly controlled by the sub-surface distribution of palaeo-channel (PC) aquifer (depth < 70 m; As-polluted), Last Glacial Maximum Palaeosol (LGMP), at depths variable between 34 and 36 m, and palaeo-interfluvial (PI) aquifer (depth 36–70 m; As-free). The groundwater on the western side of the river is generally thought to be As-free, except in few local pockets. The present study area is one such pocket and has been reported to abstract sizeable quantity of groundwater for drinking and agricultural purposes. This study, for the first time, tries to understand whether As-pollution in the western bank of the river is similarly controlled by the distribution of PC and PI aquifers. The model is tested by mapping the colour of the well completion of 148 wells, drilling 22 boreholes up to depth of 46 m and testing 141 groundwater samples collected from depth range of 10–75 m in field and laboratory for As, Fe and Mn. The well-completion colour survey and field and laboratory testing of As reveal that wells with black stains on well completion are As-free while those with red stains on well completion are As-polluted. Drilling indicates that the shallow palaeo-channel (SPC, depth < 29 m) aquifer and deep palaeo-channel (DPC, 29–70 m) aquifer consisting of grey sands have >10 μg/L As, and PI aquifer made of brown sands have <10 μg/L As. Chemical analysis of As, Fe and Mn in groundwater samples reveals that PI aquifer and deep aquifer (>70 m

U. Ghosal · S. Basu · U. Dasgupta
Department of Geology, Presidency University, Kolkata, India

Department of Environment Management, Indian Institute of Social Welfare and Business Management, Kolkata, India

P. K. Sikdar (✉)
Department of Environment Management, Indian Institute of Social Welfare and Business Management, Kolkata, India

S. Hazra et al. (eds.), *Geographies of the Indian Subcontinent*,
https://doi.org/10.1007/978-3-031-74813-4_2

depth) groundwaters have low concentrations, whereas palaeo-channel groundwaters have high concentrations of these elements. Therefore, use of PI and deep aquifers instead of PC aquifers would mitigate the health risks from consumption of As-polluted groundwater at an affordable cost.

1 Introduction

Groundwater is a crucial source for drinking and domestic purposes in India, with about 80% of the rural and 50% of the urban population depending on it (PHED, 1991; DPHE, 1999; van Geen et al., 2003a; Jakariya et al., 2007; Nickson et al., 2007; Fendorf et al., 2010). Although it is observed to be potable in general, reports of poor groundwater quality have become a major source of concern in recent time. One of the most serious issues is the rise in incidences of high levels of arsenic (As) in the eastern part of the country, particularly the Bengal Basin, west of river Hugli (Chakraborti et al., 2009; BBS/UNICEF, 2011). Arsenic has been known to have adverse impacts on human health, leading to the World Health Organisation (WHO) setting a guideline value of 10 μg/L for it in drinking water. The problem of As-enriched groundwater is even more critical for the Bengal Basin, where it is estimated that about 60 million people use it for drinking purposes (Chakraborti et al., 2009; BBS/UNICEF, 2011). It is seen that high concentrations of As in groundwater (ranging up to several hundred units in some places) are typically confined to a depth of around 30 m below ground level (bgl) (Karim et al., 1997; DPHE, 1999).

In spite of this knowledge, controlling widespread usage of As-polluted groundwater is difficult in the region due to various factors. They are:

(a) In rural communities, well screens are typically placed at the shallowest depth possible in order to minimise the cost of well installation. The screens of these wells are placed at depths generally ranging between 20 and 30 mbgl in As-rich aquifers comprising grey sands (DPHE, 2001; Pal et al., 2002; van Geen et al., 2003a, b; McArthur et al., 2004; Zheng et al., 2005; von Brömssen et al., 2007; Stollenwerk et al., 2007).

(b) The shallow tubewells are present ubiquitously in every household, making it almost impossible to monitor As in all the tubewells of a region using standard analytical methods. This causes problem in finding a feasible solution for the issue.

(c) Most of the commercially available field test kits for measuring arsenic in tubewell water are based on the principle of generating arsine gas, which has its own health hazards even at trace levels (Pakulska & Czerczak, 2006). Moreover, the process involved in As measurement in a single sample by these kits takes at least 20 min, making them expensive and time-consuming for large-scale assessment.

Hence, detection and mitigation of As hazards for the rural poor using an affordable and simple method is the most important objective for most of the researchers working in this field. Keeping this in mind, McArthur et al. (2011, 2012a) proposed the colour of stain developed on the well completion (mentioned as well-completion colour assessment method henceforth) as a parameter to screen for As in tubewells installed in the shallow aquifer (<70 m). The presence of two distinct types of aquifers has been observed through this method: (a) palaeo-interfluvial (PI) aquifer with low As and high manganese (Mn) that mainly displays black stain on well completion and (b) palaeo-channel (PC) aquifer with high As and iron (Fe) that displays red stain on well completion. Hence, their study discussed the prospect of well-completion colour assessment method as a tool for screening of As and Mn in groundwater samples. The distribution of As pollution in Bengal Basin has been postulated to be strongly controlled by the sub-surface distribution of palaeo-channel (PC, <70 m depth and As-polluted), Last Glacial Maximum Palaeosol (LGMP, depths variable between 15 and 35 mbgl), and palaeo-interfluvial (PI, 35–70 m depth and As-free) aquifers, known as Palaeosol model (McArthur et al., 2011; Hoque et al., 2012; Ghosal et al., 2015). The well-completion colour assessment method and palaeosol model of arsenic pollution (McArthur et al., 2011; Hoque et al., 2012) have been demonstrated in areas east of the River Hugli in West Bengal and Bangladesh but not in the area west of the River Hugli, where sporadic occurrences of As in groundwater have been reported (PHED, 1991; Acharya & Shah, 2007).

These models are gaining importance in the recent times as they show that the mitigation of the widespread As-pollution in the shallow aquifers of the region can be mitigated by the exploitation of deep Pleistocene aquifers, especially where a PI aquifer exists at an accessible depth under the protection of LGMP or other aquicludes. Since the drillers in these communities have some basic understanding of the local stratigraphy, educating them about how the stratigraphy relates to As pollution is likely to be an easy and cost-effective mitigation strategy for a pressing problem that exists across the Bengal Basin.

1.1 Study Area

Based on the surface network system, the study area is a part of the Bengal Basin and is located in the northern part of Balagarh block, Hooghly District, West Bengal. It comprises a 12 km NNE–SSW traverse across 11 villages under the Charkrishnabati, Guptipara-I and Guptipara-II gram panchayets (village-governing institute) and is situated about 50 km NW of the areas studied previously (McArthur et al., 2004, 2008, 2010, 2011, 2012a, b; Ghosal et al., 2015), between latitudes 23.159° and 23.198°N and longitudes 88.374° and 88.476°E (Fig. 2.1). Geologically, there is an Archaean crystalline basement followed upwards by Gondwana sediments, Rajmahal volcanics of Mesozoic age, Tertiary formation and huge thickness of Quaternary alluvium. The generalised stratigraphic sequence of the Bengal basin

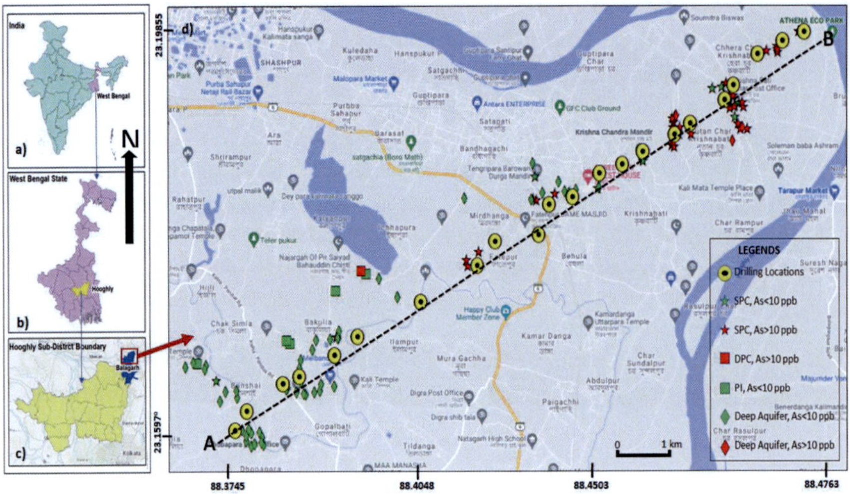

Fig. 2.1 Locations of the study area (**a**) India, (**b**) West Bengal, (**c**) Hooghly district and the small red square in the north of Balagarh block. (**d**) Sampling location distribution and the line A–B indicates the 12 km traverse along which 22 boreholes were drilled to understand the subsurface lithology. In the legend, SPC is shallow palaeo-channel; DPC is deep palaeo-channel and PI is palaeo-interfluve

as suggested by Roy and Chatterjee (2015) is given in Table 2.1. These above-mentioned studies indicate subsurface lithologies down to a depth of 150 m from the surface consist mainly of clay, silt and sand of different grades varying from fine to coarse, occasionally mixed with fine gravel, which forms a rich repository of groundwater. Each aquifer system consists of two or three aquifers separated by thin clay layers, which are not regionally extensive (Acharya & Shah, 2007).

Across the study area, groundwater is the main source of water for drinking and domestic use, tapped by hand-pump fitted tubewells of variable depths. For irrigation, groundwater is tapped through motor-fitted shallow tubewells, to grow paddy, mustard, potato, brinjal, peas, cauliflower, tomato, etc.

The main problem of the area is water quality, which is contaminated with arsenic. Many shallow irrigation and drinking water wells have arsenic concentration >10 μg/L. The groundwater of this area also contains high iron (Acharya & Shah, 2007).

2 Methods

For mapping the distribution of PI and PC aquifers, four methods adopted in previous studies in the Bengal basin (McArthur et al., 2011; Hoque et al., 2012, 2014; Ghosal et al., 2015) are used.

Table 2.1 The generalised stratigraphic sequence of the Bengal basin as suggested by Roy and Chatterjee (2015)

Age	Lithology	Formation
Recent to Pleistocene	Gravel, Sand, Silt, Clay	Bengal Alluvium
Plio-Pleistocene	Dominantly silty with minor sandstone, Clay stone and calcareous bands	Debogram/Ranaghat
Mio-Pliocene	Siltstone with minor sandstone and clay stone	Pandua/Malta
Oligocene	Sandstone, Siltstone, Carbonaceous shale, Lignite, Shale	Burdwan/Memari/Diamond Harbour
Unconformity		
Mid-Eocene to Upper Eocene	Grey shales Calcareous shale, Limestone	Sylhet limestone/ Kopili
Upper Jurassic to Lower Eocene	Sandstone, Shale, Lignite, Limestone bands	Bolpur/Ghatal
Unconformity		
Jurassic	Basalt Flow	Rajmahal trap
Unconformity		
Permo-carboniferous	Carbonaceous Shale, Coal and Sandstone	Gondwana
Unconformity		
PreCambrian	Gneisses with dolerite silts and dykes	Basement

First, the distribution, including the boundary between the PC and PI settings, is predicted by documenting the colour of stain on well-completion and utensils at 141 well sites in this study area (first developed by McArthur et al. (2011)) (Fig. 2.1). Different colours of stain on well-completions are shown in Fig. 2.2. The locations of these wells are recorded by handheld GPS, referenced to the WGS-84 coordinate system along with their depth and well-completion colour.

Second, As concentration in shallow and deep groundwater is measured in the field using a Wagtech International Digital Arsenator, which is a portable field testing kit for As in water samples, with a detection limit <1 μg/L. This kit is accurate, easy to use and environment friendly. To understand the effectiveness and validate the well-completion colour as a tool for screening As in tubewells, four indices, namely sensitivity, specificity, efficiency and predictive values are calculated, with respect to WHO drinking water guidelines (Biswas et al., 2012). Sensitivity is defined as the ability of well-platform colour to identify the wells with As concentrations above the WHO (2019) drinking water guideline. Specificity is defined as the ability of the well platform colour to identify the tubewells with As concentrations below WHO (2019) value. Predictive value measures the number of platform colours showing the true value, which may be positive predictive value (PPV) or negative predictive value (NPV). Efficiency is the percentage of cases for which tubewell completion colour has the correct value in comparison to the total number of cases. Prevalence is the percentage of contaminated tubewells among the

Fig. 2.2 Typical well-completion colours observed in the study area: (**a**) red, (**b**) black, (**c**) mixed and (**d**) undetermined well-completion colour

Table 2.2 Assessment of well-completion colour as a screening tool for arsenic in tubewells

Indices for validation of well Well-completion colour tool (%)	Basis for calculation (formula × 100%)	Estimated value of arsenic WHO guideline (10 μg/L)
Sensitivity	A/(A + C)	71.19
Specificity	D/(B + D)	69.03
Efficiency	(A + D)/(A + B + C + D)	67.37
Positive predictive value (PPV)	A/(A + B)	70.05
Negative predictive value (NPV)	D/(C + D)	72.73
Prevalence	(A + C)/(A + B + C + D)	47.32

tested tubewells at a particular time. The formula for calculating the indices is given in Table 2.2.

Third, these 141 groundwater samples are collected in 15 mL polythene tubes from well sites, which include 81shallow private wells (11–61 mbgl), 60 deep municipal wells (90–182 mbgl), which are used as proxies for PI, PC and deep aquifer settings. The samples are filtered through 0.22-μm membrane filters only when distinctly turbid. The age, depth, well-completion colour and the name of the owner for each of these wells are recorded during the field survey. Laboratory analysis for groundwater for total As and Fe is done by ICP-MS and for Mn by ICP-AES with a detection limit around 0.1 and 1 μg/L respectively. Bundles of three samples were bracketed by a blank and a standard in order to correct for instrumental drift.

Fourth, the distribution of PI and PC aquifers predicted by the above three methods is then tested by drilling boreholes up to 46 m depth at 22 locations along the NNE-SSW traverse mentioned above (Fig. 2.1d) to reveal sub-surface sedimentology.

3 Results

3.1 Well-Completion Colour and Field Analysis of Arsenic

The primary survey reveals that amongst the 83 shallow tubewells and 65 deep tubewells, 100 are observed to have red stains on the well completions, 11 black and 12 mixed, while the remaining 25 did not have any discernible stain, as the colour is disguised by rust from the pump, algal growth, regular cleaning to remove colour or coloured completions (e.g. red brick) (Fig. 2.2).

On the basis of their location, depth and well-completion colour, among 148 tubewells, 141 are further selected for analysis of As concentration in the field (Fig. 2.3). The wells are selected in a way so as to cover representative samples of all four well types, based on well-completion colour—black stained (11), red stained (95), mixed coloured stains (11) and stains with unidentified colour (24) (Fig. 2.4).

The result of this analysis shows that among selected tubewells ($n = 141$), 57.5% ($n = 81$) exceed the WHO drinking water guideline value of 10 μg/L for As (WHO, 2019). The well-completion colour-wise distribution indicates that 72.73% ($n = 8$) of the analysed tubewells with black-coloured well-completion ($n = 11$) have safe level of As, while 67.37% ($n = 64$) of all tubewells with red coloured

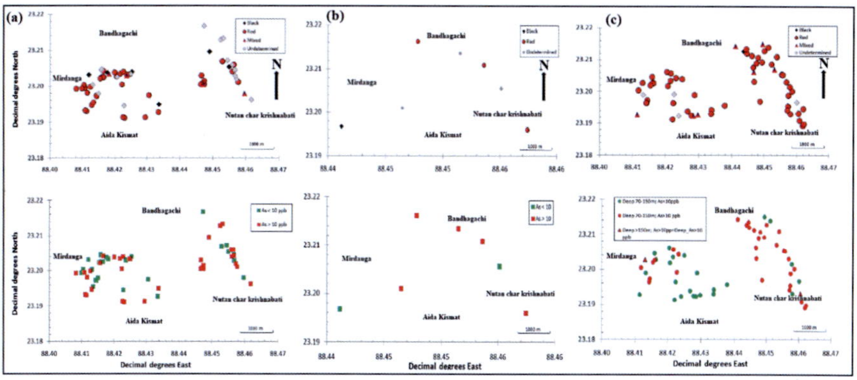

Fig. 2.3 Distribution of black, red, mixed, undetermined wells on top and wells with As < 10 ppb and As > 10 ppb at the bottom, (**a**) in shallow aquifer (<29 m) (**b**) in DPC aquifer (29–70 m) and (**c**) in deep aquifer (>70 m)

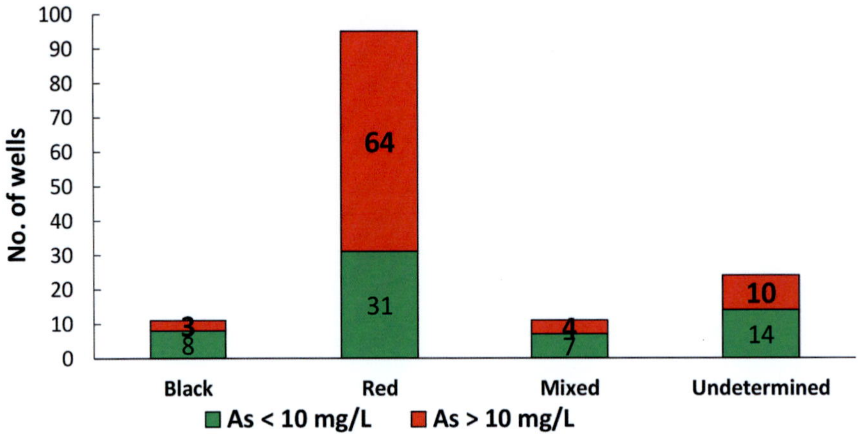

Fig. 2.4 Number of wells of each type of well-completion colour with As concentration

well-completion ($n = 95$) have As > 10 µg/L (Fig. 2.5a). In 27.27% ($n = 3$) cases, black-coloured well completion indicates that tubewells have As concentration >10 µg/L. Further, when tubewells with mixed coloured well-completion are considered, 63.64% well water ($n = 7$) contains As > 10 µg/L, and 36.36% well water ($n = 4$) contains <10 µg/L As.

Considering the aquifers <70 mbgl, it is observed that out of 56 wells, 75% ($n = 6$) of the tubewells with black coloured well-completion have safe level of As, while75% ($n = 36$) of the analysed tubewells with red coloured well-completion have As levels beyond WHO guideline value (Fig. 2.5b).

On the other hand, out of 50 deep wells (>70 mbgl) with red colour well-completion, only 58% ($n = 29$) have As values beyond WHO guideline, while the only deep well ($n = 1$) with black coloured well-completion show high As level, thereby giving a result that is contradictory to general assessment (Fig. 2.5c).

The well-completion colour to be effective as a screening tool for As in tubewell water, it should have high probability of true-positive and negative values and reasonably low probability of false-positive and negative values with respect to a particular drinking water standard. The sensitivity, specificity, efficiency, predictive values (positive and negative) and prevalence of the tool are summarised in Table 2.2. The data shows that as per WHO guidelines, the positive predictive value (PPV) of red coloured well-completion is 70.05%, and the negative predictive value (NPV) of black coloured well-completion is 72.73%. The respective efficiency, sensitivity and specificity of the tool are 67.37%, 71.19% and 69.03%.

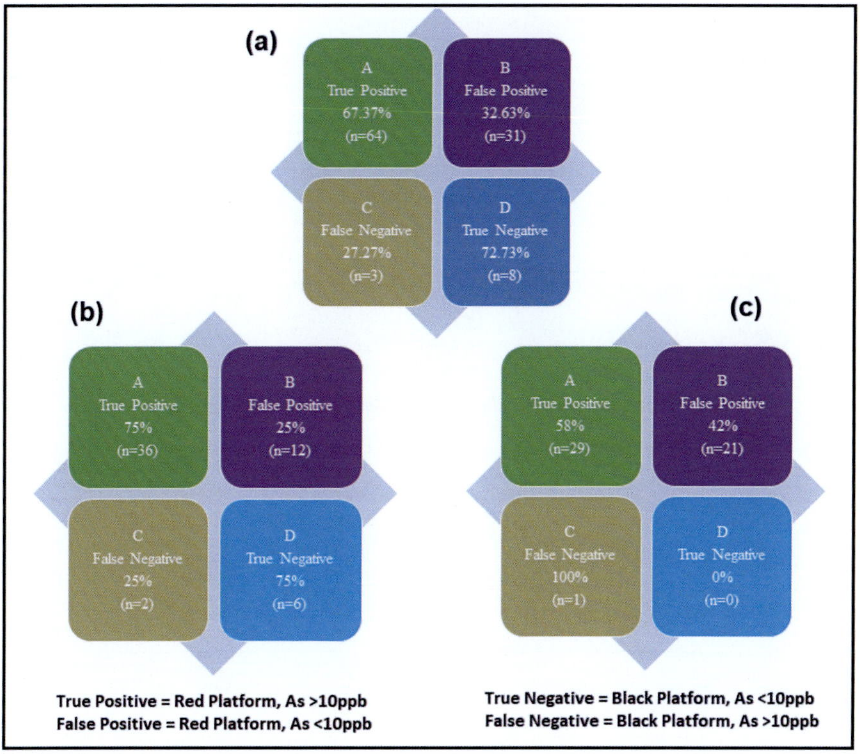

Fig. 2.5 Validation of well-completion colour as a screening tool for arsenic as per WHO guideline (10 μg/L (**a**) for combined shallow and deep aquifers) (**b**) for shallow aquifer only and (**c**) for deep aquifer only

3.2 *Drilling*

Based on the depth, well-completion colour and As concentration of wells, drilling at 22 locations are carried out to investigate the sub-surface lithology and locate palaeo-channels and palaeo-interfluves. Each of these locations is at nearly 500 m interval, along a traverse of 12 km length (Fig. 2.6a, b). Across the area, the surficial cover comprises an upper aquitard of light brown silty clay or sandy silt. The thickness of this cover is typically 3–6 m, but exceptions occur. Beneath this surficial upper aquitard, mainly four types of sedimentary sequences are recognised on the basis of the occurrence of, and depth to, brown sands and grey sands; the presence or absence of the LGMP; and the stratigraphic order of these lithologies. The sequences recognised are as follows:

(a) *Palaeo-channel*, FM4 (Fig. 2.7a): A (1.5 m) thick weathered oxidised brownish grey clay is present at the upper part of the sub-surface geological column, followed by grey clay up to a depth of (13.7 m). This is underlain by fine to medium grey sand up to a depth of (38 m). Below this there is again grey coarse

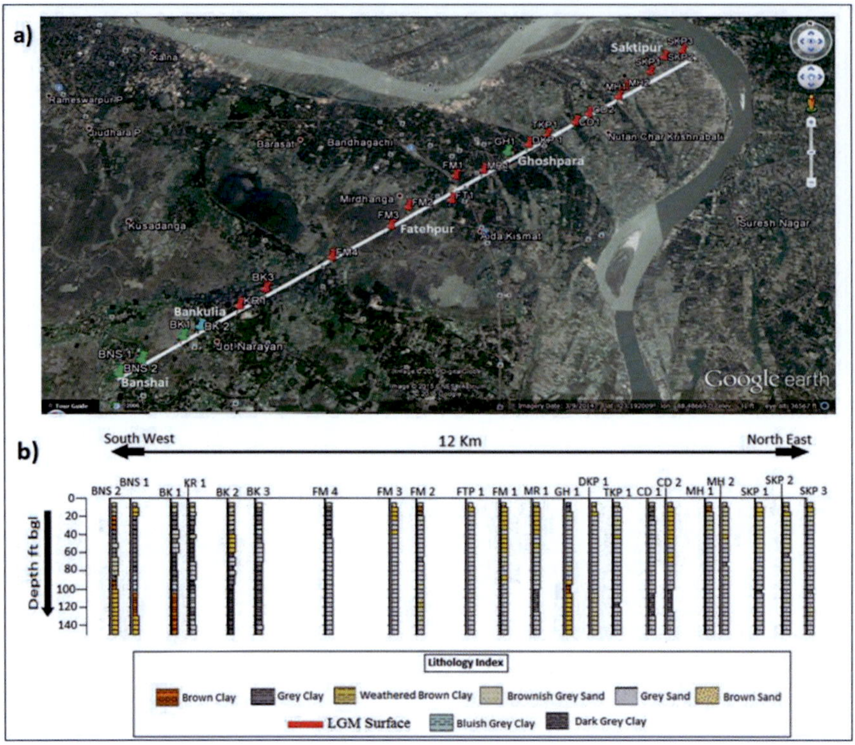

Fig. 2.6 (**a**) The line A–B indicates the 12 km traverse along which 22 boreholes were drilled to understand the subsurface lithology; (**b**) strip logs showing palaeo-channel and palaeo-interfluvial boreholes along the line of section

sand up to the drilling depth of 70 m. The upper 34 m of sand is termed the shallow palaeo-channel (SPC) aquifer and below that is termed the deep palaeo-channel (DPC) aquifer.

(b) *Palaeo-interfluve under SPC*, GH-1 (Fig. 2.7b): Brown sands are overlain by the LGMP between 29 and 32 m. Together, these units comprise the palaeo-interfluvial (PI) sequence. The LGMP consists of brown, sometimes mottled, stiff clay with carbonised rootlets, calcium carbonate and iron nodules (Fig. 2.7b). An SPC aquifer of brownish grey sand between 3 and 29 m overlies the LGMP, constituting the first local aquifer with many domestic and irrigation wells installed in it.

(c) *Palaeo-interfluve under floodplain clays*, BNS1 (Fig. 2.7c): Silty clays overlie a palaeo-interfluvial sequence of oxidised palaeosol over brown sand. Grey clay extends up to a depth of 21 mbgl, which is followed by 10.7 m thick bluish grey clay. The LGMP lies between 32 and 39.6 m. Below the LGMP, brown fine sand occurs atleast up to a depth of 45.7 mbgl.

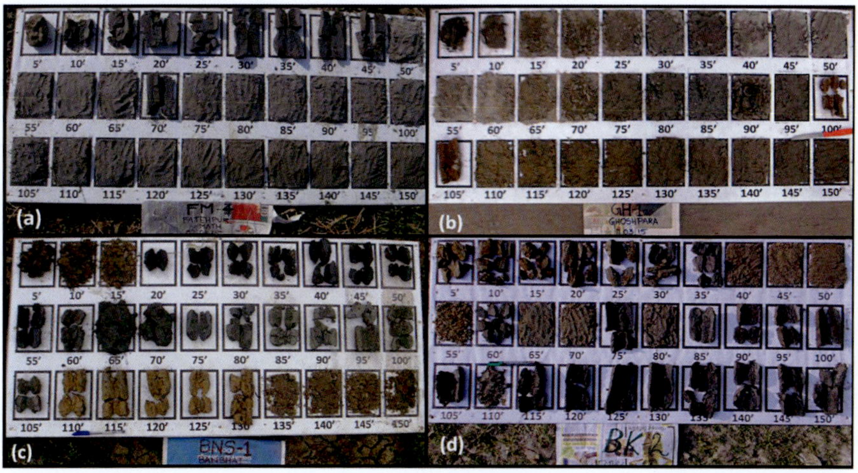

Fig. 2.7 (**a**) Borehole-FM4. This represents a typical palaeo-channel sequence where the LGM palaeosol is absent, (**b**) Borehole-GH1. This represents a typical palaeo-interfluvial sequence where the LGM palaeosol is present, (**c**) Borehole-BNS1. This represents a palaeo-interfluvial sequence under floodplain clays, and (**d**) Borehole-BK2. This represents a typical floodplain sequence

(d) *Floodplain*, BK2 (Fig. 2.7d): Dark grey clay lies throughout the sequence intercalated with brown clay from 1.5 to 11 m, coarse brown sand from 11 to 17 m and from 18 to 21 m.

3.3 Groundwater Composition

The SPC as well as DPC groundwaters contain high As and Fe, but the PI groundwaters are low in As and Fe. The concentration of As is also dependent on the depth below ground level. The depth-wise distribution of As, Fe and Mn in wells indicates that these concentrations are higher in the aquifers up to a depth of 70 m (Fig. 2.8). The deep aquifer (>70 m) generally have <10 μg/L As, low Mn, but high Fe (Table 2.3; Fig. 2.8).

4 Discussions

4.1 Well Completion Colour as a Screening Tool

In order to be an effective screening tool for As contamination of groundwater, the well-completion colour model should have a high probability of true-positive (red well-completion, As > 10 μg/L) and true-negative (black well-completion,

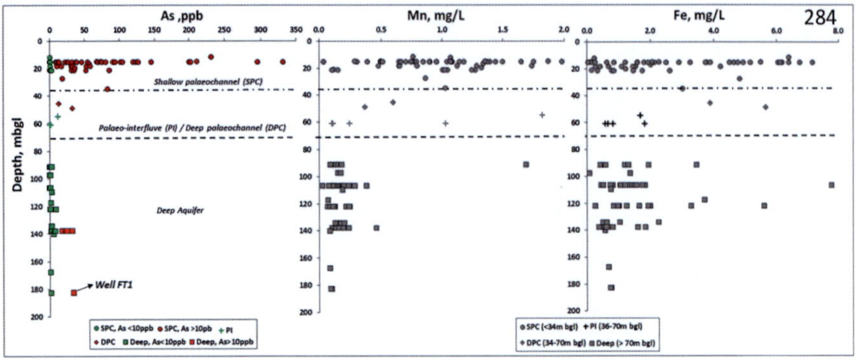

Fig. 2.8 Depth-wise distribution of different chemical parameters in the wells

Table 2.3 Concentration of elements in different aquifers

	As (μg/L)			Fe (mg/L)			Mn (mg/L)		
Type of aquifer	Max	Min	Mean	Max	Min	Mean	Max	Min	Mean
SPC	332	0	67	19	0	3.2	2	0	1
DPC	32	13	22	5.7	3.9	4.8	1	0.4	0.5
PI	11	0	3	1.8	0.6	1.1	1.9	0.1	0.7
Deep	34	0	4	7.8	0.1	1.4	1.7	0.03	0.2

As < 10 μg/L) values and a reasonably low probability of false-positive (red well-completion, As < 10 μg/L) and false-negative (black well-completion, As > 10 μg/L) values (Biswas et al., 2012). The majority of the cases for wells <70 mbgl show black well-completion with As concentration < 10 μg/L, and red well-completions show considerable numbers with As > 10 μg/L (Fig. 2.5). The values of sensitivity, specificity, efficiency, predictive values (positive and negative) and prevalence of the tool of the entire aquifer system (shallow and deep aquifers) (Table 2.2) indicate that the red stain on well-completions can be used to screen wells tapping water likely to be contaminated by As and the black stain on well-completions is likely to indicate wells to be As free with equal precision. Again, if we compare the sensitivity, specificity, efficiency, predictive values separately for the shallow and deep aquifers (Table 2.4), they show that the model is more applicable for wells in shallow aquifers rather than deep aquifers. Therefore, the well-completion colour model is an effective screening tool for As contamination in aquifers <70 m deep.

4.2 Aquifer Setting Deduced from Sediments and Groundwater Composition

Based on the sub-surface lithological disposition, a tentative evolution of the study area and adjoining areas can be outlined. During the late Pleistocene epoch, when the

Table 2.4 Assessment of well-completion colour as a screening tool for arsenic in tubewells for shallow and deep aquifer

Indices for validation of well Well-completion colour tool (%)	Basis for calculation (formula × 100%)	Estimated value of arsenic	
		Shallow aquifer (<70 mbgl)	Deep aquifer (>70 mbgl)
Sensitivity	A/(A + C)	75	36.71
Specificity	D/(B + D)	75	0
Efficiency	(A + D)/(A + B + C + D)	75	29
Positive predictive value (PPV)	A/(A + B)	75	58
Negative predictive value (NPV)	D/(C + D)	75	0
Prevalence	(A + C)/(A + B + C + D)	50	79

Last Glacial Maximum was recorded (c.20 ka), the sea level was about 120–130 m lower than the present. Erosion of interfluves between 125 and 20 ka created a broad coastal platform with intervening channels, at a depth of 35–40 m below the present land surface. The interfluves were then weathered, which allowed a red soil to develop on the interfluves at 29–40 m below the present ground level (brown stiff clay at GH1 and BNS1), which is the Last Glacial Maximum Palaeosol (LGMP), a key stratigraphic marker (Goodbred Jr & Kuehl, 2000a, b). The LGMP formed discontinuously over the basin, which was preserved in the interfluvial areas. The interfluvial sands beneath this soil were also weathered and oxidised to brown colour. The degree of oxidation of the brown sand is judged by the intensity of colour, which decreases with depth. The maximum depths of flushing of the ancient palaeo-interfluves are marked by the change of colour of sands from brown to grey, which in the present study area is between 67 and 73 mbgl.

The palaeo-interfluve and the LGMP represent an old landscape, now buried, but dissected and weathered prior to the post-LGM transgression. Later, sea level rose rapidly, forming estuaries along the incised pre-LGM channels that were subsequently in-filled mainly by grey sand. The rising sea level inundated the interfluvial platforms, allowing sediment to be deposited widely across these platforms. The sediment supply increased as Himalayan glaciers retreated and monsoonal conditions intensified from the early Holocene to the climatic optimum at 8 ka, when rainfall, temperature and river flow were much higher than now (McArthur et al., 2008). This strong pulse of increased water and sediment supply rejuvenated local channels (the post-LGM palaeo-channels), resulting in channel migration and channel deposition. Subsequent floodplain deposition by the post-LGM palaeo-channels, consisting of grey fine sand with clay-silt units, buried both palaeo-interfluves and the pre-LGM palaeo-channels (Fig. 2.7c, d; BNS1 and BK2).

The modification of these deltaic landforms produced an aquifer system in the area, where post-LGM palaeo-channel grey sand and palaeo-interfluvial brown sand lie juxtaposed in the shallow sub-surface, covered by a post-LGM palaeo-channel sediments consisting of fine grey sand and clay (Fig. 2.7a, b; FM4 & GH1). The fine-to-medium sand with brownish tinge indicates the downward movement of dissolved organic carbon into the pre-existing brown sand, driving redox reaction

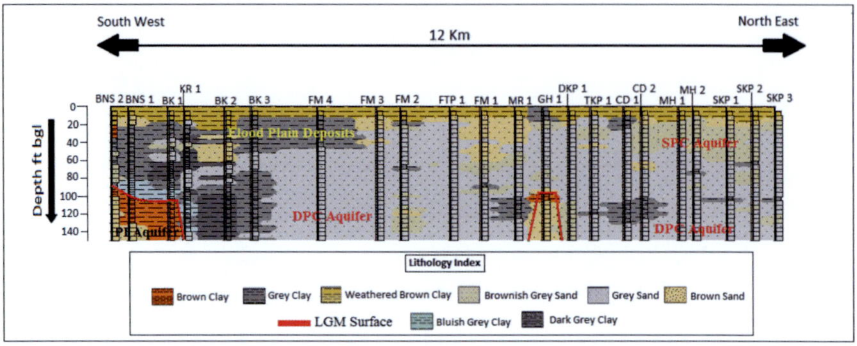

Fig. 2.9 Sedimentary sequences across the study area showing palaeo-channel and palaeo-interfluvial zones along the line of section shown in Figs. 2.1e and 2.6

to proceed in the aquifer, stripping most of the iron content of the brown sand under reducing condition and converting it to grey sand.

Pre-LGM palaeo-channel brown sands are physically separated by overlying Holocene post-LGM palaeo-channel grey sands where the LGMP is present (Figs. 2.7b, c and 2.9). On the other hand, palaeo-channel deposits lacking the LGMP are recharged by downward percolation. The LGMP prevents downward recharge into palaeo-interfluvial aquifers, which are therefore, mainly recharged by lateral flow from adjacent post-LGM deep palaeo-channels (DPC; 29–70 m). Minor recharge also takes place in palaeo-interfluvial aquifers by groundwater from post-LGM As-polluted shallow palaeo-channel above the palaeosol that moves laterally above it to the nearest palaeo-interfluve, where it rolls over the edge of the palaeosol (under pumping conditions) to contribute to recharge. The post-LGM palaeo-channel sediments constitute the first local aquifer (shallow palaeo-channel; SPC; $<$29 m depth) in and around the study area, where innumerable domestic and agricultural wells have been installed. The SPC and DPC sediments constitute the unconfined aquifer across the entire area. It is always polluted by As. In our study area, the palaeo-interfluvial aquifer, which is As-free is mostly not tapped by the local people because of the lack of knowledge of the presence of such As-free aquifer.

Therefore, the above subsurface geological model validates the McArthur et al. (2008) palaeosol model and also explains that the subsurface distribution of As-pollution to be strongly controlled by the subsurface distribution of palaeo-channel (As-polluted) and palaeo-interfluvial (As-free) aquifers in the area west of River Hugli.

Groundwater in the SPC and DPC aquifers shows a range of concentrations of Fe, Mn, and As but provides a clear picture of difference in character when PI aquifer is compared to DPC aquifer alone. SPC and DPC groundwater with high As ($>$10 μg/L) concentration have high amount of Fe and indicates that As in Bengal Basin is derived from reductive dissolution of sedimentary iron oxide (FeOOH). In PI groundwaters, the concentration of As and Fe are low because the water is in an insufficiently reducing environment for Fe to liberate into solution (Nickson et al.,

1998, 2000). Figure 2.6 also show As concentrations in our groundwater from depth >70 mbgl relate to areas of palaeo-interfluve and palaeo-channel. Of our 58 groundwater sampled at depths >70 mbgl, 39 have As concentrations <10 μg/L and are screened beneath regions of palaeo-interfluve. The remaining 19 samples containing >10 μg/L of As are screened beneath deep palaeo-channels. The highest As concentration (34 μg/L) in our deep groundwaters occurs in Well FT1 (depth 183 m), which is beneath a deep palaeo-channel.

Our results delineated a small PI aquifer within the central part and a large area of PI aquifer towards the south-western side of our study area, hosting low-As groundwater (≤10 μg/L As). The depth to the top of the PI aquifers in our area ranges mostly from 34 to 36 mbgl (Fig. 2.9). Previous studies (Biswas et al., 2014; Ghosal et al., 2015; McArthur et al., 2011) have reported that the top of the PI aquifer, on the eastern side of the River Hugli, lies at a depth between 30 and 38 mbgl. Therefore, to abstract low As (<10 μg/L) groundwater in our study area, the top of the screen of the wells should be placed in the late Pleistocene PI brown sands or grey sands below the LGMP.

Exploitation of aquifers of brown and grey sands, below the LGMP, has been suggested as an As-mitigation strategy. It relies on the sorptive capacity of the brown sands for As (DPHE, 1999; Zheng et al., 2004, 2005; Stollenwerk et al., 2007; von Brömssen et al., 2007, 2008; McArthur et al., 2008; Biswas et al., 2014).

The reason for the underexploitation of the PI aquifers is that these aquifers may be vulnerable to downward migration of As-polluted groundwater from overlying SPC aquifers (e.g. Stollenwerk et al., 2007), although local protection by intercalated clays has also been noted (Swartz et al., 2004; Zheng et al., 2005). The concern about the downward migration of As is real. The vertical thickness of the brown sands, i.e. the oxidised part of PI aquifers, is typically only a few tens of metres. The sorption capacity of underlying grey sands is not known. Where SPC grey sand aquifers sit directly on PI brown sand aquifers without the intervening LGMP (e.g. at Moyna, about 90 km south-east of the present study area), groundwater moving downward has created a reduction front where As is released from the sediment into the groundwater. This front has burned 7–8 m downward into the brown sands and the As released has migrated another 5 or so metres downward (McArthur et al., 2008).

It is therefore unwise to assume brown sands will prevent downward migration of As into PI aquifers in the medium or long term, largely because of the limited thickness of the brown sands. The PI brown sands must be separated from any overlying SPC aquifer by a LGMP confining aquiclude, or an equivalent in order to ensure long-term sustainability of the low-As source.

5 Conclusion

The study conducted on the western side of the River Hugli revealed the presence of several domestic and irrigation wells. The screens of these wells tap groundwater from As-polluted SPC aquifer (depth < 29 mbgl) made of Holocene grey sands, As-polluted DPC aquifer (depth 29–70 mbgl) made of grey sands, As-free PI aquifer at depths below LGMP (varying between 29 and 32 mbgl) made of late Pleistocene brown sands and deep aquifer (depth > 70 mbgl) made of late Pleistocene grey sands. It is difficult for the villagers to identify and demarcate As-polluted wells and As-free wells. Field investigation of the well-completion colour indicates that it is an effective first-level screening tool for As contamination in aquifers <70 m depth. Red stains on well-completion usually indicate presence of Fe and As in well water, and black stains on well-completion indicate the presence of Mn and that the water is As-free. Mixed coloured stains on well completions indicate change in the redox level within the aquifer sediments and may be As-polluted.

Drilling revealed the presence of palaeo-channel as well as palaeo-interfluvial sequences with an intervening LGMP at places. Chemical analysis revealed that As concentration is dependent on the lithology as well as the depth of the well screen. The Holocene SPC sediments deposited above the LGMP and DPC sediments have As-polluted groundwater whereas the Late Pleistocene PI sediments and deeper sediments below it have As-free groundwater. Thus, it can be concluded that the depositional history of the sediments and As distribution pattern is similar on both sides of the River Hugli. Therefore, to abstract low As (<10 μg/L) groundwater for the foreseeable future, the screen of the wells should be placed in the late Pleistocene PI brown sands or grey sands below it underneath the LGMP cover.

Acknowledgements The work was supported by DST, India grant SR/S4/ES-399/2009 to P. K. Sikdar (P. K. S.) and NERC grant NEG016879/1 to JM McArthur and P. K. S. The authors thank Prof. JM McArthur for his many discussions and carrying out the chemical analysis of the water samples in his laboratory in University College, London. We also thank Shri Mrityunjoy Das and his drill team and Joton Mahato for their assistance in the field. We thank them for their care and forbearance.

References

Acharya, S. K., & Shah, B. A. (2007). Arsenic-contaminated groundwater from parts of Damodar fan-delta and west of Bhagirathi River, West Bengal, India: Influence of fluvial geomorphology and Quaternary morphostratigraphy. *Environmental Geology, 52*, 489–501.

BBS/UNICEF. (2011). *Bangladesh national drinking water quality survey 2009*. Bangladesh Bureau of Statistics and UNICEF.

Biswas, A., Nath, B., Bhattacharya, P., Halder, D., Kundu, A. K., Mandal, U., et al. (2012). Testing tubewell platform color as a rapid screening tool for arsenic and manganese in drinking water wells. *Environmental Science & Technology, 46*(1), 434–440.

Biswas, A., Bhattacharya, P., Mukherjee, A., Nath, B., Alexanderson, A., Kundu, A. K., Chatterjee, D., & Jacks, G. (2014). Shallow hydrostratigraphy in an arsenic affected region of Bengal Basin:

Implication for targeting safe aquifers for drinking water supply. *Science of the Total Environment, 485–486*, 12–22.

Chakraborti, D., Das, B., Rahman, M. M., Chowdhury, U. K., Biswas, B., Goswami, A. B., Nayak, B., Pal, A., Sengupta, M. K., Ahmed, S., Hossain, A., Basu, G., Roychowdhury, T., & Das, D. (2009). Status of groundwater arsenic contamination in the state of West Bengal, India: A 20-year study report. *Molecular Nutrition & Food Research, 53*(5), 542–551.

DPHE. (1999). Phase I: Rapid investigation. In *Groundwater studies for arsenic contamination in Bangladesh*. Department of Public Health Engineering (DPHE) of Government of Bangladesh, British Geological Survey (BGS) and Mott MacDonald Ltd (MML).

DPHE. (2001). *Arsenic contamination of groundwater in Bangladesh* (D. G. Kinniburgh & P. L. Smedley, Eds., p. 267). Department of Public Health Engineering (DPHE) of Government of Bangladesh and British Geological Survey (BGS) Keyworth.

Fendorf, S., Michael, H. A., & van Geen, A. (2010). Spatial and temporal variations of groundwater arsenic in South and Southeast Asia. *Science, 328*, 1123–1127.

Ghosal, U., Sikdar, P. K., & McArthur, J. M. (2015). Palaeosol control of arsenic pollution: The Bengal Basin in West Bengal, India. *Groundwater, 53*(4), 588–599.

Goodbred, S. L., Jr., & Kuehl, S. A. (2000a). The significance of large sediment supply, active tectonism, and eustasy of margin sequence development: Late Quaternary stratigraphy and evolution of the Ganges-Brahmaputra delta. *Sedimentary Geology, 133*(3), 227–248.

Goodbred, S. L., Jr., & Kuehl, S. A. (2000b). Enormous Ganges-Brahmaputra sediment supply, active tectonism, and eustasy on margin sequence development: Late Quaternary stratigraphy and evolution of the Ganges-Brahmaputra delta. *Geology, 28*, 1083–1086.

Hoque, M. A., McArthur, J. M., & Sikdar, P. K. (2012). The palaeosol model of arsenic pollution of groundwater tested along a 32 km traverse across West Bengal, India. *Science of the Total Environment, 431*, 157–165.

Hoque, M. A., McArthur, J. M., & Sikdar, P. K. (2014). Sources of low-arsenic groundwater in the Bengal Basin: Investigating the influence of the last glacial maximum palaeosol using a 115-km traverse across Bangladesh. *Hydrogeology Journal, 22*(7), 1535–1547.

Jakariya, M., Vahter, M., Rahman, M., Wahed, M. A., Hore, S. K., Bhattacharya, P., & Jacks, G. (2007). Persson LÅ Screening of arsenic in tubewell water with field test kits: Evaluation of the method from public health perspective. *Science of the Total Environment, 379*(2–3), 167–175.

Karim, M., Komori, Y., & Alam, M. (1997). Arsenic occurrence and depth of contamination in Bangladesh. *Journal of Environmental Chemistry, 7*, 783–792.

McArthur, J. M., Banerjee, D. M., Hudson-Edwards, K. A., Mishra, R., Purohit, R., Ravenscroft, P., Cronin, A., Howarth, R. J., Chatterjee, A., Talukder, T., Lowry, D., Houghton, S., & Chadha, D. K. (2004). Natural organic matter in sedimentary basins and its relation arsenic in anoxic ground water: The example of West Bengal and its worldwide implications. *Applied Geochemistry, 19*, 1255–1293.

McArthur, J. M., Ravenscroft, P., Banerjee, D. M., Milsom, J., Hudson-Edwards, K. A., Sengupta, S., Bristow, C., Sarkar, S., Tonkin, S., & Purohit, R. (2008). How paleosols influence groundwater flow and arsenic pollution: A model from the Bengal Basin and its worldwide implication. *Water Resources Research, 44*, W11411.

McArthur, J. M., Banerjee, D. M., Sengupta, S., Ravenscroft, P., Klump, S., Sarkar, A., et al. (2010). Migration of As, and 3H/3He ages, in groundwater from West Bengal: Implications for monitoring. *Water Research, 44*, 4171–4185.

McArthur, J. M., Nath, B., Banerjee, D. M., Purohit, R., & Grassineau, N. (2011). Palaeosol control of groundwater flow and pollutant distribution: The example of arsenic. *Environmental Science & Technology, 45*(4), 1376–1383.

McArthur, J. M., Sikdar, P. K., Nath, B., Grassineau, N., Marshall, J. D., & Banerjee, D. M. (2012a). Sedimentological control on Mn, and other trace elements in groundwater of the Bengal delta. *Environmental Science & Technology, 46*(2), 669–676.

McArthur, J. M., Sikdar, P. K., Hoque, M. A., & Ghosal, U. (2012b). Waste-water impacts on groundwater: Cl/Br ratios and implications for arsenic pollution of groundwater in the Bengal Basin. *Science of the Total Environment, 437*, 390–402.

Nickson, R., McArthur, J. M., Burgess, W. G., Ahmed, K. M., Ravenscroft, P., & Rahman, M. (1998). Arsenic poisoning in Bangladesh groundwater. *Nature, 395*, 338.

Nickson, R., McArthur, J. M., Ravenscroft, P., Burgess, W. G., & Ahmed, K. M. (2000). Mechanism of arsenic release to groundwater, Bangladesh and West Bengal. *Applied Geochemistry, 15*, 403–413.

Nickson, R., Sengupta, C., Mitra, P., Dave, S. N., Banerjee, A. K., Bhattacharya, A., Basu, S., Kakoti, N., Moorthy, N. S., Wasuja, M., Kumar, M., Mishra, D. S., Ghosh, A., Vaish, D. P., Srivastava, A. K., Tripathi, R. M., Singh, S. N., Prasad, R., Bhattacharya, S., & Deverill, P. (2007). Current knowledge on the distribution of arsenic in groundwater in five states of India. *Journal of Environmental Science and Health Part A – Toxic/Hazardous Substances & Environmental Engineering, 42*, 1707–1718.

Pakulska, D., & Czerczak, S. (2006). Hazardous effects of arsine: A short review. *International Journal of Occupational Medicine and Environmental Health, 19*(1), 36–44.

Pal, T., Mukherjee, P. K., & Sengupta, S. (2002). Nature of arsenic pollutants in groundwater of Bengal basin—A case study from Baruipur area, West Bengal, India. *Current Science, 82*, 554–561.

PHED. (1991). *National drinking water mission project on arsenic pollution in groundwater in West Bengal*/Final Report, Steering Committee, Arsenic Investigation Project, Public Health Engineering Department, Kolkata, Govt. of West Bengal, India, 57pp.

Roy, A. B., & Chatterjee, A. (2015). Tectonic framework and evolutionary history of the Bengal Basin in the Indian subcontinent. *Current Science, 109*(2), 271–279.

Stollenwerk, K. G., Breit, G. N., Welch, A. H., Yount, J. C., Whitney, J. W., Foster, A. L., Uddin, M. N., Majumder, R. K., & Ahmed, N. (2007). Arsenic attenuation by oxidized aquifer sediments in Bangladesh. *Science of the Total Environment, 379*(2–3), 133–150.

Swartz, C. H., Blute, N. K., Badruzzman, B., Ali, A., Brabander, D., Jay, J., Besancon, J., Islam, S., Hemond, H. F., & Harvey, C. F. (2004). Mobility of arsenic in a Bangladesh aquifer: Inferences from geochemical profiles, leaching data, and mineralogical characterization. *Geochimica et Cosmochimica Acta, 68*(22), 4539–4557.

Van Geen, A., Ahmed, K. M., Seddique, A. A., & Shamsudduha, M. (2003a). Community wells to mitigate the arsenic crisis in Bangladesh. *Bulletin of the World Health Organization, 81*(9), 632–638.

Van Geen, A., Zheng, Y., Versteeg, R., Stute, M., Horneman, A., Dhar, R. K., Steckler, R., Gelman, M., Small, C., Ahsan, H., et al. (2003b). Spatial variability of arsenic in 6000 tubewells in a 25 km^2 area of Bangladesh. *Water Resources Research, 39*(5), 1140–1155.

von Brömssen, M., Jakariya, M., Bhattacharya, P., Ahmed, K. M., Hasan, M. A., Sracek, O., Jonsson, L., Lundell, L., & Jacks, G. (2007). Targeting low-arsenic aquifers in Matlab Upazila, Southeastern Bangladesh. *Science of the Total Environment, 379*(2–3), 121–132.

von Brömssen, M., Larsson, S. H., Bhattacharya, P., Hasan, M. A., Ahmed, K. M., Jakariya, M., Sikder, M. M. A., Sracek, O., Bivén, A., Doušová, B., Patriarca, C., Thunvik, R., & Jacks, G. (2008). Geochemical characterisation of shallow aquifer sediments of Matlab Upazila, South-eastern Bangladesh—Implications for targeting low-As aquifers. *Journal of Contaminant Hydrology, 99*(1–4), 137–149.

WHO. (2019). *Preventing disease through healthy environments exposure to arsenic: A major public health concern*. WHO.

Zheng, Y., Stute, M., van Geen, A., Gavrielli, I., Dhar, R., Simpson, H. J., Schlosser, P., & Ahmed, K. M. (2004). Redox control of arsenic mobilization in Bangladesh groundwater. *Applied Geochemistry, 19*(2), 201–214.

Zheng, Y., Datta, S., Stute, M., Dhar, R., Hoque, M. A., Rahman, M. W., Ahmed, K. M., Schlosser, P., & van Geen, A. (2005). Stable isotopes (^{18}O, ^{2}H) and arsenic distribution in the shallow aquifers in Araihazar, Bangladesh. *Eos, Transactions American Geophysical Union, 86*(52). Fall Meeting Supplement, Abstract H31B-1305.

Chapter 3
Assessing the Ecosystem Services of Wetlands: A Case Study from North 24 Parganas District, West Bengal

Anwesha Haldar, Mousumi Paul, and Anish Bhattacharyya

Abstract Wetlands are specific regions where water is present either at or near the surface of the soil or covers it all around the year or for varying periods of time during the year. Wetland ecosystem services (WES) are ranked the first among all kinds of natural ecosystems. The value of WES is expected to be 47% of the total value of the global ecosystem, yet the Ramsar Convention reports degrading conditions of wetlands worldwide, creating greater economic losses. The Millennium Ecosystem Assessment (MEA, Ecosystems and human well-being: wetland ecosystem services. Island Press, 2005) classifies provisioning, regulating, cultural, and supporting services of the ecosystems, and thus the wetland values affect three levels of biological hierarchy to include population, eco-system, and global. The Wetland (Conservation and Management) Rules 2017, formulated by the Ministry of Environment, Forest and Climate Change (MoEFCC), Government of India, prescribe various measures to improve the wetlands across the country. However, with rapid urbanization, these wetlands are fast degenerating, depleting, and engulfed into urban land use. An effective way to create awareness regarding their restoration is by estimating the cost-benefit analysis of the WES. This chapter elaborates the techniques of calculating ecosystem services, its benefits, and challenges when computing in wetlands of Lower Gangetic Delta region. Moreover, it also assesses its relevance to the wetlands commonly called "Baorh" in North 24 Pargana district of West Bengal, India. The environmental changes and the ecosystem services of the wetlands have been studied at a regional level with special focus on the Kankona Baorh, which have been studied using both primary and secondary data, photographs, and interviews along with geospatial mapping techniques. In this study, both primary and secondary data were collected, like fishing data, photographs, and local

A. Haldar · M. Paul
Department of Geography, East Calcutta Girls' College, Kolkata, West Bengal, India

A. Bhattacharyya (✉)
Department of Biotechnology, Adamas University, Kolkata, West Bengal, India
e-mail: anish.bhattacharyya1@adamasuniversity.ac.in

S. Hazra et al. (eds.), *Geographies of the Indian Subcontinent*,
https://doi.org/10.1007/978-3-031-74813-4_3

resident interviews along with published references. The gradual degradation of the lake can be observed from the study; however, assessing the importance, the need for conservation and proper management through the government and community participation is of utmost necessity.

1 Introduction

Wetlands are a critical component of the global ecosystem and provide numerous services that are essential to human well-being. In India, wetlands play a significant role in supporting the diverse ecosystems, agriculture, and fishing industries and provide important ecosystem services (ES) such as water storage, flood control, and carbon sequestration. Wetlands are areas of land that are saturated with water, either permanently or seasonally. These ecosystems play a crucial role in providing a range of ecosystem services to the people in India. Ecosystem service is a concept developed to support our understanding of human uses and management of natural resources. Ecosystem services refer to the benefits that human populations derive from natural ecosystems, such as clean water, climate regulation, and biodiversity conservation, which are fundamental for the well-being of all humans as well as maintenance and development of society on earth (Costanza et al., 1997). ES also implies quality life for mankind in terms of recreation or the cultural and inspirational values of natural wonders, which in turn improves cognitive thinking as well as emotional perspectives (West, 2015). As per the 2006 Millennium Ecosystem Assessment (MEA), ecosystem services are "the benefits people obtain from ecosystems." The MEA also delineated the four categories of ecosystem services. These services include provisioning services such as food, water, and medicine; regulating services such as air and water purification, soil and climate regulation, and pest control; cultural services such as recreation and aesthetic enjoyment; and supporting services such as water and nutrient cycling, floral growth, and soil formation (Tallis & Polasky, 2011). The wetland ecosystem services (WES) per hectare have been ranked first among all kinds of natural ecosystems (Meng & Dong, 2019). The value of WES is expected to be 47% of the total value of the global ecosystem (Xu et al., 2020), and yet, as per the Ramsar Convention reports, the conditions of wetlands worldwide are fast depleting, creating greater economic losses. While coastal or tidal wetlands are closely linked to the estuaries of West Bengal, inland or non-tidal wetlands are most common on floodplains along rivers and streams (riparian wetlands) all through the eastern states of India.

An estimation of ecosystem services is important for understanding the value of these services to humans and for making informed decisions about land use and resource management (De Groot et al., 2010). This would involve gathering data on the physical and ecological characteristics of the wetland and conducting studies to quantify the benefits that these services provide to the local community and the region as a whole. The whole process was summarized by Potschin and Haines-Young (2016) as (a) identification of related and significantly influencing environmental attributes, (b) the management practices and activities involved in that ecosystem; both of these may be facing environmental stress or pressures and hence need proper location-specific policy interventions to continue offering the services; (c) the natural as well as anthropogenic processes operative; and (d) finally the services it provides overcoming the environmental limitations to either in situ beneficiaries, remote benefits, or globally influencing services.

There are several methods for calculating the value of ecosystem services. Each of these methods has its strengths and limitations, and the choice of the method will depend on the specific ecosystem services being valued and the questions being asked (Tallis & Polasky, 2011). In order to ensure that the estimates of ecosystem services are robust and accurate, it is recommended to use a multi-disciplinary approach that integrates the expertise of biologists, economists, and social scientists. These approaches can help to address the challenges of data availability and quality, as well as the complexity of valuing ecosystem services that are non-market goods (Kaya et al., 2019). There are also software-based calculations of ecosystem services such as Artificial Intelligence for Ecosystem Service (ARIES) designed for integrated socioeconomic and environmental modeling, which can accommodate a range of different users and user needs, including scenarios, spatial assessment and economic valuation of ES, optimization of payments for ecosystem services programs, and spatial policy planning (Villa et al., 2014); Integrated Valuation of Ecosystem Service and Tradeoffs 3.4.2 (InVEST) for mapping and quantifying ES in biophysical or economic terms under different scenarios (e.g., policy or management option) (Sharp et al., 2014; Hamel et al., 2020; Cong et al., 2020). Toolkit for Ecosystem Service Site-based Assessment v.2.0 (TESSA) generates information that can be used to influence decision-making. It does not require computer modeling, but it does require stakeholder participation and encourages primary data collection using the methods provided through guidance and low-cost methods to evaluate the benefits people receive from nature at particular sites (Peh et al., 2013); and Protected Areas Benefits Assessment Tool (PA-BAT) is a rapid, workshop-driven, and standardized assessment of different stakeholders' perceptions about ES benefits from protected and other areas but requires modeling or other computer skills (Neugarten et al., 2018; Ivanić et al., 2020).

Some common methods used till date to quantify the ecosystem services of wetlands include:

1. Ecological footprint analysis: This method calculates the area of land required to produce the resources and absorb the waste generated by human activities. The total land required is divided by the total area of the wetland to determine the ecological footprint of the wetland (Daily, 1997).
2. Value transfer: This method uses the values of ecosystem services from one wetland to estimate the values of the same ecosystem services in a different wetland (Bateman et al., 2002).
3. Benefit transfer approach: This method uses data from studies on similar wetlands to estimate the values of the ecosystem services provided by a wetland (Adamowicz et al., 1998). This involves using data from similar studies

conducted in other regions to estimate the value of an ecosystem service in a new location. This method can be useful when conducting a full-scale valuation study is not feasible or when data is limited. Additionally, the BT approach can provide valuable information for decision-making processes, such as evaluating the costs and benefits of a particular land use change or development project (Loomis & White, 1996). However, the BT approach also has some limitations. The results of the BT approach may not accurately reflect the true value of an ecosystem in the new context, as the environmental and economic conditions may differ from those in the original study (Loomis & White, 1996). Additionally, the results of the BT approach may not be representative of the preferences and values of individuals in the new context, as the sample population and survey design may differ from those in the original study (Loomis & White, 1996).

4. Choice experiments: Natural ecosystems provide not only services that have value in the market but also non-market-value benefits, and for measuring these economic values, the revealed and stated preference techniques, which measure the increase or decrease in the utility or economic value of environmental changes for individuals, are used. This method, just like benefit transfer method, is the choice modeling approach, which involves presenting individuals with a series of hypothetical choices and using their responses to estimate the values of the services provided by the ecosystem (Kaya et al., 2019). A choice experiment is a survey method that involves asking people to state their preference for hypothetical alternative scenarios, goods, or services, which are combinations of attribute levels generated by the experimental design. Each alternative "good" is described by several attributes in terms of different attribute levels (Petcharat et al., 2020). It includes surveying individuals about their willingness to pay for specific ecosystem services (Loomis & White, 1996).
5. Direct use values: This method calculates the direct use values of ecosystem services based on the direct benefits that people receive from them, such as the value of fish caught from a lake or the value of timber harvested from a forest, recreation, water supply, etc. (Daily, 1997; Dudley et al., 2008). It determines the market price of the services or goods provided by an ecosystem, such as the value of fish caught from a lake or the value of timber harvested from a forest, and thus, the value of the ecosystem service is calculated as the market value of the goods or services minus the costs of harvesting or providing those services (Costanza et al., 1997). However, this method has some limitations. For example, it does not take into account the indirect benefits of an ecosystem service, such as the value of water purification provided by a wetland, or the value of carbon sequestration provided by a forest.
6. Indirect use values: This method is based on the indirect benefits that people receive from them, such as the benefits derived from the provision of habitat for wildlife, water purification, and carbon sequestration (MEA, 2005), and focuses on the indirect benefits that people receive from these services (De Groot et al., 2002). To calculate the indirect use value of an ecosystem service, one can use market prices for substitute goods or services or use revealed preference

methods, such as willingness-to-pay surveys, to estimate the value of the service to people. But it may be difficult to quantify aesthetic and recreational services as well as to find appropriate substitute goods or services to use as the basis for market prices, and the estimates may be subject to significant uncertainty or may not accurately reflect the value of the ecosystem service in question (De Groot et al., 2002).

7. Option value method: This method calculates the value of ecosystem services based on the value that people place on the possibility of using these services in the future (Daily, 1997). For example, the value of a forest as a source of future timber or the value of a wetland as a future source of water (Costanza et al., 1997; Hanley et al., 2000). This method is based on the idea that people may value the possibility of using an ecosystem service in the future, even if they do not use it currently. For example, a person may value the option of fishing in a river in the future, even if they do not fish there now, and economists often use the contingent valuation method, which asks people directly how much they would be willing to pay to preserve the option of using the service in the future (Hanley et al., 2000). However, the Option Value method may be subject to biases, such as hypothetical bias, which occurs when people give different answers to survey questions depending on the context in which they are asked (Adamowicz et al., 1994). Additionally, it can be difficult to accurately predict how people will value the option of using an ecosystem service in the future (Hanley et al., 2000).
8. Passive use value method: This method calculates the value that people place on simply knowing that these services exist, even if they never directly use those (Daily, 1997). To calculate the passive use value of an ecosystem service, economists often use the stated preference method, which asks people directly how much they value knowing that the ecosystem exists, but stated preference surveys may be subject to biases, such as hypothetical bias, which occurs when people give different answers to survey questions depending on the context in which they are asked (Aylward & Albin, 2006). Additionally, people may not accurately report their values for passive use, as they may not be aware of the benefits that an ecosystem provides or may not fully understand the concept of passive use.
9. Total economic value method: This method combines the values obtained from the direct use, indirect use, option, and passive use value methods to arrive at a total estimate of the value of ecosystem services (Costanza et al., 1997). The Total Economic Value (TEV) method, a comprehensive approach to calculating the value of ecosystem services, can be particularly useful in decision-making processes, such as when evaluating the costs and benefits of a particular land use change or development project. The strength of the TEV method is that it provides a comprehensive estimate of the value of ecosystem services, taking into account all of the different types of values; however, it can be complex and time-consuming, as it requires the use of multiple methods and data.

10. Contingent valuation method (CVM): This most commonly used method involves surveying individuals to determine their willingness to pay for a particular ecosystem service, and the results of these surveys can then be used to estimate the total value of the service to the wider population (Mitchell & Carson, 1989). The CVM is considered a useful tool for estimating the value of ecosystem services because it can capture both the direct use values and the non-use values, such as the option and passive use values, of an ecosystem. Additionally, the CV method can provide valuable information for decision-making processes, such as evaluating the costs and benefits of a particular land use change or development project (Mitchell & Carson, 1989). However, the CV method also has some limitations. Care must be taken so as not to be influenced by factors such as survey design and sample selection, which can impact the validity of the results.

The use of biophysical models to estimate the provision of ecosystem services has also become more common in India. For example, the Integrated Assessment Model (IAM) method involves creating a mathematical model of an ecosystem and its components, such as the flow of energy, nutrients, and water, and the interactions between different species (de Groot et al., 2002). However, it requires a high level of technical expertise and specialized software, which may not be available in all cases and may not accurately reflect the true value of ecosystem services, as they may not account for all the direct and indirect benefits that an ecosystem provides. Others, like Wetland Ecosystem Service Assessment Model (WESAM) and Soil and Water Assessment Tool (SWAT), have been used to estimate the contribution of wetlands to water provision and flood control. The use of economic valuation techniques, such as input-output analysis (IOA), involves analyzing the flow of goods and services by quantifying the interdependencies between different economic sectors and the ecosystem services in an economy to estimate the contribution of a particular sector, such as agriculture, to the overall economy (Rao et al., 2014).

There is no one specific formula for calculating the ecosystem services of a wetland as the method of estimation can vary depending on the particular wetland and the specific ecosystem services being analyzed and will depend on its unique physical and biological characteristics. It is also important to note that when calculating the ecosystem services of a pond, it is necessary to consider the context in which the pond is located and the specific characteristics of the ecosystem. Hence the estimations may be based on perceptions and rankings of the beneficiaries or may be quantified based on economic or monetary terms. Economic valuation (EV) of wetland ES facilitates efficient management and sustainable use of environmental resources by assigning quantitative values to wetland goods and ES in the absence of any market price for the same with the willingness to pay model or perceived benefit model (Mitra et al., 2022). Additionally, the values of ecosystem services may change over time as a result of changes in the pond's physical and biological conditions or as a result of changes in the way the ecosystem is used by people. It is also important to recognize that the calculation of ecosystem services is just one step towards management and conservation of these important ecosystems. Once the values of the ecosystem services have been estimated, it is necessary to consider the trade-offs between different uses of the pond and to make informed decisions about how best to manage and conserve the ecosystem to ensure that its services are maintained into the future.

2 Objectives and Methodology

This chapter is based on the methodologies of estimation of ecosystem services of water bodies, especially of the wetlands in the Lower Ganga Delta (LGD) region. The popular methods have been evaluated to understand the most suitable assessment technique in estimating the valuation of the services specific to the wetlands of this region. For this, a case study area has been selected, the Kankona Baorh and its adjacent canal Ratna Khal. It also evaluated why the benefit analysis is more appropriate rather than having just an economic approach. Exploring beyond traditional methods, due to a lack of reliable assessments of practical changes to services, an undue focus on monetary valuation and insufficient integration of social and policy research is proposed to be eliminated.

The work was conducted by structured questionnaire surveys of the local stakeholders surrounding the water bodies and semi-structured site-based expert interviews using snowball sampling method. There were 38 local experts who were either direct or indirect beneficiaries of the Baorh and Khal, selected based on their familiarity with the area and adequate background on ecological aspects aged 20–57 years residing within the locality of the wetland. The experts ranged from people using the water, be it domestic, aquaculture, or irrigation, to people viewing it for recreation and tourism, ecologists, climatologist, as well as government officials dealing with land and revenue records.

The ecosystem service assessment was computed by the choice-based experiment method discussed formerly through the Matrix Model analysis based on ranked preferences of the beneficiaries (Müller et al., 2016). One of the earliest choice experiments was conducted by Louviere and Hensher (1983) who examined the effect of attributes and ticket price on attendance at a proposed bicentennial international exposition in Australia (Snowball, 2008). It is the most practicable approach that can be used to present a more holistic overview of ecosystem services aspects in this region and allows for the estimation of the comparative prominence of multiple environmental attributes and their levels in the wetlands. This can be applied in terms of policy, institutional, and local participation in sustainable management. The basic idea of the assessment was to point out a strategic analysis to evaluate the ecosystem services in spatial manner and which of the attributes were most considered most significant by the beneficiaries. Twenty-eight sub-services or attributes were evaluated by using a six-point Likert-type scale that were scored with the ranks based on expert responses as follows: “no relevant capacity” (0), “low relevant capacity” (1), “relevant capacity” (2), “medium relevant capacity” (3), “high relevant capacity” (4), and “very high relevant capacity” (5). To understand the hydro-geomorphic character of the wetland, geospatial mapping techniques were computed. From the satellite images, Landsat 5 image dated 24.12.1987 and Landsat 8 dated 1.11.2023, the normalized difference water index (NDWI) proposed by McFeeters (1996) by the algorithm (GREEN band – NIR band)/(GREEN band + NIR band) is mapped. NDWI is used to highlight open water features in a satellite image, to detect and monitor slight changes in water content of the water bodies and allowing a water body to "stand out" against the soil and vegetation. Multidated images were used to access the changing area covered by the Kankona Baorh.

3 Assessing the Character of Kankona Baorh Wetland

The North 24 Parganas district is part of the LGD region of West Bengal, flanked by the Hugli River in the west and the Ichhamati River in the east that borders the countries of India and Bangladesh. It being the most densely populated district in the state of West Bengal, and with rapid urbanizing trends, most natural land covers have been anthropogenically replaced by the use for settlement and livelihood options. An investigation on the wetlands or as locally called *Baorhs* in North 24 Parganas district highlights numerous scattered paleo-channels, swamps, and low-lying areas in between the tidal rivers of Hugli and Ichhamati, which are intermittently connected by the northern-most cross-channel of the LGD named Jamuna. Presently, the majority of land has been converted from wetland areas to agricultural and settlement plots or even into aquacultural farms, leaving few freshwater pools to be utilized for artificial groundwater recharge and rainwater harvesting. This study has been conducted in the surrounding area with special focus on the Kankona Baorh, a lacustrine (oxbow lake) wetland, and its surroundings, which also include the Ratna Khal, in Swarupnagar C.D. Block (Fig. 3.1). Lacustrine wetlands are swamps with lakes or reservoirs with shallow depths and growth of aquatic vegetation. These types of wetlands are inhabited by delicate aquatic flora and fauna. To the villagers of Khatura, Haidadpur, Tepul, Parui, and Deora, the lake and the channel are a boon, as many lives are dependent on this lake. The main aim of this study is to assess the ecosystem services of wetlands with reference to the trends of land use, land cover, and environmental changes in the floodplains of Ichhamati River in North 24 Parganas.

The meander geometry of the river systems of Bhagirathi and the Ichhamati in the LGD, greatly affected by the monsoonal floods and tidal events, is so complex and intricate that it was described as an "abnormality" by Hirst (1915). In this area, the river, paleo channel, and oxbow lake are spread like a network. Among them, many paleo channels are totally degraded over time. The Jamuna divides the deltaic West Bengal into two unequal parts, and the area lying south of this east-flowing river may be described as the Southern Deltaic West Bengal (SDWB). The moribund course of the Jamuna is now disconnected from its feeder near Kalyani and flows eastward for about 66 km in an extremely sinuous course to join the Ichhamati near Charghat of North 24 Parganas (Rudra, 2014). The highly populated Jamuna River basin and its

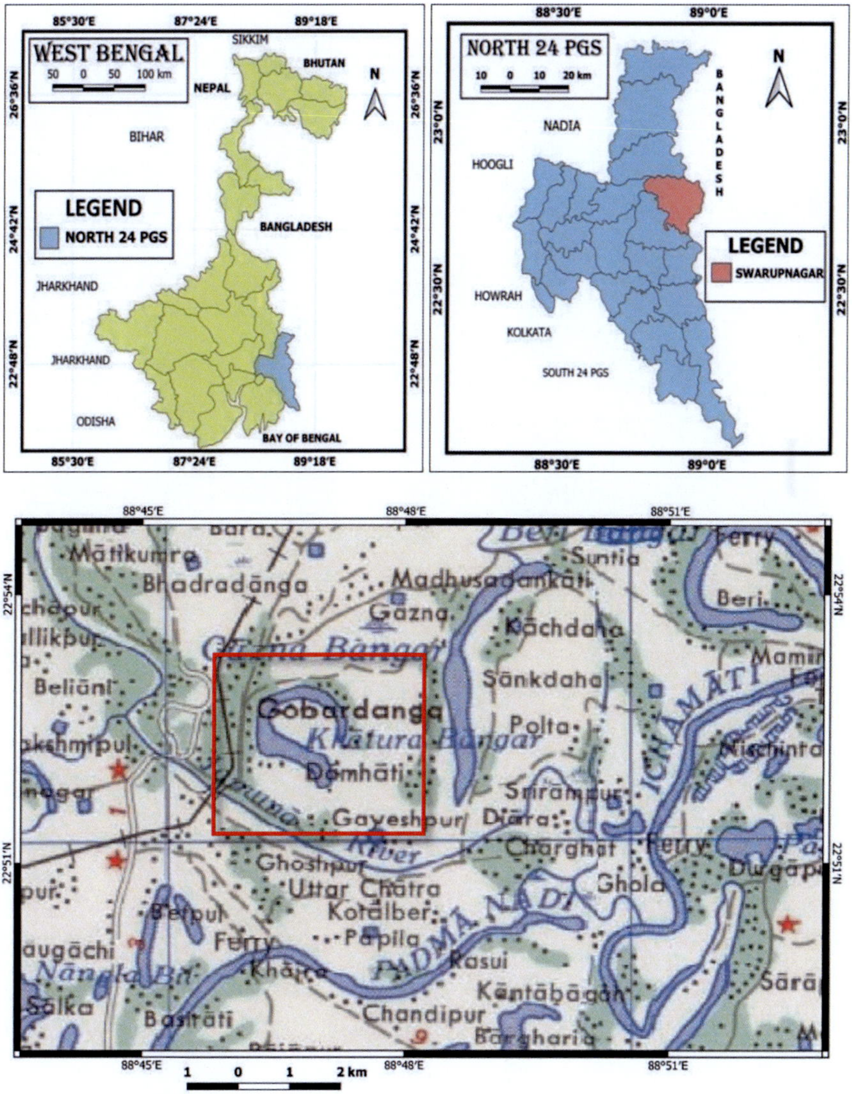

Fig. 3.1 Location map

surroundings are characterized by gentle gradient flat topography with various dominant geomorphic features like river floodplain, wetlands, meander cut offs, meander scars, swamps, old and new natural levees etc. (Sarkar et al., 2023). The Kankona Baorh, an oxbow lake, is famous among the few healthy wetlands still present, but like most wetlands in this region, shows a decreasing trend in its surface

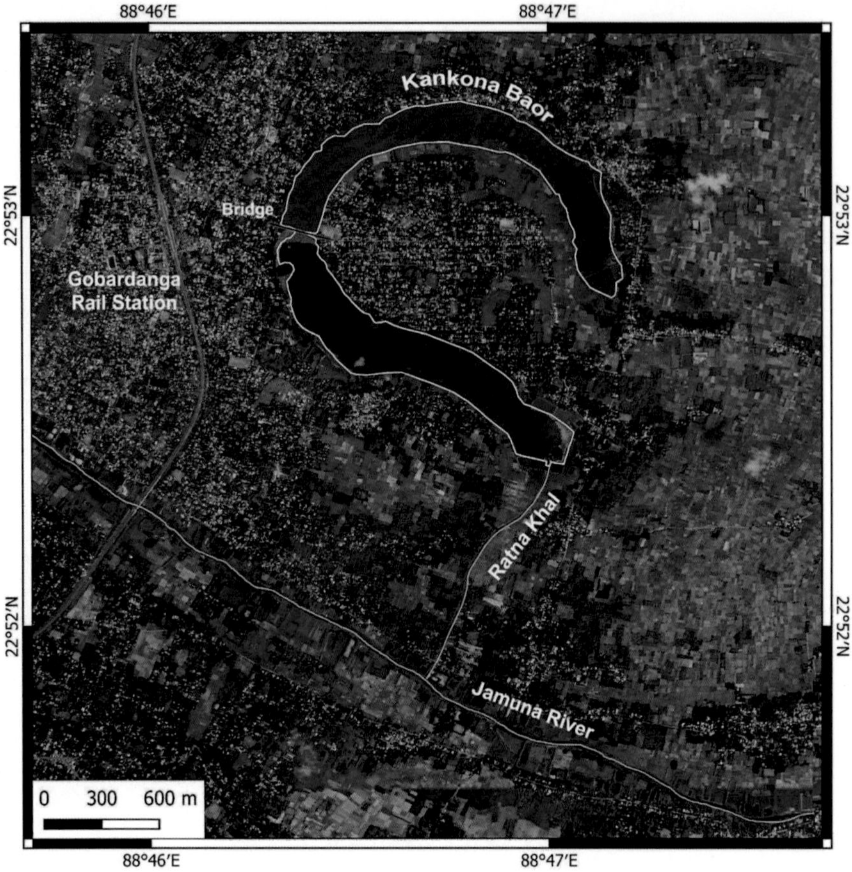

Fig. 3.1 (continued)

extensions (Fig. 3.2). Morphologically, the "inland wetland" is an abandoned channel (geographically termed oxbow-lake due to its crescent shape) of the river Jamuna on its right bank. This alluvial plain wetland is considered "Permanent Freshwater Lake." The left arm of this "open type" lake is connected with Ratna Khal, and the Ratna Khal is connected with the Jamuna River near Gobardanga Burning Ghat, which flows to join the Ichhamati River, that forms the south-western boundary between India and Bangladesh. Kankona Baorh (locally known as Mediya Baorh) is situated on the tract of the Bengal deltaic region, at the Swarupnagar Block in North 24 Parganas districts of West Bengal (22°52′38″N, 88°46′57.33″E and 22° 52′02.12″N, 88°46′50.58″E). Covering a total water area of 967,711 sq.m. in 2021, reduced to 929,088 sq.m in 2024, Kankona Baorh is an open-mouthed bangle-shaped water body that gives it such a name. The flood frequency is more or less a 5 year interval, as responded by the local people.

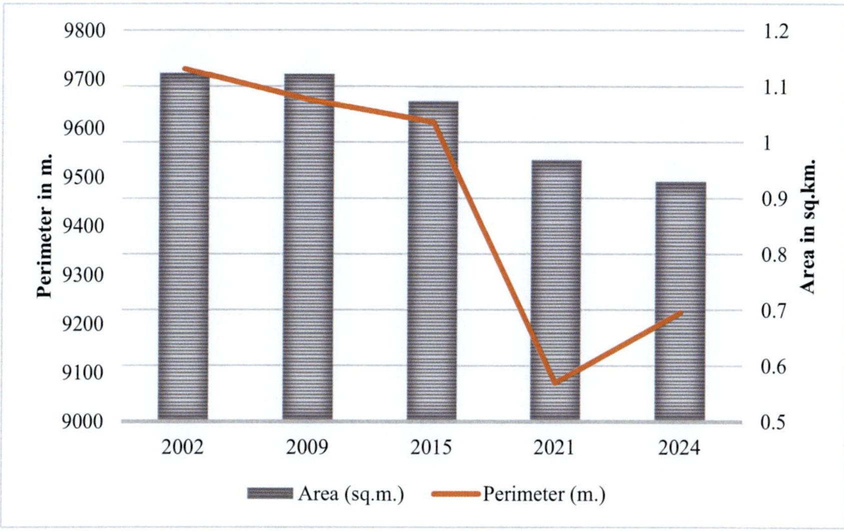

Fig. 3.2 Measurement of spatial-temporal changes of Kankona Baorh

The geomorphological character of the area, much like the rest of the LGD region, has a high moisture index as calculated by Normalized Difference Water Index (NDWI) but shows a decreasing trend. The NDWI makes to possible to map the areas of the water-logged locations. The NDWI values may be interpreted using the following ranges: 0.2 to 1 as water surfaces, 0.0 to 0.2 as waterlogged humid conditions, −0.3 to 0.0 as moderate drought, non-aqueous surfaces, −1 to −0.3 as non-aqueous surfaces (EOS, 2024). From the following the two maps (Fig. 3.3) it can be deduced that owing to the high sedimentations and degradation of the wetlands due to their highly seasonal nature now, the NDWI values are depreciating.

4 Evaluating the ES of Kankona Baorh and Ratna Khal

Climate change is a major determinant of wetland ecosystem services, as it can affect the hydrological regime, vegetation cover, and habitat conditions of wetlands (Davidson et al., 2018). The hydrological regime of Kankona Baorh, like most wetlands, is a key factor of the ecosystem services, as it influences water availability, water quality, and habitat conditions for aquatic and terrestrial species. During periods of heavy flooding, they are useful to mitigate flood and to trap suspended solids and sequester numerous nutrients (Bhattacharya, 2014). They retain water during dry periods, thus keeping the water table high and relatively stable. Vegetation cover in wetlands plays an important role in maintaining ecosystem services, such as carbon sequestration, nutrient cycling, and habitat provision for wildlife (Liu et al., 2018). The ecological health of the lake depends upon its biodiversity richness and nutrient cycling (Mitsch & Gosselink, 2015).

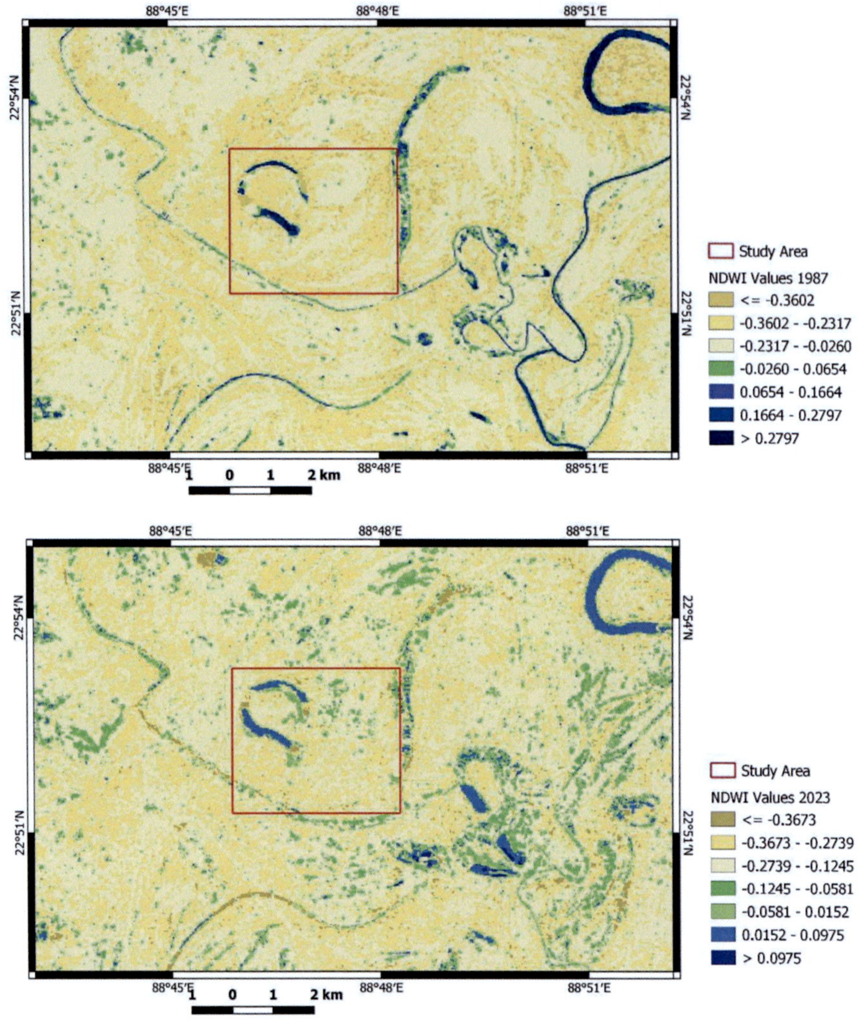

Fig. 3.3 NDWI Map 1987 and 2023

There are two canals generating from the horns of this oxbow lake, the Ratna Khal and the Tepul canal. The Tepul canal has almost degenerated, though during monsoons it diverts the excess water from the aquacultural farms and cultivable lands into the Baorh. The Ratna Khal supplies water from the Jamuna channel into the Baorh and vice-versa to keep the equilibrium conditions of the water bodies. While assessing the land use and management practices in the surrounding catchment area, it was observed that the water from the Kankona Baorh wetland was not majorly used for irrigation but runoff from the surrounding areas mixed with the water. Agricultural practices that involve the use of fertilizers and pesticides can lead to water pollution and eutrophication, reducing the capacity of wetlands to support

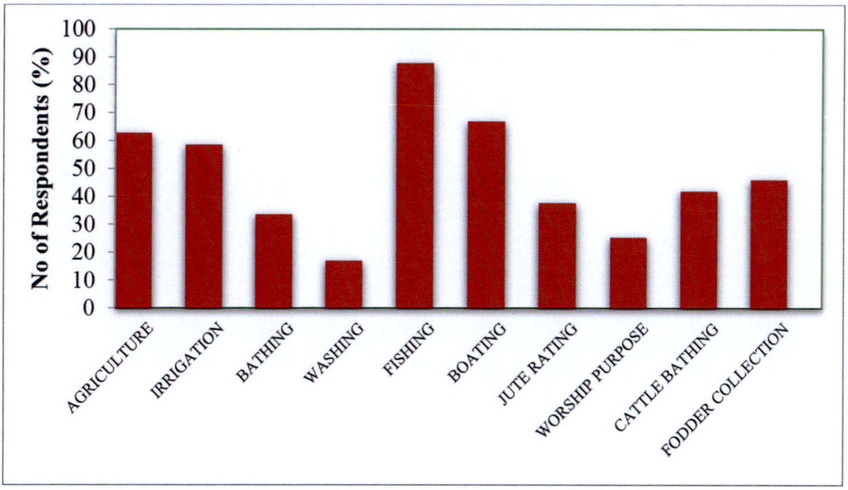

Fig. 3.4 Water use purposes of Kankona Baorh

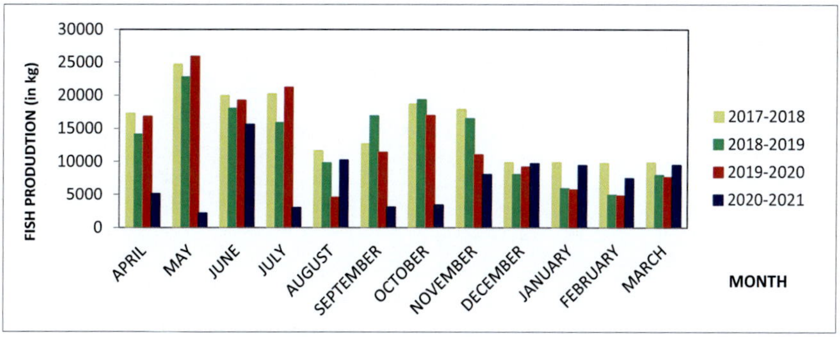

Fig. 3.5 Annual fish catch (2017–2021)

fisheries and other aquatic services (Mitra et al., 2018). Socio-economic factors, such as population growth, urbanization, and poverty, can also influence the provision of wetland ecosystem services (Islam & Haque, 2017). The wetland provides a number of provisional services, which are mostly domestic, while some are for livelihood, such as fishing and aquacultural farming, irrigation etc. and a few other cultural services are also present (Fig. 3.4).

Inhabitants of the Mediya village surrounded by the Kankona Baorh engage in aqua farming for a living. There is a local fishermen's association named "Mediya Government Colony Udbastu Matshyajibi Somobai Samiti." It was established in 1959 with 23 members but today stands with 235 members. Every year on annual fishing day (15th August), the association distributes fishing tickets to all those who are interested, whether or not involved with the Matsyajibi Somobai Samiti (Fig. 3.5). Two types of fishing methods are common here: (a) *Kochal* fishing and

Table 3.1 Species of "Raja and Rani Mach" in Kankona Baorh

Rani Mach (common names)	Scientific name	*Raja Mach* (common names)	Scientific name
Shol	*Channa striata*	Rui	*Labeo rohita*
Taki	*Channa punctate*	Catla	*Catla*
Shing	*Heteropneustes fossilis*	Silver carp	*Hypophthalmichthys molitrix*
Magur	*Clarias batrachus*	Grass carp	*Ctenophoryngodon idella*
Koi	*Anabas testudineus*	Mrigal	*Mylopharyngodon piceus*
Punti	*Puntius spp*	Black carp	*Cyprinus carpio*
Maya	*Amblypharyngodon mola*		
Kholisha	*Trichogaster app*		
Bele	*Glosspogobius giuris*		
Chingri	*Macrobrachium malcpolmsonii*		
Khoyra	*Gonialosa manmina*		
Baim	*Anguilla bengalensis*		
Chanda	*Chanda nama*		

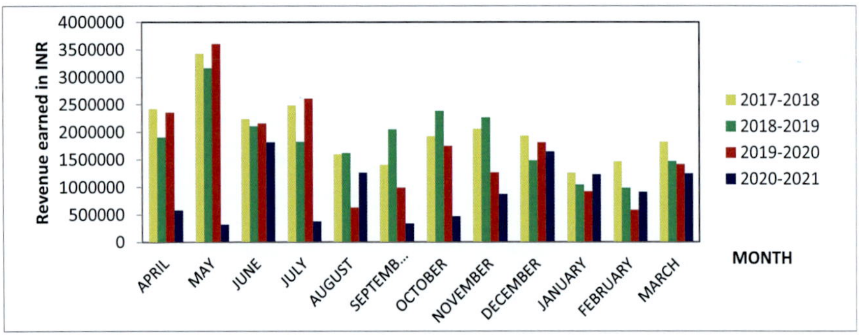

Fig. 3.6 Annual distribution of income from fishing (2017–21)

(b) *Komor* fishing. According to last 5 years production data and Mediya Government Colony Udbastu Matshyajibi Somobai Samiti information, it was found that six types of carp species and non-stocked fishes (locally known as "Raja Mach") were used to culture (fry released and then harvested when these are of the marketable size) in Kankona Baorh (Table 3.1). Among the other common fish varieties locally known as "Rani Mach," the highest production is Khayra (approx. 76%). Generally, Raja Mach is harvested twice a year (November–December and May–June), and Rani Mach was harvested all around the year from Kankona Baorh. The fishing season is at its peak from May to July every year, where most of the earnings are made (Fig. 3.6). Fishing dates are fixed by the Baorh authority. After fixing the date,

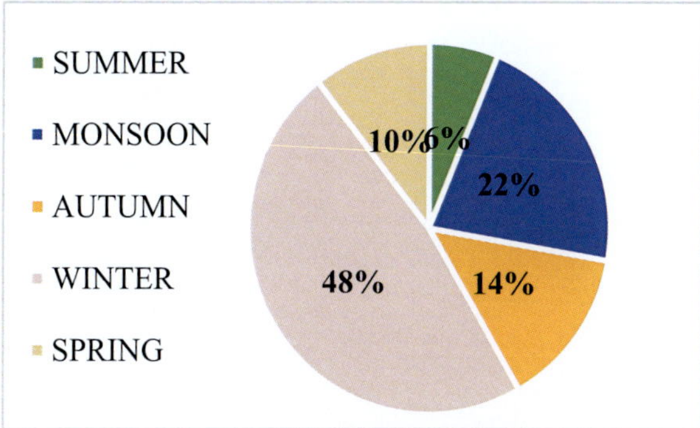

Fig. 3.7 Annual distribution of tourist

Fig. 3.8 Kankona Baorh photographed at 22°52′ 45.9952″N and 88°46′ 9.00584″E on 28.12.2023

a fishing schedule is prepared for fish harvesting, which is approved by the project director to prevent over exploitation of aquatic resources.

The area has been recently encouraging tourism, mainly day trips and for picnic purposes around the wetland that peaks during the winters when the migratory birds flock (Fig. 3.7). There is also a park named Kankona Binodon Park.

Other than ecotourism and fishing services, this lake provides many other ecosystem services like agriculture, irrigation, boating, jute retting, and cattle fodder supply that adds to the local livelihood, revenue diversity, and economic enhancement.

There is a bridge across the Baorh, established on 31st December, 1965, which causes an ecosystem fragmentation as the pillars collect sedimentation and waste. This bridge is the only way to cross this lake from one side to another side. With the growing water hyacinth (*Eichhornia crassipes*) concentrations, one of the floating aquatic weeds that is harmful for the living organisms in this lake, associated activities like boating, fishing, etc. are getting hampered, and also the utilizable water supply is degrading. The accumulation of water hyacinth has been found to be considerably more in the side of the bridge that connects the main land with the land chunk surrounded by the oxbow lake, which ultimately is increasing the sedimentation rate as well as initiating habitat fragmentation. Sedimentation from the dead and decomposed water hyacinths has been identified as the prime cause of the decrease in lake depth. The weed also disturbs solar light penetration into water bodies, reduces dissolved oxygen concentration through decomposition, alters the

Fig. 3.9 Ratna Khal sluice gate photographed at 22°51′ 30.7703″N and 88°46′ 28.3922″E on 28.12.2023

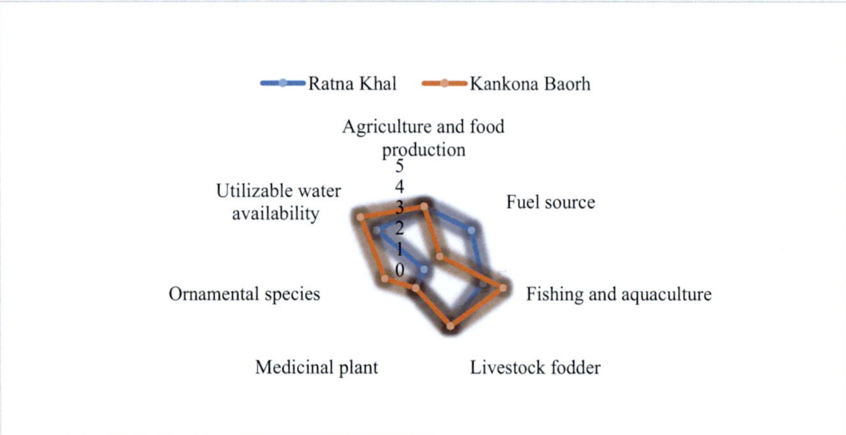

Fig. 3.10 Provisional ecosystem services

chemistry of water, and substantially increases the evapotranspiration of this wetland. Yet contrastingly, this bridge plays an important role as maximum tourists gather here for the scenic view of surroundings (Fig. 3.8). The lock gate at the junction of the oxbow lake and Ratna Khal canal regulates the flow of water as well as suspended solids or wastes (Fig. 3.9) whereby ensuring improved water quality both in the canal and the oxbow lake.

The utilization of the Kankona Baorh with respect to its ecosystem services is substantial because of its location, accessibility, sustaining of biodiversity, and vastness. The services received from Ratna Khal are higher only in a few cases, especially in jute retting, providing fuel and fodder, bioremediation, and for religious purposes as it is beside a very significant temple of the region due to its longer path that covers large agricultural and aquacultural sections. The ecotone area of the Baorh with the surrounding vegetated areas is also larger, leading to the diversity of flora and fauna. Human association with the water bodies in the area is also very high and prominent in the form of diverse activities, for example, domestic use, boating, fishing, bird watching, photography, picnics, and other recreational activities. However, responsible tourism, especially ecotourism, has not developed due to lack of

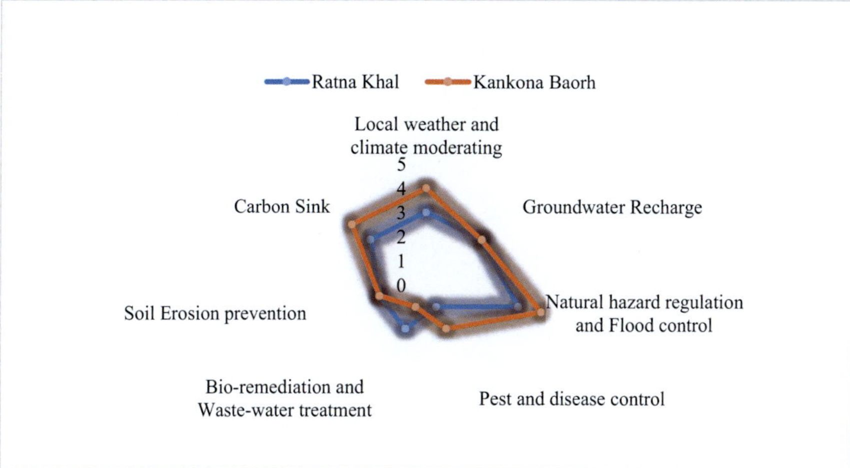

Fig. 3.11 Regulating ecosystem services

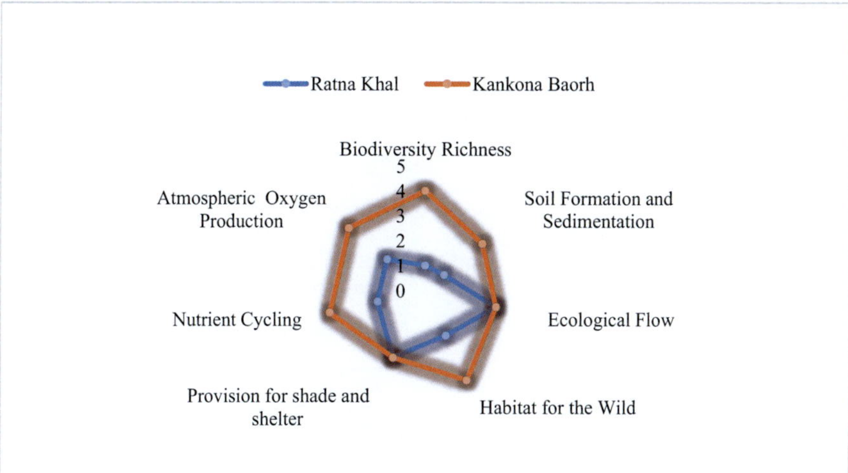

Fig. 3.12 Supporting ecosystem services

planning and investment, but the opportunities are plenty. The location of the Baorh is very advantageous for the development of an eco-tourism hub since the place is close to Gobardanga town, which is well connected by road and rail routes and can be reached within 2 h from Kolkata.

In the cartograms given below (Figs. 3.10, 3.11, 3.12, and 3.13) four types of ecosystem services derived from wetlands, based on matrix assessment, were mapped. This was done with an average of the perceptions of the direct and indirect local beneficiaries that were ranked out of five. Even though they may not always yield economic benefits, and more so in a district where water bodies take up a large

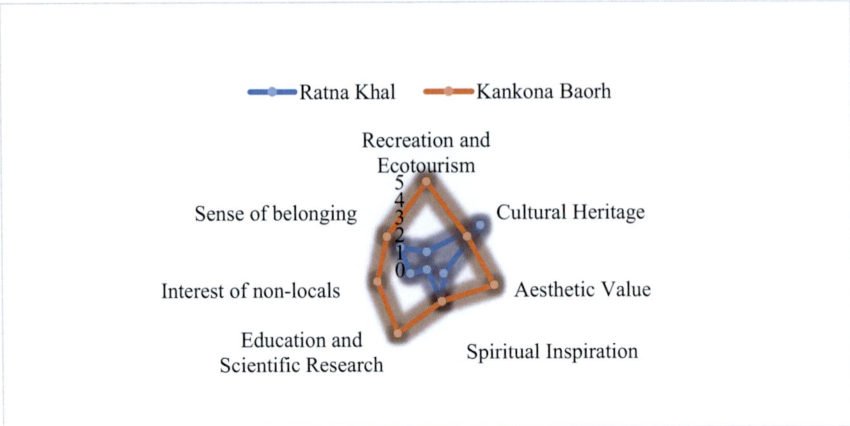

Fig. 3.13 Cultural ecosystem services

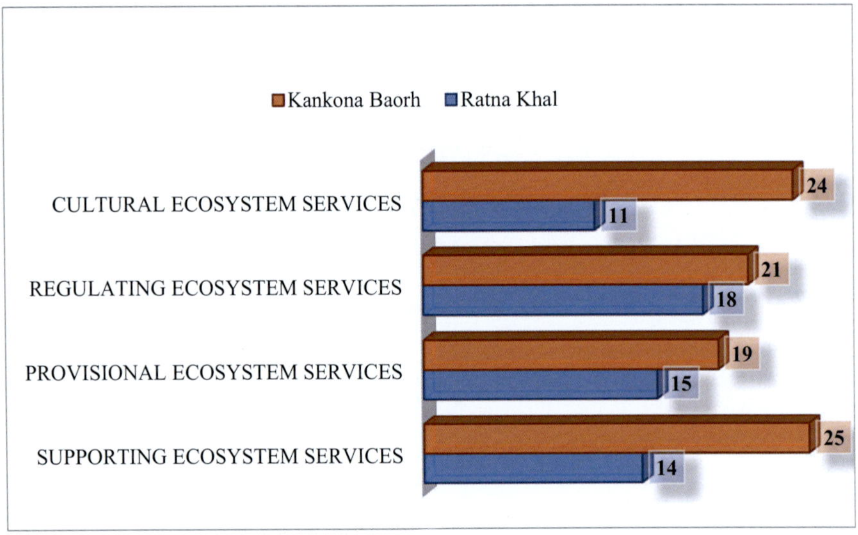

Fig. 3.14 Total ecosystem services

percentage of the total area, it can be inferred that the choice-based qualitative method prioritizes the need for conservation of wetlands.

Analyzing the provisional and regulating services, it can be observed that both Ratna Khal and Kankona Baorh facilitate the beneficiaries equally, whereas Kankona Baorh provides much greater services from the perspective of cultural and supporting aspects (Fig. 3.14). The Ratna Khal carries out its usual functions by providing more regulating and provisional services, much unlike the Kankona Baorh according to the locals; the highest ranked among them were fuel wood, fishing and aquaculture, flood control, etc. and showed no signatures of degeneration.

From the decision-making perspective or policy documentation, it can be inferred that supporting ecosystem services were most preferred to be developed, as ecological and environmental consideration seemed to be of highest concern among the learned respondents of the area around the Kankona Baorh. Local residents feel that wetlands do help in conserving the ecological diversity and health in this region. This was closely followed by the cultural services, as this ecosystem serves as the recreational hub, tourism destination, and cultural as well as religious linkages among the locals as well as people from other districts of West Bengal. While regulating services showed little significance for the Baorh, it had a much greater importance in case of Ratna Khal. It was interesting to observe that even though Kankona Baorh had the least relevance to generate significant livelihood opportunities in the form of provisional services, yet people desired to expand and develop the Baorh area in future.

5 Conclusion

This study highlights the simplicity of the choice experiment method in evaluating the significance of the ecosystem services provided, especially by that particular wetland. The parameters considered here are in sync with the local processes and uses of wetlands in the LGD regions; hence they can be standardized for evaluation of other wetlands in this region. However, it was observed that of the four types of ecosystem services, only the provisioning services can be expressed in terms of TEV as in the direct-use method of ES estimation. It is also inferred that, as actual valuation or pricing differs from location to location, service categories cannot be standardized on financial quotients. From the study, it can be concluded that the two types of wetlands assessed in this region show higher significance towards environmental concerns and development of a sociocultural attraction site. Thus, the beneficiaries emphasized government intervention or funding and providing impetus towards perceiving wetlands, not just for receiving any direct economic gains but for sustaining the ecology of the region and development of ecotourism to highlight the scenic beauty and natural biodiversity of the Kankona Baorh. Therefore, the authors propose the application of choice-based qualitative analysis and scoring of perceptions as the best possible way to assess the ecosystem service valuation to the stakeholders. This in turn would help in understanding the importance of wetlands, based on nonfinancial gains, whereby taking serious initiatives to conserve them.

References

Adamowicz, W. L., Louviere, J. J., & Williams, M. (1994). Combining stated and revealed preference methods for valuing environmental amenities. *Land Economics, 70*(2), 165–175.

Adamowicz, W. L., Louviere, J. J., & Williams, M. (1998). Combined direct and indirect approaches to non-market valuation with example applications to recreational fishing. *Journal of Environmental Economics and Management, 36*(3), 271–292.

Aylward, B., & Albin, J. (2006). Passive use values of the National Wildlife Refuge System. *Environmental Science & Policy, 9*(2), 125–135.

Bateman, I. J., Anderson, B., & Muir-Wood, R. (2002). Valuing the indirect use of wetland ecosystems. *Ecological Economics, 43*(1), 35–47.

Bhattacharya, P. (2014, December 15–16). *A study to investigate the importance of an Oxbow Lake in West Bengal, India.*

Cong, W., Sun, X., Guo, H., & Shan, R. (2020). Comparison of the SWAT and InVEST models to determine hydrological ecosystem service spatial patterns, priorities and trade-offs in a complex basin. *Ecological Indicators, 112*, 106089.

Costanza, R., d'Arge, R., de Groot, R., Farber, S., Grasso, M., Hannon, B., Limburg, K., Naeem, S., O'Neill, R. V., Paruelo, J., Raskin, R. G., Sutton, P., & van den Belt, M. (1997). The value of the world's ecosystem services and natural capital. *Nature, 387*(6630), 253–260.

Daily, G. C. (1997). *Nature's services: Societal dependence on natural ecosystems*. Island Press.

Davidson, N. C., Finlayson, M., van der Valk, A. G., & Zahawi, R. A. (2018). Climate change impacts on inland wetlands: A review of potential threats and mitigation options. *Wetlands, 38*(3), 361–383.

De Groot, R., Wilson, M. A., & Boumans, R. M. J. (2002). A typology for the classification, description and valuation of ecosystem functions, goods and services. *Ecological Economics, 41*(3), 393–408.

de Groot, R. S., Alkemade, R., Braat, L., Hein, L., & Willemen, L. (2010). Challenges in integrating the concept of ecosystem services and values in landscape planning, management and decision making. *Ecological Complexity, 6*, 453–462.

Dudley, N., Stolton, S., Sukhdev, P., & von Heland, M. (2008). *Wetlands: Water, life, and culture*. IUCN.

EOS. (2024). *Normalized difference water index (NDWI)*. https://eos.com/make-an-analysis/ndwi/#:~:text=The%20NDWI%20index%20was%20proposed,bodies%20in%20a%20satellite%20image

Hamel, P., Valencia, J., Schmitt, R., Shrestha, M., Piman, T., Sharp, R. P., et al. (2020). Modeling seasonal water yield for landscape management: Applications in Peru and Myanmar. *Journal of Environmental Management, 270*, 110792.

Hanley, N., Wright, R. E., & Adamowicz, W. L. (2000). *Valuing environmental goods: An assessment of the contingent valuation method*. Edward Elgar Publishing.

Hirst, F. C. (1915). *Report on the Nadia Rivers*. Reprinted by Gazetteer Department, Government of West Bengal, Kolkata, 3–61.

Islam, M. M., & Haque, M. A. (2017). Changes in wetland ecosystem services in Bangladesh and their valuation. *Ecosystem Services, 26*, 274–281.

Ivanić, K. Z., Stolton, S., Figureueroa Arango, C., & Dudley, N. (2020). *Protected Areas Benefits Assessment Tool+(PA-BAT+): A tool to assess local stakeholder perceptions of the flow of benefits from protected areas*. IUCN.

Kaya, Y., Burak, A. E., & Kibar, Y. (2019). Ecosystem services of wetlands: A review of recent developments. *Environmental Science and Policy, 92*, 123–134.

Liu, W., Chen, J., Zheng, H., Zhang, Q., & Ouyang, Z. (2018). Effects of vegetation cover on ecosystem services of wetlands: A case study in the Sanjiang Plain, China. *Ecological Indicators, 93*, 802–812.

Loomis, J., & White, D. (1996). Economic benefits transfer: A review. *Agricultural and Resource Economics Review, 25*(1), 49–59.

Louviere, J. J., & Hensher, D. A. (1983). Using discrete choice models with experimental design data to forecast consumer demand for a unique cultural event. *Journal of Consumer Research, 10*(3), 348–361.

McFeeters, S. K. (1996). The use of the Normalized Difference Water Index (NDWI) in the delineation of open water features. *International Journal of Remote Sensing, 17*(7), 1425–1432.

Meng, L., & Dong, J. (2019). LUCC and ecosystem service value assessment for wetlands: A case study in Nansi Lake, China. *Water, 11*(8), 1597.

Millennium Ecosystem Assessment (MEA). (2005). *Ecosystems and human well-being: Wetland ecosystem services*. Island Press.

Mitchell, R. C., & Carson, R. T. (1989). *Using surveys to value the benefits for public goods*. Resources for the Future.

Mitra, A., Banerjee, K., Saha, S., & Chowdhury, R. (2018). Ecological services of the Sundarbans: Opportunities for sustainable livelihood. *International Journal of Environmental Studies, 75*(1), 78–94.

Mitra, P., Haldar, A., & Banerjee, P. (2022). Public participation in restoration and sustainable use of wetland ecosystem services in India. In *Handbook of research on monitoring and evaluating the ecological health of wetlands* (pp. 280–303). IGI Global.

Mitsch, W. J., & Gosselink, J. G. (2015). *Wetlands*. Wiley.

Müller, F., Burkhard, B., Hou, Y., Kruse, M., Ma, L., & Wangai, P. (2016). Indicators for ecosystem services. In *Routledge handbook of ecosystem services* (pp. 157–169). Routledge.

Neugarten, R. A., Langhammer, P. F., Osipova, E., Bagstad, K. J., Bhagabati, N., Butchart, S. H., et al. (2018). *Tools for measuring, modelling, and valuing ecosystem services* (p. 70). IUCN, International Union for Conservation of Nature.

Peh, K. S. H., Balmford, A., Bradbury, R. B., Brown, C., Butchart, S. H., Hughes, F. M., et al. (2013). TESSA: A toolkit for rapid assessment of ecosystem services at sites of biodiversity conservation importance. *Ecosystem Services, 5*, 51–57.

Petcharat, A., Lee, Y., & Chang, J. B. (2020). Choice experiments for estimating the non-market value of ecosystem services in the Bang Kachao Green Area, Thailand. *Sustainability, 12*(18), 7637. https://doi.org/10.3390/su12187637

Potschin, M., & Haines-Young, R. (2016). Defining and measuring ecosystem services. In *Routledge handbook of ecosystem services* (1st ed., pp. 25–44). Routledge.

Rao, K. P., Vasudeva Rao, P., & Srinivas, K. (2014). *Input-output analysis of ecosystem services in Karnataka, India*.

Rudra, K. (2014). Changing river courses in the western part of the Ganga–Brahmaputra delta. *Geomorphology, 227*, 87–100. https://doi.org/10.1016/j.geomorph.2014.05.01

Sarkar, S., Bera, B., Bhattacharjee, S., & Shit, P. K. (2023). Large-scale stresses, decay of River Jamuna and resilience strategies in the Anthropocene. *Quaternary Science Advances, 12*, 100111.

Sharp, R., Tallis, H. T., Ricketts, T., Guerry, A. D., Wood, S. A., Chaplin-Kramer, R., et al. (2014). InVEST user's guide. *The Natural Capital Project*, 306.

Snowball, J. D. (2008). The choice experiment method and use. In *Measuring the value of culture: Methods and examples in cultural economics* (pp. 177–215). Springer.

Tallis, H., & Polasky, S. (2011). Natural capital and ecosystem services informing decisions: From promise to practice. *Annual Review of Environment and Resources, 36*, 6–23.

Villa, F., Bagstad, K. J., Voigt, B., Johnson, G. W., Portela, R., Honzák, M., & Batker, D. (2014). A methodology for adaptable and robust ecosystem services assessment. *PLoS One, 9*(3), e91001.

West, A. (2015). Ecosystem services. *Proceedings of the National Academy of Sciences, 112*(24), 7337–7338.

Xu, X., Chen, M., Yang, G., Jiang, B., & Zhang, J. (2020). Wetland ecosystem services research: A critical review. *Global Ecology and Conservation, 22*, e01027.

Chapter 4
Surface Water Vulnerability of the Urban Fringe Area in Kolkata Metropolitan Development Authority (KMDA), West Bengal, India

Debdip Bhattacharjee

Abstract Change in land use and land cover is an inevitable phenomenon in urban areas. The urban dynamic shows impacts on its fringes. Edge urban transformation is inextricably linked to the shrinkage of surface water bodies. The changes in surface water bodies as well as the vulnerability status in the urban fringes of the Kolkata Metropolitan Development Authority (KMDA), India region, have been the focus of this chapter. Basic approaches have been marked in this paper. In the first part, the author has prepared a comprehensive and reliable methodology to delineate the urban fringe area using digital image processing techniques based on the "fringe effect" theory in landscape ecology. The existing land cover information entropy model incorporating scale theory, the Shannon Diversity Index (SHDI), is significantly used to generalize the actual mixed pixels (i.e. fringe) in satellite image. The election of the optimal size of grid for the calculation of SHDI is selected using the following two algorithms: value stability and spatial heterogeneity preservation. Ultimately, the selection of an optimal threshold value to delineate the fringe area is marked by applying Otsu's thresholding method. The second part dealt with the changing surface water area during the last 30 years (1989–2019) in the urban fringe areas of KMDA. Water quality parameters such as turbidity, total dissolved solid, electronic conductivity, temperature, and pH have been examined through satellite image band ratio, and accuracy assessment is done by collecting sample water from different surface water bodies of the fringe area. The water collection follows the following steps like spot analysis and sample preservation to process it for further analytical computation. The enumerated analysis of water quality of the fringe portion of KMDA is found to be a great concern, and it is felt that there should be a separate administrative body to plan uniquely only for fringe areas to dwindle the effects of environmental disparities.

D. Bhattacharjee (✉)
Department of Geography, Sukanta Mahavidyalaya, Jalpaiguri, India

S. Hazra et al. (eds.), *Geographies of the Indian Subcontinent*,
https://doi.org/10.1007/978-3-031-74813-4_4

1 Introduction

The transitional portion of the landscape between urban and rural areas is termed as fringe. Fringe is depicted as mixture of both rural and urban landscapes. People who reside on the fringe can experience both urban and rural nature. Bird's eye view can separate the changing characters of land use and land covers of this dynamic environment. People seek to get all facilities of urban areas as well as rural nature of environment. Fringe is a good place to fulfill all desires. The land use and land cover of the fringe region are dynamic and sensitive because of the rapid use of these zones for the sake of new developmental activities. Detection of fringe regions is beneficial to delineate the area of interest when answering a wide range of environmental research questions related to the role of urbanization in climate, biogeochemistry, and hydrological cycles (Schneider et al., 2010). However, many attempts have been made to depict out the best method to extract the fringe area. This chapter will produce a strong methodology to extract the exact fringe area by applying Otsu's algorithm along with geospatial techniques.

Over the main portion of the twentieth century, development in Indian urban communities remained to a great extent restricted to city limits (Brush, 1968). Indeed, even as late as the 1970s, Shafer and Brush (1977: 64) noticed that in India, as diverged from the west, "a great part of the populace development and relocation to urban communities has been obliged by swarming an ever increasing number of individuals into existing urban zones as opposed to by extension of urban areas into rural areas and periphery regions." This example of development and populace retention has, nonetheless, changed fundamentally in the course of the last few decades, and the change has been the most honed in the biggest metropolitan urban areas. Littler metropolitan urban areas in India despite everything keep on encountering in-filling of settlements inside city limits, while the enormous metros for the most part show declining development in the center and keep on extending outward, immersing numerous towns and littler towns in the encompassing territory. The outward extension of the biggest metros has implied expanding and progressively complex communications with the encompassing provincial zones and steady changes in their property uses and occupations, changing them into semi-urban or "peri-urban" territories. Such regions have been concentrated before, especially as far as their monetary and social linkages with the city (Ramachandran, 1989). In an early investigation by Nangia (1976), the metropolitan area of impact was delineated based on streams of products and individuals. The positive parts of these streams and collaborations at the country's urban periphery are additionally caught in McGee's idea of the "desakotta" where the two districts gain, the rustic zones through expanded income and bigger markets, and the urban territories through reserve funds on lodging costs and less blockage in the developed regions (Ginsburg et al., 1991). It carries an ignorance of the techniques to recognize the periphery regions of enormous metropolitan zones like Kolkata, which is the eighth biggest urban agglomeration on the planet and third biggest in India, after Mumbai and Delhi (2011). Additionally, it is the biggest and the main metropolitan territory in eastern

India. It is almost certainly true that an enormous urban territory like this could never stop its dynamic development inside a specific limit, and that is the means by which it makes inquiries upon the ecological supportability. Since the 1990s, be that as it may, concerns have been raised about the conceivable negative effect of spreading urbanization, and this has originated from researchers taking a shot at the natural effect of the spreading city and its impacts on the peri-urban (Bentinck, 1996). These impacts are the aftereffect of the procedures of progress that such regions involved in spreading urbanization, and they remember changes for land use from farming to private and modern/business, changes in the utilization of common assets, for example, deforestation, water consumption and contamination, land leveling in light of inordinate quarrying, land corruption due to block ovens, and expanded strong and fluid squanders. By and by, the significance of recognizing the periphery district is commonly ignored (Jindrich, 2010). It is along these lines basic to have a target and reliable strategy to depict the urban periphery region to help with overseeing and relieving the potential unfriendly outcomes of this scene adjustment. Portraying the urban periphery zone stays an emotional procedure. In the early time, Hezber Louis (1936) proposed the idea of the "Urban periphery zone" ("Stadtrand zonen") in 1936. A few endeavors were accounted for in which an assortment of socio-practical parameters and measurements were received (Desai & Gupta, 1987; Friedberger, 2000; Hall et al., 2006; Lesage & Charles, 2008; Russwurm, 1975). The vast majority of the above endeavors depended primarily on evaluation and study information. While this information gives the greater part of the information on the financial condition, they don't take into consideration inclinations of qualities and experience the ill effects of spatial confuse between the working spots and homes for a great many people (Weeks, 2010). The possibility of a continuum proposes that urban and country territories are, truth be told, the parts of the bargains, as opposed to speaking to a division (Weeks, 2010). Remote detecting offers the chance of joining physical and morphological information with segment and financial information in one application (Banzhaf et al., 2009). Land spread change data speaks to the interface between biophysical conditions and anthropogenic impacts through time (Yang et al., 2013). Various examinations on this point utilizing remote detecting strategies have been accounted for. Zhang et al. (1999) indicated an edge of land spread data entropy (otherwise called Shannon Diversity Index) as the urban periphery. Wilson et al. (2004) likewise built up an urban development model that could be used to depict the urban periphery zone. Heavenly attendant et al. (2007) proposed a standard of the level of developed pixels to outline the urban periphery. Significant outcomes were gotten from these models. Be that as it may, the examination scale and depiction edge are resolved discretionarily. It is obvious, consequently, that the degree of reproducibility among various mappers and for various districts is restricted. As a crucial issue in geosciences, the impacts of scale are all around perceived as the essential property and "one of the most basic parts of any exploration" (Peterson & Parker, 1998; Quattrochi & Goodchild, 1997). This express accentuation on spatial heterogeneity requires the thought of the impact of scale in characterizing the spatial degree of urban periphery (Corry & Nassauer, 2005; Wu, 2013). Later on, Huang et al. (2016) has shown the importance of scale and significance of threshold values to delineate urban fringe areas in the Guangzhou-Foshan metropolitan area of China, which efficiently got approval to clip the transition between rural and urban zones.

Therefore, this study aims to establish a strong methodology to delineate the actual edge urban pixels of a satellite image to depict the real picture of urban fringe region. To go through this analysis, the author has taken a metropolitan unit, i.e., Kolkata Metropolitan Development Authority (KMDA) of India, as an area of interest. Moreover, this study has another focus on the water qualities of the depicted fringe pixels. The whole analysis has been carried out through digital image processing techniques followed by laboratory-based water quality analysis.

2 Objectives of the Study

The objective of this chapter is to study the quality of surface water body in urban fringe of KMDA. To fulfill this intension, the author has to go through four separate sections.

- Detection of fringe region of KMDA (2017)
- Extraction of surface water mask of the fringe region to show the spatial change for a period of 28 years (1989–2017)
- Study of fringe area's surface water quality status in 2017
- Accuracy assessment of the verified results through laboratory-based water quality samples.

3 Study Area

Calcutta Metropolitan District (CMD), which was along these lines known as Kolkata Metropolitan Area (KMA), was portrayed by the Calcutta Metropolitan Planning Area Control Act 1964, for urban foundation advancement is taken as territory of enthusiasm for this chapter. The Kolkata Metropolitan Area is a conurbation of urban communities, towns, and connecting provincial zones spread over almost 100 km along the two banks of the River Hooghly in the southern piece of the territory of West Bengal, India. The examination region falls between scopes 22°20′ N to 23°00′ N and longitudes 88°07′ E to 88°35′ E (Survey of India geological guide numbers 79B/1, 79B/2, 79B/3, 79B/5, 79B/6, 79B/7, 79B/9, 79B/10 and 79B/11). KMA establishes 41 legal towns, 80 enumeration towns, 16 outgrowths, one cantonment board, and 456 towns (according to plan distributed in Kolkata Gazette dated 26th June 2001 and according to Govt. Warning number 3543-UD/O/M/SB/S-4/98 dated 6.9.2001 and notice number 1820-40/UD/O/M/SB/S–2/97 dated 29.05.2003-Population Profile of KMA, 2001, KMDA) from 23 blocks of six areas—Kolkata, North and South 24 Parganas, Howrah, Hugli, and Nadia (mostly

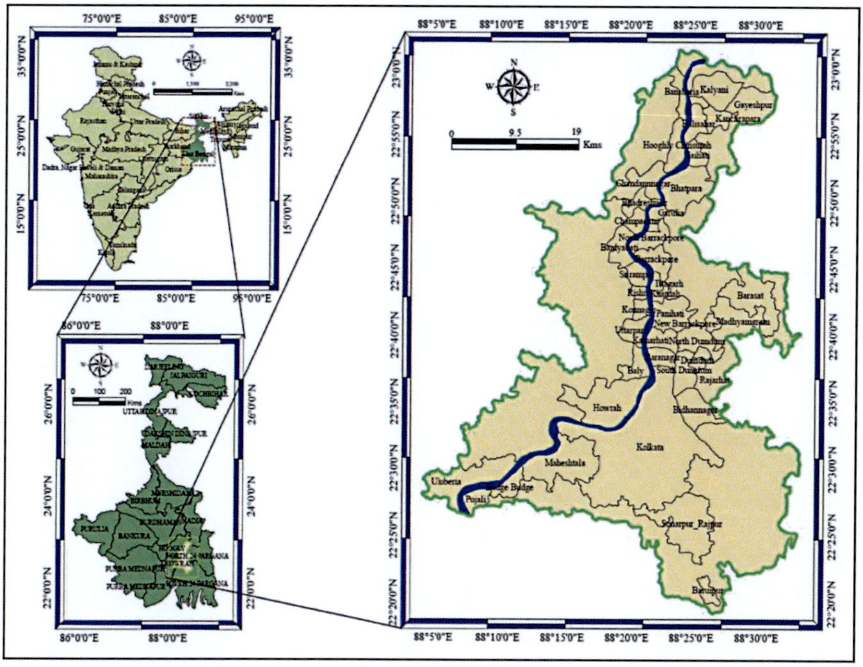

Fig. 4.1 Location map of Kolkata Metropolitan Development Authority (KMDA), West Bengal, India. (Source: Map Prepared by the author)

or completely included inside KMA) of West Bengal. The metropolitan region covers a region of 1819 km^2 (barring the zone of the River Hooghly) and falls under the ward of Kolkata Metropolitan Development Authority (KMDA). KMDA is the organization answerable for arranging, advancing, and building up the KMA. With a populace of around 16 million individuals, the thickness of populace of the metropolitan zone is more than 8700 people/km^2 (Census, 2011) (Fig. 4.1).

4 Data Used

Two scenes of predominantly cloud-free Landsat Thematic Mapper (TM) and Operational Land Imagery (OLI) of 1989 and 2017 were used for this study. Various image preprocessing techniques such as subset, atmospheric correction and radiometric correction were done. The TM image of 1989 was only used to retrieve the water mask of that time in fringe area and OLI image of 2017 was used in the actual fringe area as well as to perform various image band ratios for water quality mapping. Apart from this image analysis, this work is also related to laboratory-based water quality data. In total, 16 samples were collected from different parts of the urban fringe region. The maximum likelihood classifier was applied to classify the image and prepare the land use land cover map of the KMDA. There were five categories, i.e., urban, settlement with vegetation, open space, green space, and Water bodies. The training sets were taken from GPS survey and covered most of the study area. The overall accuracy of this classification achieved a level of 99.47% along with 99.18% of Khat statistic.

5 Methodology

5.1 *Land Cover Information Entropy Model to Analyze the "Fringe Effect"*

Journey through urban landscape toward rural is not always totally easy, as one should realize the change in land use and land cover properly. There is always a transition between two landscapes where a tourist can be confused in detecting proper urban or proper rural. Here comes the concept of ecotone, generally used to detect the transition zone between two biomes. An ecotone, a transition area between two biomes or different patches of the landscape (Senft, 2009), presents a tendency of increased biodiversity referred to as the "fringe effect" since it provides greater spatial and temporal variation in resources (Graves, 2016). The Shannon Diversity Index (SHDI) is significantly used to generalize the actual mixed pixels in satellite image. This index produces the information entropy values, which are widely applied in the detection of fringe regions with the application of remote sensing and GIS techniques (Eq. 4.1). This information entropy can be shown as a gradient of changing dynamics of land use and land cover pixels. Certain portion of gradient is clipped to get a view of real edge of urban region, which is called the fringe.

$$H = -\sum_{i=1}^{n} P_i \ln P_i \tag{4.1}$$

In Eq. (4.1): H is the pixel diversity index or information entropy value. P_i means that particular area classified as Land Use Land Cover (LULC) class i. It computes the diversity of LULC types based on two parameters: the number of different LULC types (separability) and the proportional area distribution among land cover types (specificity). As a general rule, the index (H) increases when the number of LULC classes increases and when the proportion of small classes approaches that of large classes in a given unit of measurement. One of the basic assumptions of this study is that a higher pixel diversity index (information entropy value), which corresponds to a diversified LULC status, designates an urban fringe, while a lower diversity index (information entropy value) denotes a perfect urban area or rural area.

5.2 Selection of Optimal Size of Grids for the Calculation of Shannon Diversity Index (SHDI)

Selection of an optimal scale and calculation of SHDI according to that applied to be the next step of this analysis. With reference to past literature, cells of different sizes (1 km^2, 5 km^2, and 10 km^2) are created to calculate SHDI. Then a transect profile was made to examine the gradient of SHDI in different cells or grids. This 50 km long west-east directional transect was made joining two extreme points of KMDA (East point: 88°7′40″ E and 22°31′12″ N to West point: 88°32′13″ E and 22°28′55″ N). This transect passes through urban, rural, and mixed pixels (fringe), and later the profile is used to detect the clear pictures of gradients, which helps to analyze the fringe regions of the study area.

The appropriate scale of grid is optimized from the following two algorithms: value stability and spatial heterogeneity preservation (Huang et al., 2016).

Value Stability

Stable values are optimized by computing the average information entropy pixels (variable x) of the cells on the transect profile, followed by a second-degree polynomial fit for all x (Eq. 4.2).

$$f(x) = ax_i^2 + bx_i + c \quad (i = 1, 2, 3) \tag{4.2}$$

When $f(x) = 0$, its corresponding stationary point of the curve is S_0 (x_0, y_0). In Eq. (4.3), variable S_i is scale (cell/grid size), which has three possible values: S_1, S_2, and S_3. This part is done to identify the most stable scale of cells/grids (S_a), which is most adjacent to stationary point S_0 (Eq. 4.3).

$$S_a = [S_1 / \min / S_i - S_0 /, \quad i = 1, 2, \text{and}\, 3] \tag{4.3}$$

Spatial Heterogeneity Preservation

Significant heterogeneity transect is an example of both convex and concave regions. As the transect crosses both rural and urban pixels and mixed pixels as well, it looks like a topsy turvy curve. Convex region indicates higher entropy value, which significantly designates the mixed pixels, i.e., fringe region. Concave portion of curve indicates evenness of entropy value, which ideally indicates either urban or rural region.

Further analysis of this chapter will be based on this optimal scale or cells/grids obtained from Eqs. (4.2) and (4.3).

5.3 *Selection of Optimal Threshold Value to Delineate the Fringe Area Applying Otsu Thresholding Method*

Otsu's thresholding method involves iterating through all the possible threshold values and calculating a measure of spread for the pixel levels each side of the threshold, i.e., the pixels that either fall in foreground or background. The aim is to find the threshold value where the sum of foreground and background spreads is at its minimum.

After getting the optimal scale or grid/cell, the author calculated the thresholding value using the SHDI image of the study area. The entire SHDI image was classified into six categories containing two series of data. Then, within class variance (σ_{w}^2) or threshold value was calculated, followed by the computations of weight (W_{b}), mean (μ_{b}), and variance (σ_{b}^2) of both series.

The Eqs. (4.4, 4.5, 4.6, and 4.7) used to solve these algorithms area:

$$\text{Weight } (W_{\mathrm{b}}) = (G_1 + G_2 + G_3)/\text{Total number grid of the image} \tag{4.4}$$

$$\begin{aligned} \text{Mean } (\mu_{\mathrm{b}}) = (0 \times G_1) + (1 \times G_2) \\ +(2 \times G_3)/\text{Total number of grid of the series} \end{aligned} \tag{4.5}$$

$$\begin{aligned} \text{Variance}\left(\sigma_{\mathrm{b}}^2\right) = & \left[\left\{(0-\mu_{\mathrm{b}})^2 \times G_1\right\} + \left\{(1-\mu_{\mathrm{b}})^2 \times G_2\right\} + \left\{(0-\mu_{\mathrm{b}})^2 \times G_1\right\}\right. \\ & \left. + \left\{(1-\mu_{\mathrm{b}})^2 \times G_3\right\}\right] /\text{Total number grid of the image} \end{aligned} \tag{4.6}$$

$$\text{Within class variance } \left(\sigma_{\mathrm{w}}^2\right) \text{ or threshold value: } W_{\mathrm{b}_1} \times \sigma_{\mathrm{b}_1}^2 + W_{\mathrm{b}_2} \times \sigma_{\mathrm{b}_2}^2 \tag{4.7}$$

whereas, G_1, G_2, and G_3 denote the number of grids of back ground and fore ground images.

5.4 *Extraction of Surface Water Masks*

After clipping the mixed pixels (fringe region) through Otsu's thresholding algorithms, the next steps were to examine the surface water qualities of this edge urban region. To extract the water masks of 1989 and 2017, Modified Normalized Difference Water Index (MNDWI) was used (Eq. 4.8). The values of MNDWI containing band ratio, which is greater than 0, denote water pixels. In both 1989 and 2017, values greater than 0 were extracted to prepare the water masks (Xu, 2006).

$$\text{MNDWI} = \frac{\text{Green} - \text{SWIR}}{\text{Green} + \text{SWIR}} \tag{4.8}$$

5.5 *Surface Water Quality Analysis*

The main objective of this paper was to examine the surface water quality status of fringe region of KMDA. In total, 16 water samples were collected from various parts of identified fringe regions of KMDA. Parameters that were tested in laboratory are: turbidity (NTU), electronic conductivity (μs), pH, total dissolved solid (ppt) and temperature (°C). Same parameters were also extracted from image band ratios of the entire fringe water mask of 2017 (Eqs. 4.9, 4.10, 4.11, 4.12, and 4.13) (Abdelmalik, 2018). The year of water quality collection and used OLI scene are both of 2017. The details of band ratios used in this analysis are:

$$\text{Turbidity} = \frac{\text{NIR} \times \text{Red}}{\text{Green}} \tag{4.9}$$

$$\text{EC} = \frac{\text{SWIR}}{\text{NIR}} \times \text{SWIR} \tag{4.10}$$

$$\text{pH} = \frac{\text{SWIR 1}}{\text{SWIR 2}} \tag{4.11}$$

$$\text{TDS} = \text{NIR} \tag{4.12}$$

$$\text{Temperature} = \frac{1}{2} \times \text{TIR} - [(\text{TIR} + \text{TIR})/\text{TIR}] \tag{4.13}$$

5.6 *Accuracy Assessment of the Water Quality Parameters Retrieved from Satellite Image Band Ratios*

For assessing the accuracy of the water quality parameters retrieved from satellite image band ratios, which are turbidity, electronic conductivity, pH, total dissolved solids, and temperature, the laboratory tested 16 sample data were used as reference (Tables 4.2, 4.3, 4.4, 4.5, and 4.6). Both the data sets were normalized to get values between 0 and 1; otherwise they could not be compared with each other (Eq. 4.14). The accuracy had been assessed in terms of Maximum Absolute Error (MAE), Mean Error (ME), and Root Mean Square Error (RMSE) and was computed using the following Eqs. (4.15), (4.16), and (4.17).

$$\frac{\text{Actual value} - \text{Minimum value}}{\text{Maximum value} - \text{Minimum value}} \tag{4.14}$$

$$\text{Absolute error} = \text{Laboratory tested value} - \text{Band ratio value} \tag{4.15}$$

Mean Absolute Error

$$= \sqrt{\sum_{i=0}^{n} \text{Laboratory tested value} - \text{Band ratio value}/n} \tag{4.16}$$

Root Mean Square Error

$$= \sqrt{\sum_{i=0}^{n} (\text{Laboratory tested value} - \text{Band ratio value})^2/n} \tag{4.17}$$

Where, n is the number of samples used.

6 Results

6.1 Classification of Land Use Land Cover and Shannon Diversity Index

Satellite image of KMDA was classified applying supervised classification algorithm and categorized into five classes, i.e., built-up, settlement with vegetation, water body, green space, and open space (Fig. 4.2). The next intention was to find out the land use and land cover diversity of the region. The LULC map is nothing but the association of ample number of polygons, each of which represents single or multiple landscapes. Urban core is generally the agglomeration of impervious pixels or concrete space. Lots of buildings, parking lots, and metal roads used to make this area impervious. Because of this reason, urban cores show minimum amount of landscape diversity, which could be analyzed by low amount of SHDI values. Rural area is also the integration of similar kinds of LULC pixels, which results in decreasing SHDI values. Fringe areas show higher values of SHDI. Here in this analysis, the author feels to show this SHDI diversity along transects, which are discussed later in this chapter. To show the SHDI values in grids, the LULC polygons were merged with three different-sized fishnets, i.e., 1 km^2, 5 km^2, and 10 km^2. Now, which grid size would produce the maximum interpretability? To get these answers, three different maps were prepared with varying grid sizes (Figs. 4.3, 4.4, and 4.5). Then five transects were made to get the best grid size for this analysis (Fig. 4.2).

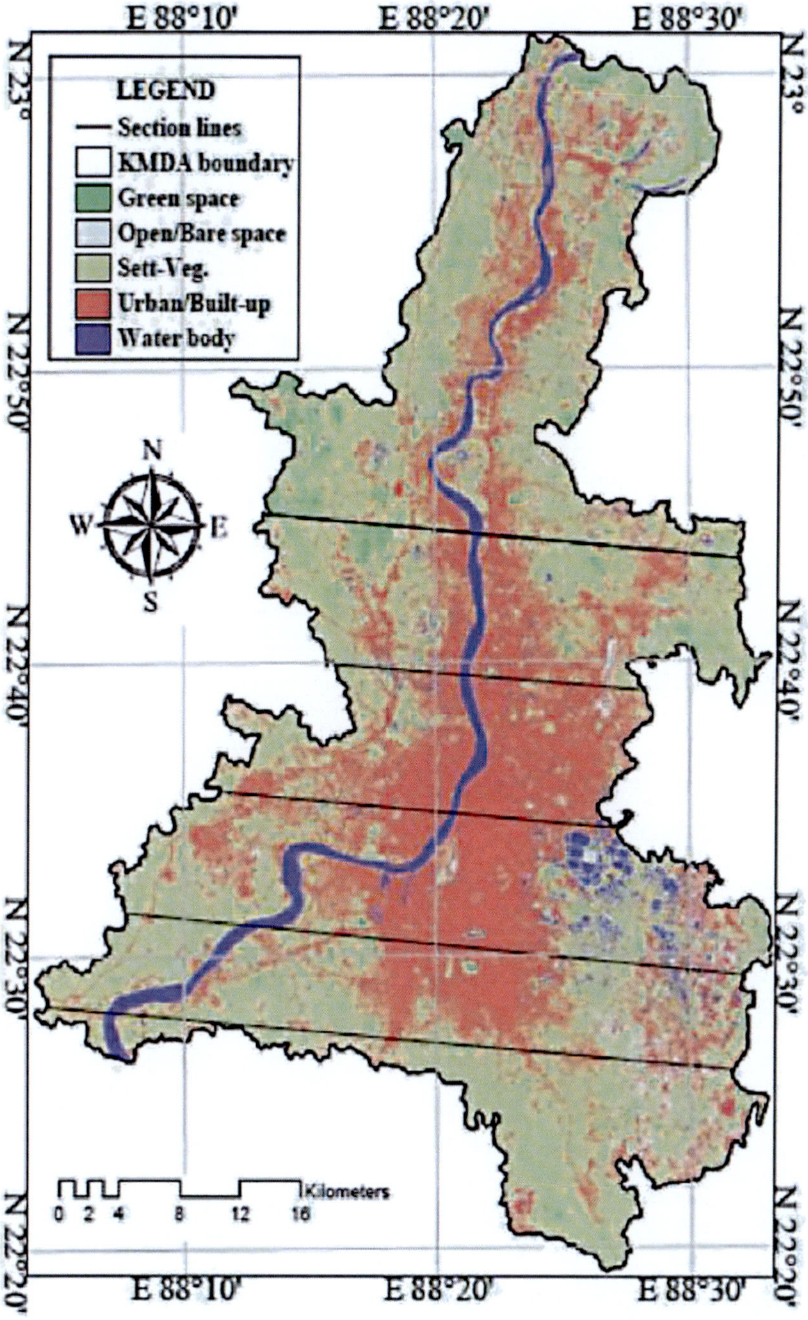

Fig. 4.2 LULC of KMDA (2017) and locations of five transects. (Map is prepared by the author using Arc GIS Professional environment)

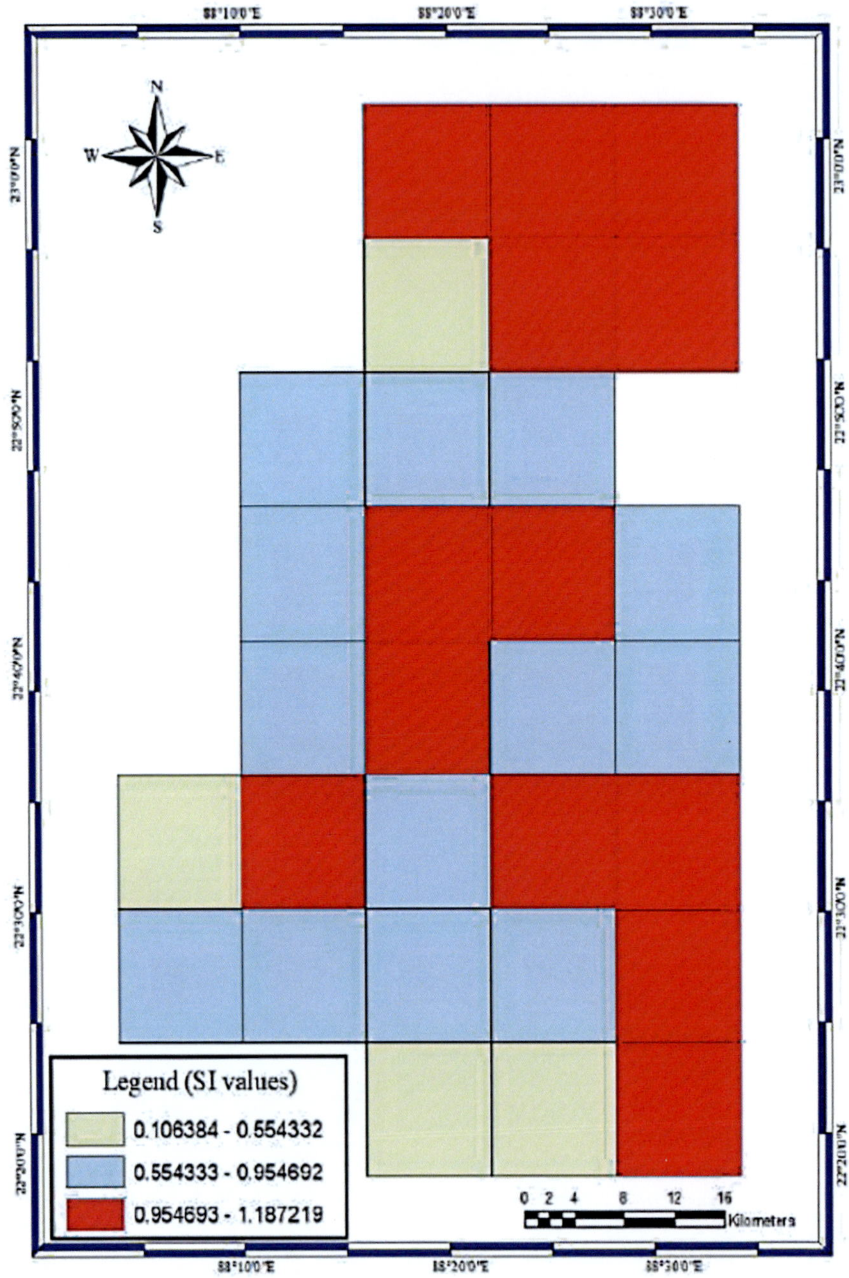

Fig. 4.3 SHDI image of KMDA (10,000 m grid). (Map is prepared by the author using Arc GIS Professional environment)

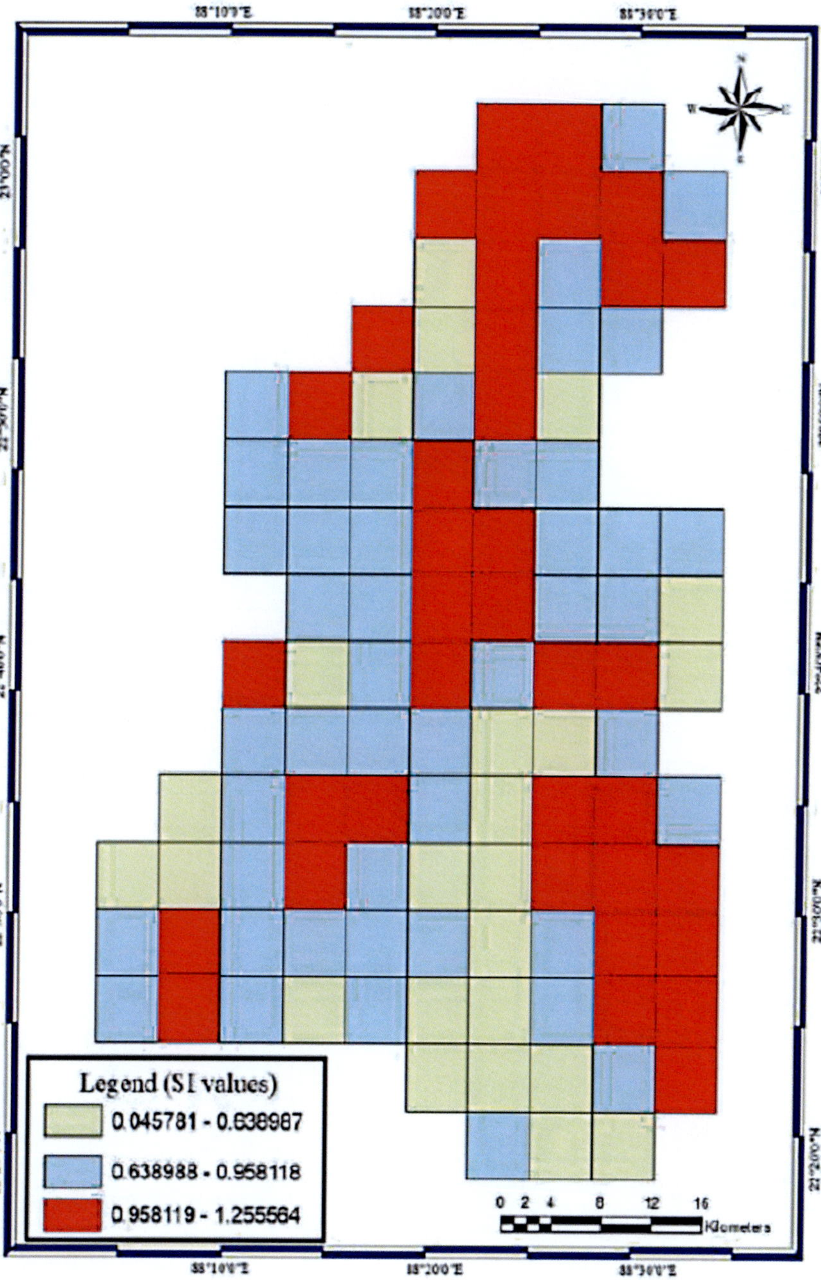

Fig. 4.4 SHDI image of KMDA (5000 m grid). (Map is prepared by the author using Arc GIS Professional environment)

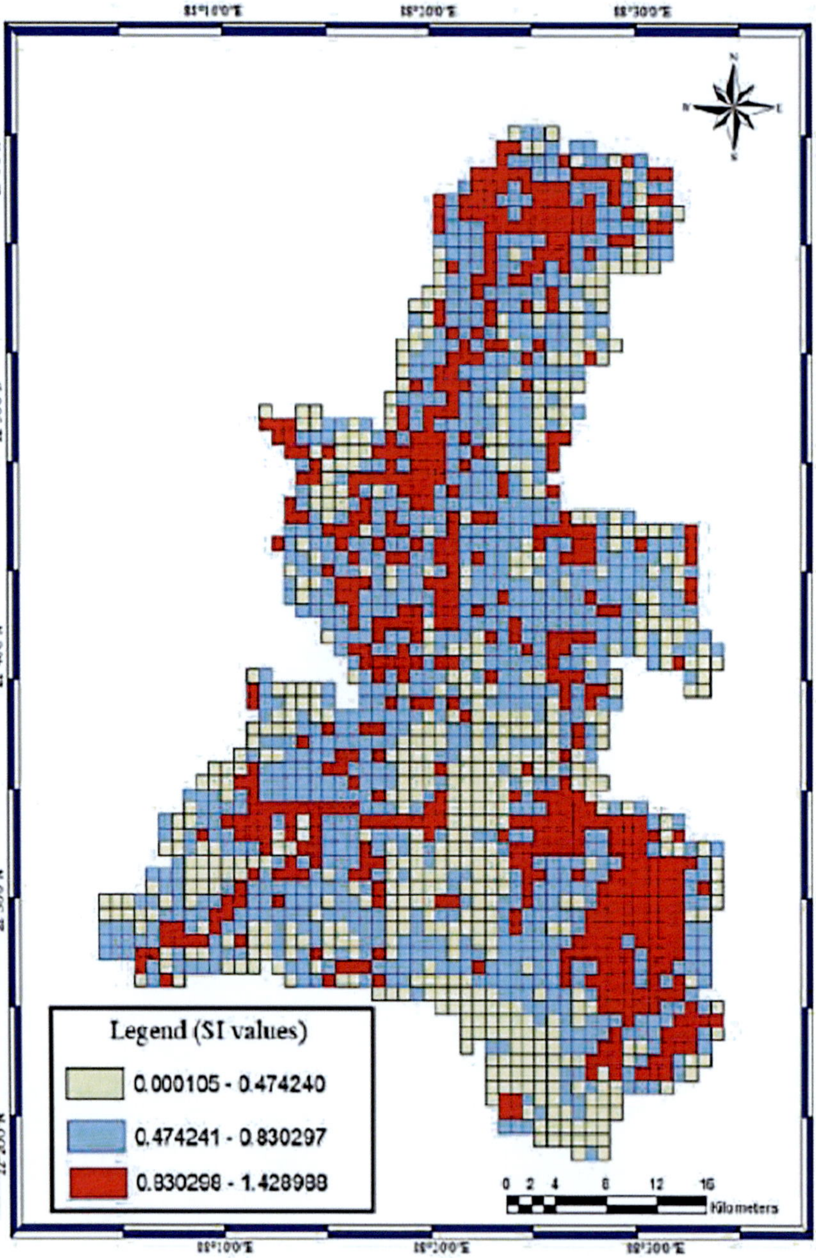

Fig. 4.5 SHDI image of KMDA (1000 m grid). (Map is prepared by the author using Arc GIS Professional environment)

6.2 Variation of Average SHDI in Different Scales and Selection of the Perfect One

The responses of average entropy value on different scales are shown here (Fig. 4.6a–e). There are five transects where the average values of SHDI are calculated on 1000 m, 5000 m, and 10,000 m sized grids. In every case, it is found that the exponential fitting curve is increased initially from 1000 to 5000 m grid and then declined toward 10,000 m grid.

According to Eqs. (4.2) and (4.3), 1000 m (S_1) is the scale that is most adjacent to stationary points S_0, which was obtained when $f(x) = 0$. Therefore, 1000 m was found as the optimal scale to compute the information entropy. The change of entropy value is most stable at this stationary point.

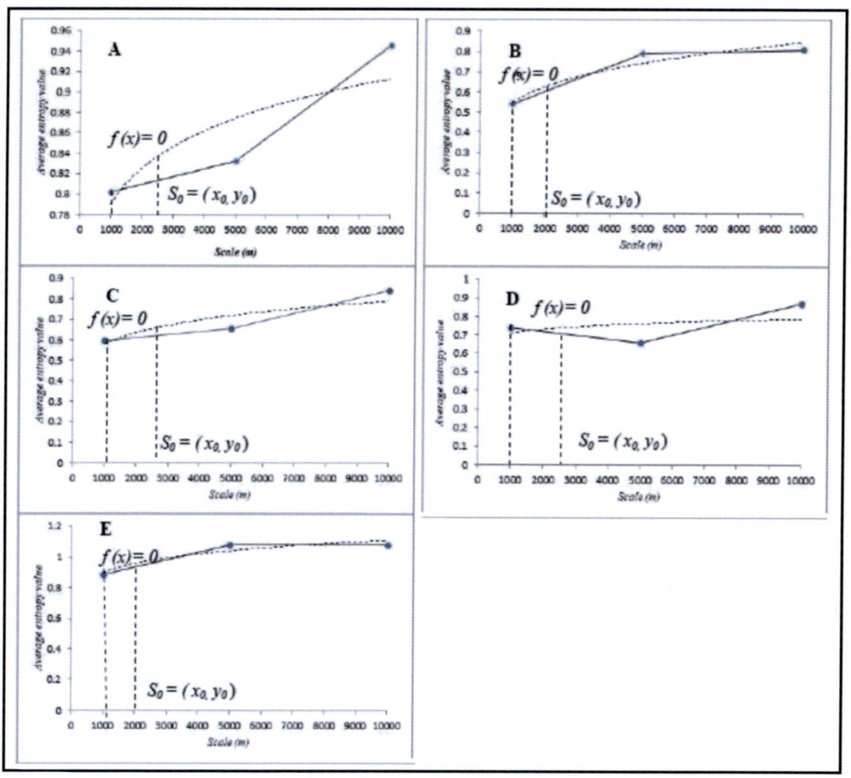

Fig. 4.6 Variation of average Shannon's entropy value with different grid size along different transects, i.e., (**a**) Transect 1, (**b**) Transect 2, (**c**) Transect 3, (**d**) Transect 4, and (**e**) Transect 5. (Source: Prepared by the author)

6.3 Heterogeneous Profiles in Different Scales

Figure 4.7a–c represents the pattern of information entropy value at various scales, i.e., 1, 5, and 10 km on 50 km. The pattern of transects are highly influenced by the variation of scale. In cases of 5 and 10 km scales, the profile mostly presents generalized views so that data loss creates problems in finding the perfect transects

However, data redundancy increases as the cell size gets smaller. Therefore, 1 km was chosen (Profile A) as the optimal scale due to spatial heterogeneity preservation and minimum data redundancy. All these transects were drawn based on the fourth transect (toward north to south in Fig. 4.7) from west to east direction, having a length of 50 km.

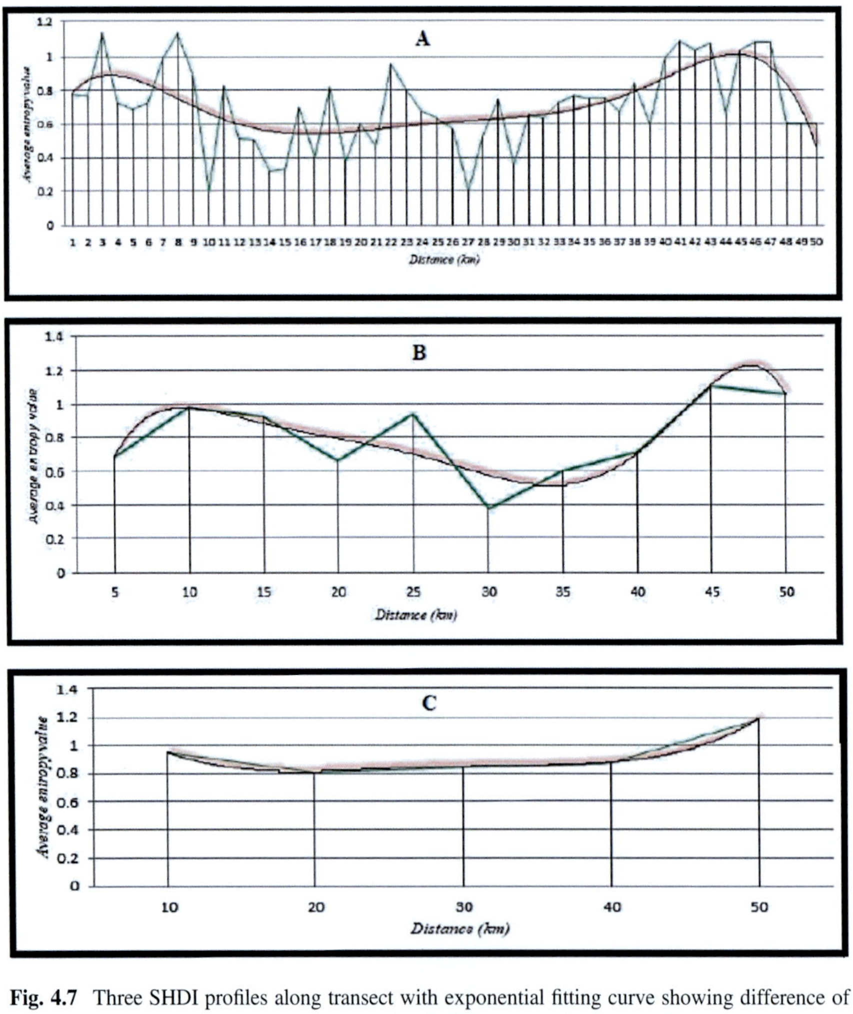

Fig. 4.7 Three SHDI profiles along transect with exponential fitting curve showing difference of data redundancies and demarcation of the optimal grid to extract fringe area of KMDA ((**a**) 1000 m grid, (**b**) 5000 m grid and (**c**) 10,000 m grid). (Source: Prepared by the author)

6.4 Retrieval of Fringe Area of KMDA

Now everything is set as the best scale (1000 m) is already decided. There are 2004 cells covering the whole KMDA with this grid size. To calculate the image in two categories in a most distinguishing manner, there was a need for a threshold value. It was clear from Fig. 4.2a that the transect is characterized by two distinct looks, i.e., convex and concave. Increasing SHDI value indicating higher mixing of LULC is the fringe portion. Lower SHDI having concave region is actually either rural or urban region. To segregate the SHDI image in most convincing manner, Otsu thresholding techniques were applied with the equations as written in Eqs. (4.4, 4.5, 4.6, and 4.7). Table 4.1 is the outcome of these applications.

After calculating weight, mean, and variance and applying Otsu thresholding algorithms, author computed the within variance value to segregate proper back ground and fore ground of the SHDI image to find out the fringe area of KMDA (Fig. 4.8). Before applying Otsu thresholding method, the SHDI image was normalized with the help of Eq. (4.14). We know that Census of India has its own parameters to treat an area as urban or rural. Here in this analysis, several parts of villages, municipal towns, cantonment towns, outgrowths, census towns, and marginal parts of municipal corporations are retrieved as fringe area of KMDA (Fig. 4.9). For examples parts of municipal towns like Bansberia, Kalyani, Gayeshpur, Kanchrapara, Halisahar, Hooghly, Small part of Chinsura, marginal parts of Naihati, Bhatpara, Chandannagar, Bhadreshwar, Garulia, Champadani, North Barrackpore, Baidyabati, Parts of Barrackpore, Srirampur, Titagarh, Rishra, Khardah, Konnagar, Panihati, Marginal part of Barasat, Madhyamgarm, New Barrackpore, North Dum Dum, small pathes of Dum Dum, marginal part of Rajarhat, very small pathes in Bidhannagar, Baranagar, Karhati Uttarpara, Baly, Sonarpur-Rajpur, Baruipur, Mahestala, small pathes in Budge Budge, Pujali, Uluberia area there in fringe region of KMDA. Marginal parts of three municipal

Table 4.1 Calculation of within class variance value after Otsu's thresholding algorithm

SHDI categories	Number of cells		Weight	Mean	Variance	Within class variance
0.000105–0.238252	222	SERIES 1 (Back ground)	0.5494	0.72752	0.6015	0.4565
0.238253–0.476400	357					
0.476401–0.714547	522					
0.714548–0.952694	598	SERIES 2 (Fore ground)	0.450599	1.63787	0.27971	
0.952695–1.190841	283					
1.190842–1.428988	22					
Total cells	2004					

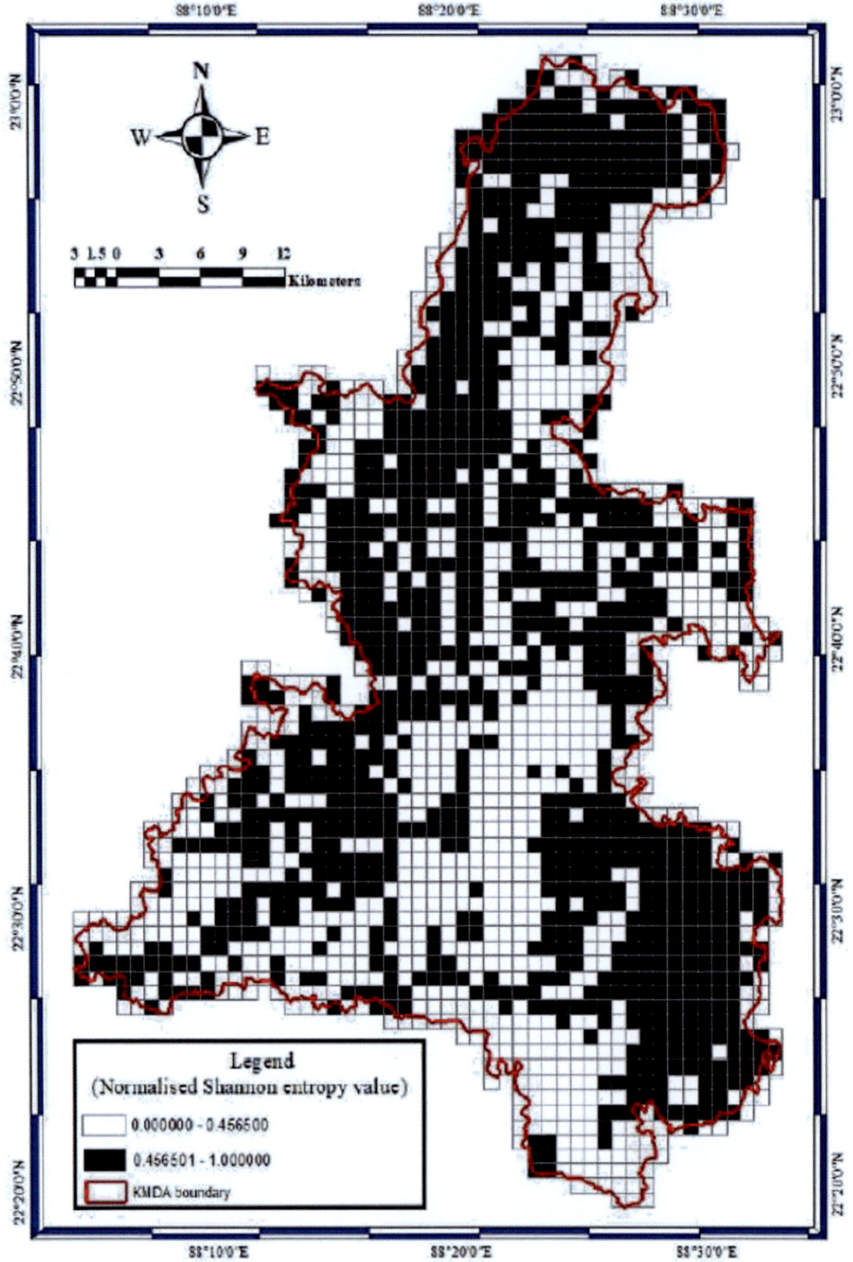

Fig. 4.8 Separation of back ground and foreground of SHDI image of KMDA after Otsu thresholding algorithm. (Source: Prepared by the author)

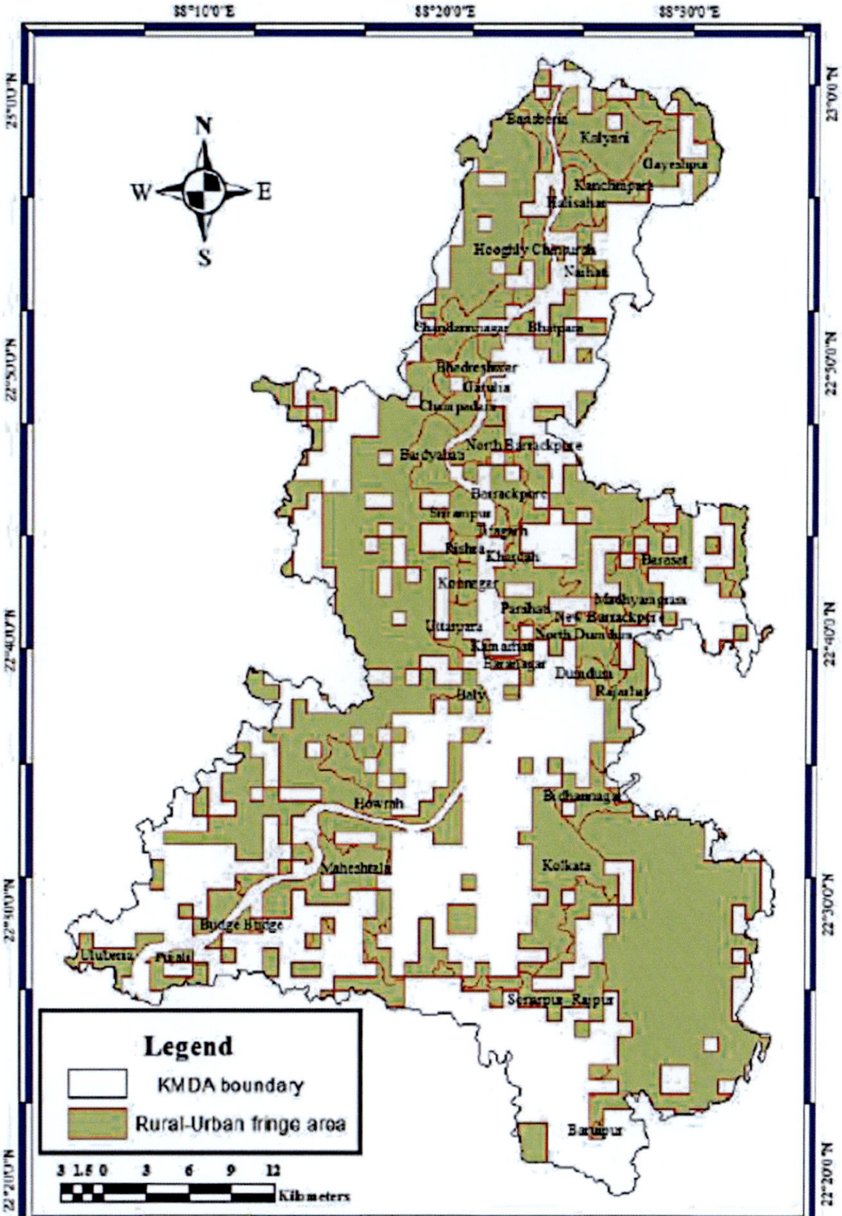

Fig. 4.9 Extracted fringe area of KMDA (2017). (Source: Prepared by the author)

corporations Howrah, Kolkata and Bidhannagar are also having some mixed pixels as fringe. Not only that, noticeable portions of villages both from east and western sides of river Hooghly are treated as fringe area here in this analysis.

6.5 *Surface Water Quality Status of Fringe Landscape in KMDA*

Once the fringe area was retrieved, the next intension was to create the water masks of 1989 and 2017. Positive values of Modified Normalized Difference Water Index (MNDWI) were used to extract the water mask of two images. Water mask of 1989 was superimposed on the detected fringe area of 2017. Water bodies with an area of 184.91 km^2 (1989) are reduced to 89.10 km^2 in 2017 (Fig. 4.10). The red-colored symbols on 2017 map indicate the locations of sample water collection, which were later tested in laboratories to compare with the results of image band ratios. There are 16 water samples, which were compared to get the accuracy of image band indices.

The applied equations (band ratios) for each water quality parameter were applied to LANDSAT OLI data in order to establish a spatial distribution map for each

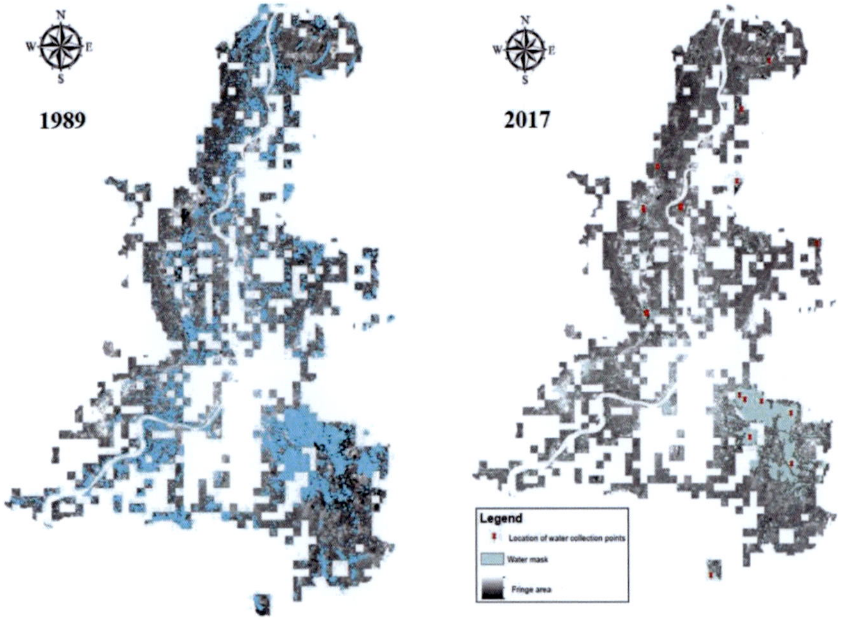

Fig. 4.10 Extracted water masks of 1989 and 2017 and superimposed on fringe area of KMDA (2017); Red dots indicating the water sample collecting sites. (Source: Prepared by the author)

studied parameter. Figures 4.11, 4.12, 4.13, 4.14, and 4.15 illustrate the produced maps for the five parameters used. The spatial distribution of normalized turbidity generally shows an increase beside the Kolkata Metropolitan Corporation (KMC) region. This region is called East Kolkata Wetland (EKW) region, and there is a decrease of turbidity toward the marginal part of fringe area. Northern paleo-channels are with low turbid level (Fig. 4.11). Generally, along the boundary portions of any water body are more turbid than the middle portions.

The higher values of Electronic Conductivity (EC) values are found at the central part and along the river Hooghly with densely settled area of KMDA (Fig. 4.12), decreasing with the distance from urban core.

As seen earlier, pH level is high toward some parts of East Kolkata Wetland at the south-eastern portion of KMDA, though the large wet land shows less pH values (Fig. 4.13).

Figure 4.14 clearly illustrates the same spatial distribution pattern for both EC and TDS respectively. The concentration of ions in the middle part of the water body may be the main reason for this pattern. This is closely related to the distribution of EC (Fig. 4.12), as the maximum values are found in the middle of the water bodies. This is due to ion mobility, whereas conductivity is also affected by heavy metal ions released into water by acid drainage. This is obvious, as recorded pH in this region shows minimum values due to the absence of buffering materials that neutralize the slightly acidic inflow.

Higher temperature in surface water is found mainly beside the river Hooghly and in some of the parts which is close to urban settlements. It decreases with the increasing distance from municipalities and corporations (Fig. 4.15).

6.6 Accuracy Assessment

Equations (4.16) and (4.17) are applied to calculate the mean error (ME) and root mean square error (RMSE) of the band indices and laboratory-based data. Each parameter was tested like wise to get the accuracy levels of band indices. In case of turbidity, ME and RMSE were found as 0.15 and 0.19 (Table 4.2). Electronic Conductivity (EC) achieved ME and RMSE values of 0.23 and 0.16 respectively (Table 4.3). In case of pH, both the values found to be not so satisfied, which are 0.61 and 0.72 (Table 4.4). Further test and convincing results are needed for this parameter in future research. On the other hand, Total Dissolved Solid (TDS) got more convincing results where ME and RMSE are 0.26 and 0.18 (Table 4.5). Water temperature data was also tested. The ME and RMSE of this parameter achieved values of ME as 0.16 and RMSE as 0.19, which is again a convincing result (Table 4.6).

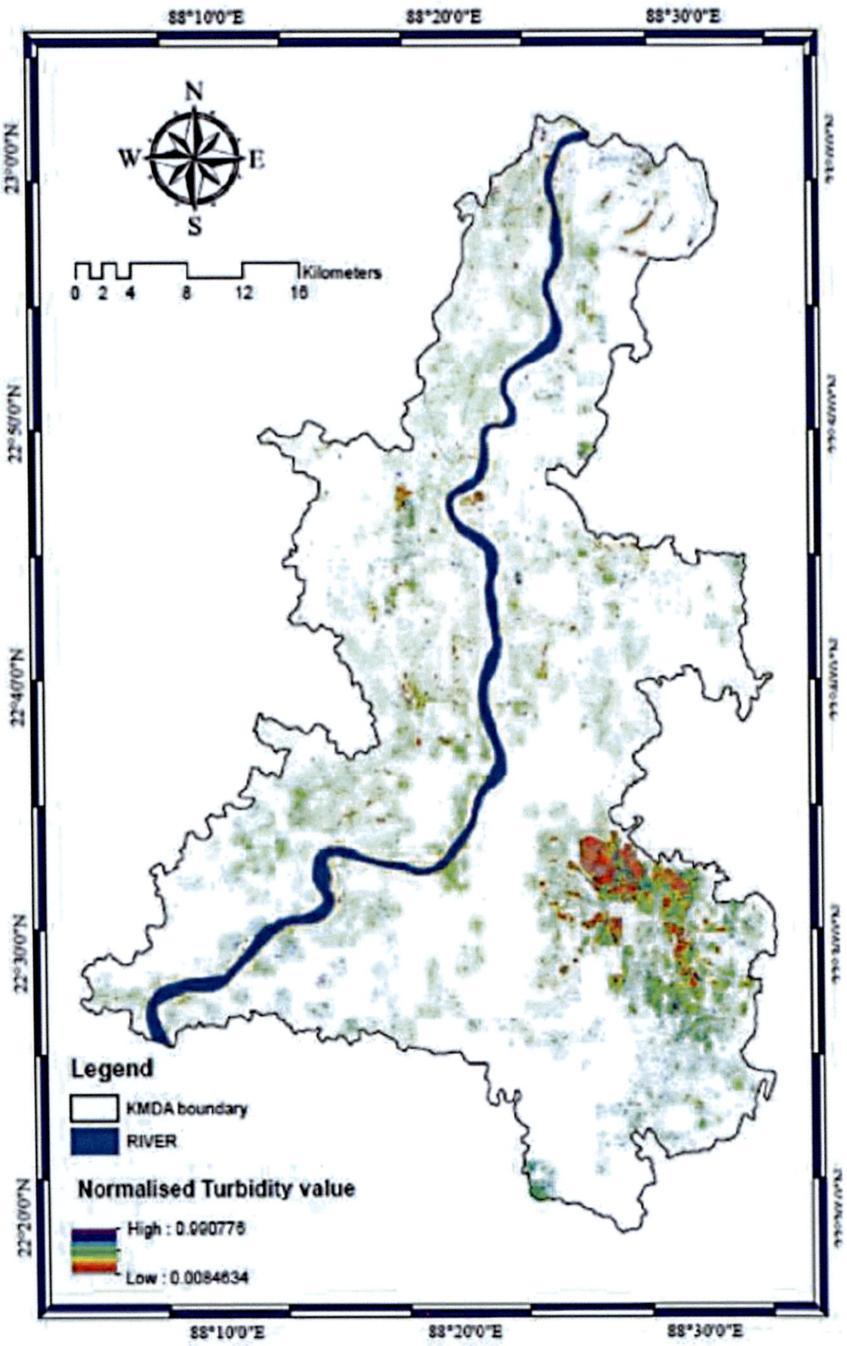

Fig. 4.11 Normalized Difference Turbidity Index (NDTI) image. (Source: Prepared by the author)

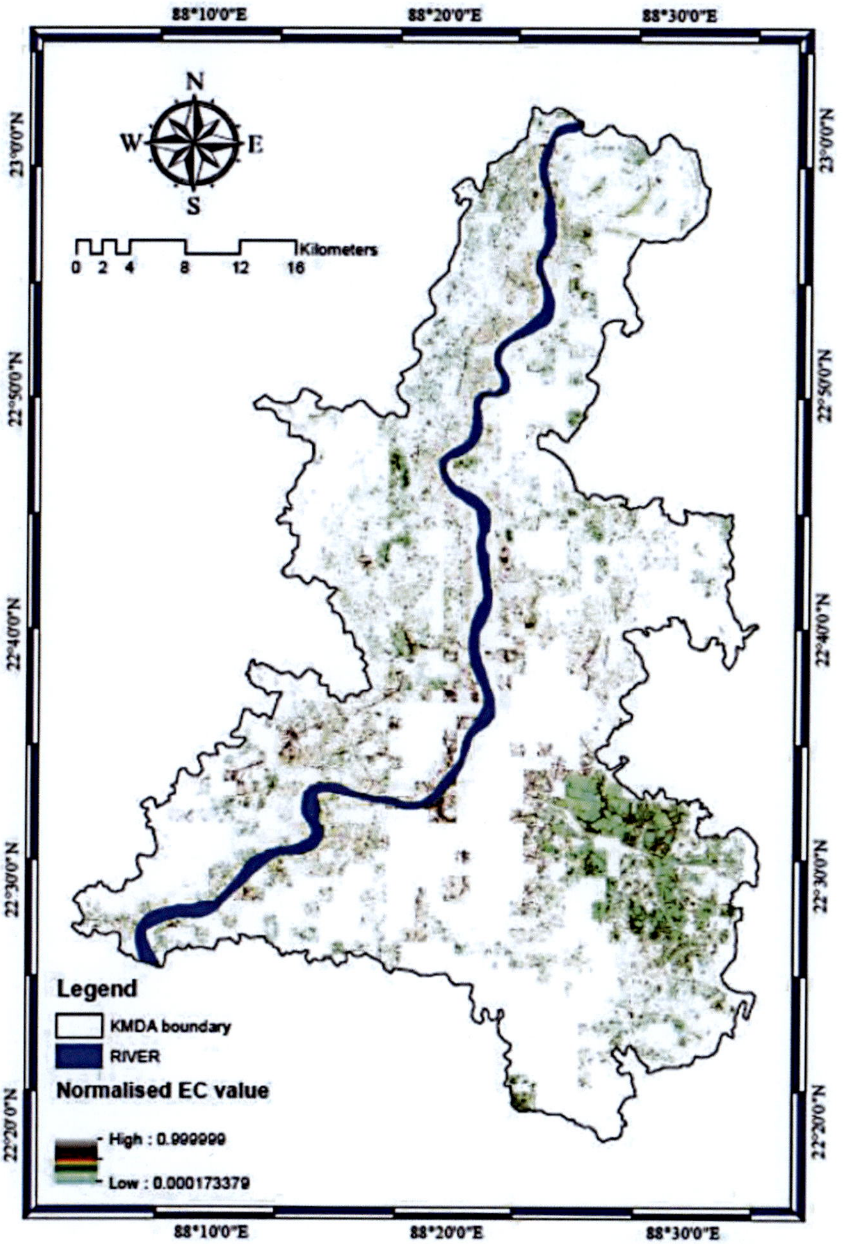

Fig. 4.12 Electronic Conductivity (EC) of surface water. (Source: Prepared by the author)

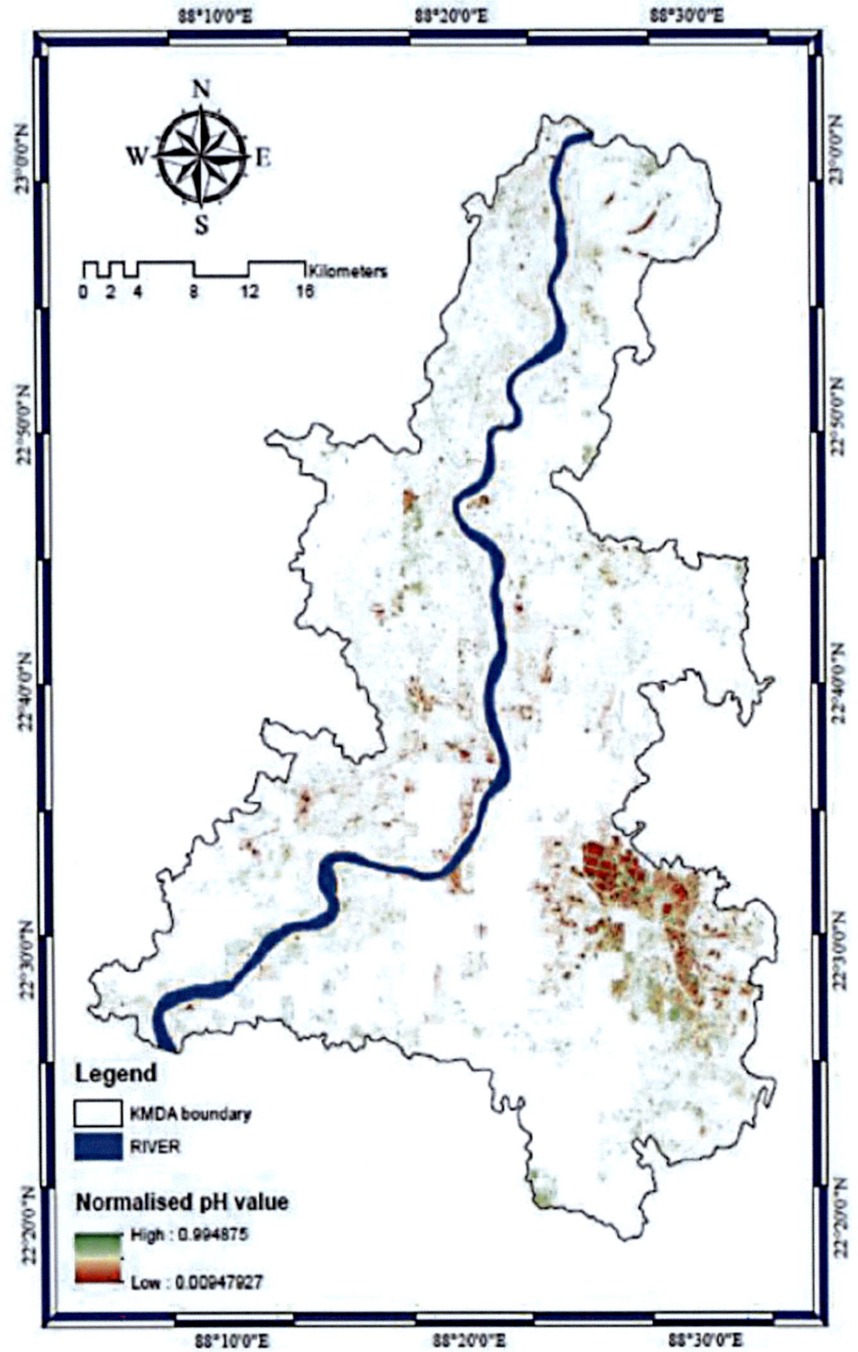

Fig. 4.13 pH distribution image

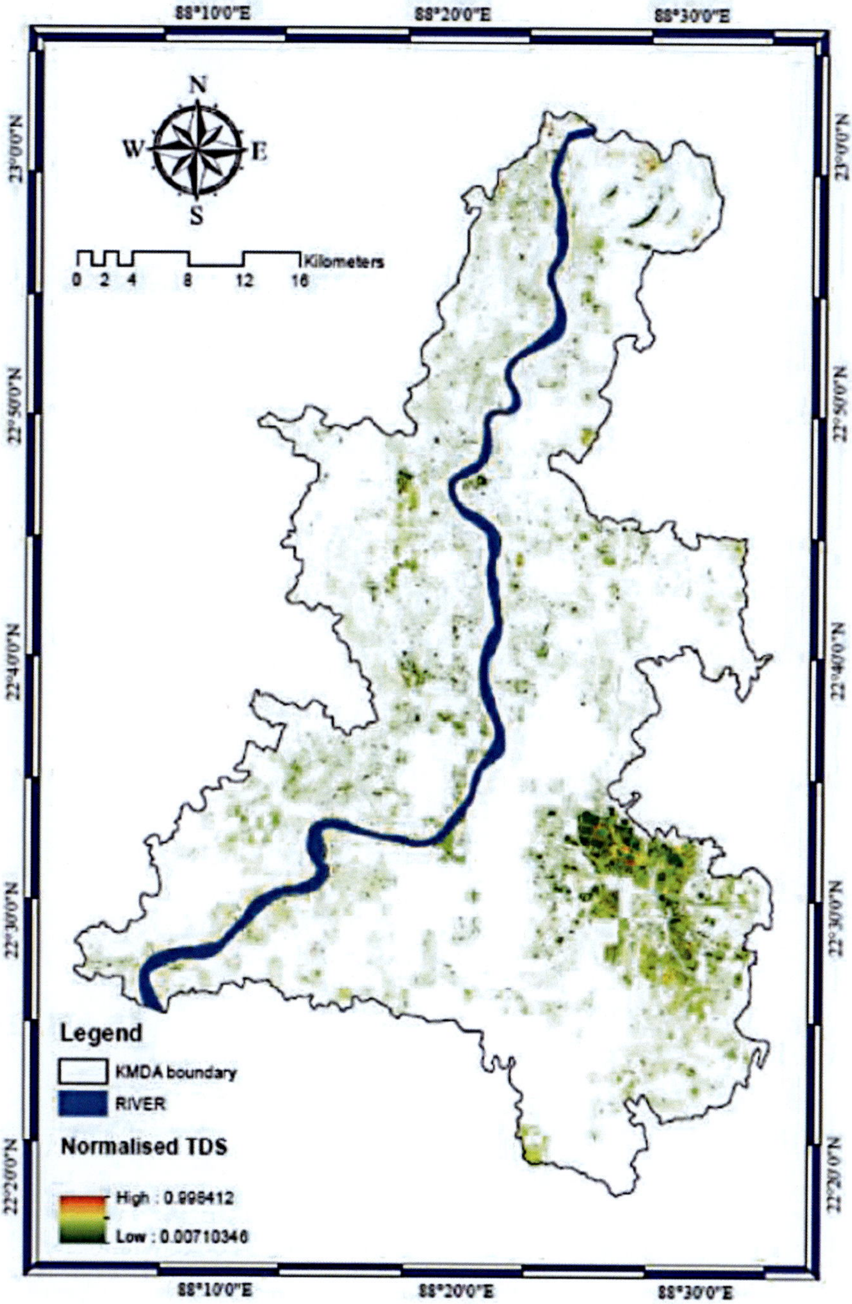

Fig. 4.14 Total Dissolved Solid (TDS) distribution image

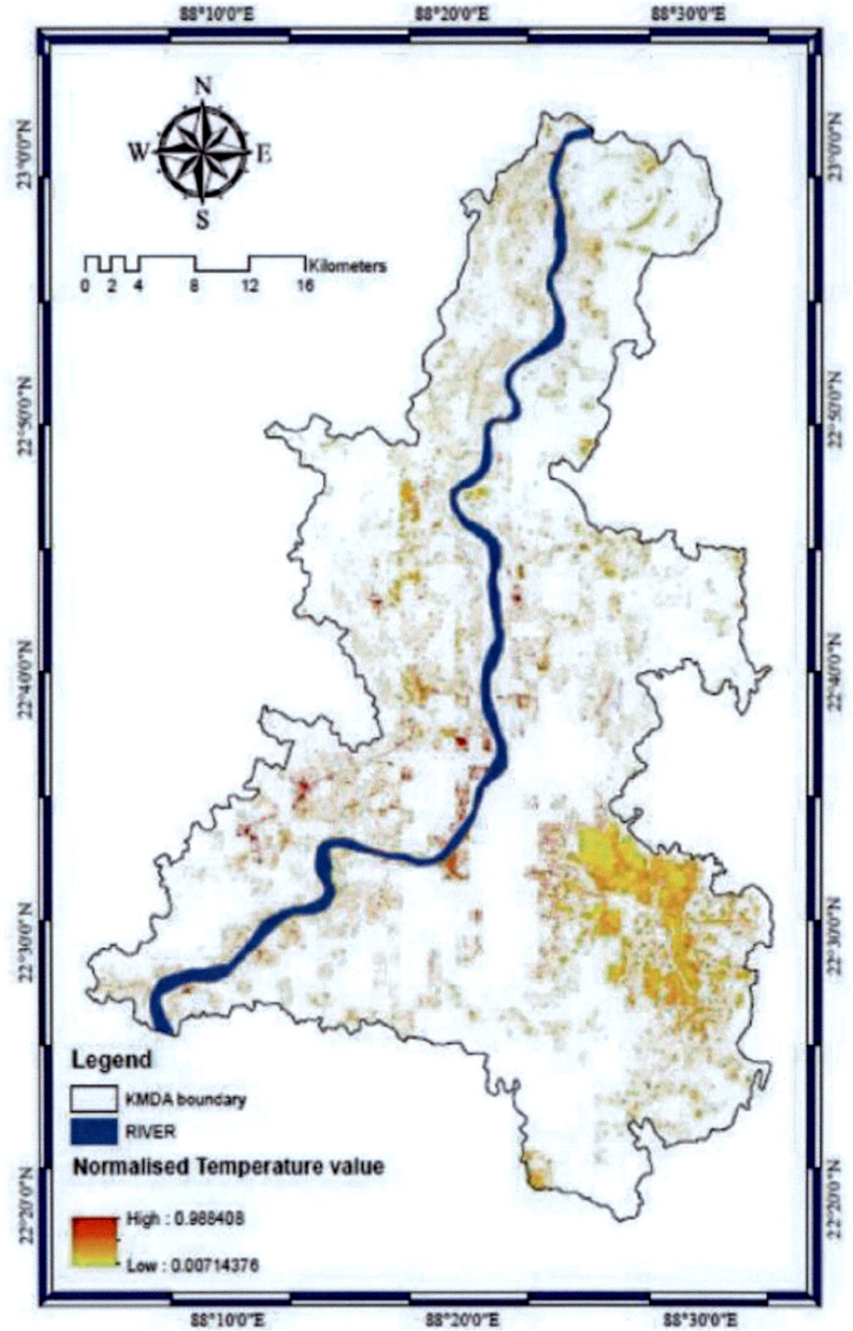

Fig. 4.15 Surface water temperature distribution image. (Source: Map prepared by the author)

Table 4.2 Calculation of ME & RMSE in Turbidity image

Turbidity	Normalized	Turbidity (NTU)	Normalized lab.	Error	Modulus	Error^2	
0.11	0.12	33.60	0.20	0.08	0.08	0.01	
0.27	0.42	36.10	0.22	−0.20	0.20	0.04	
0.41	0.68	141.50	0.86	0.18	0.18	0.03	
0.05	0.02	9.30	0.06	0.03	0.03	0.00	
0.05	0.01	8.80	0.05	0.04	0.04	0.00	
0.38	0.62	163.00	0.99	0.38	0.38	0.14	
0.04	0.00	15.80	0.10	0.10	0.10	0.01	
0.07	0.05	0.20	0.00	−0.05	0.05	0.00	
0.10	0.10	2.70	0.02	−0.09	0.09	0.01	
0.27	0.42	41.50	0.25	−0.17	0.17	0.03	
0.06	0.03	26.40	0.16	0.13	0.13	0.02	
0.23	0.34	27.00	0.16	−0.18	0.18	0.03	
0.31	0.50	47.00	0.29	−0.21	0.21	0.05	
0.36	0.59	26.40	0.16	−0.43	0.43	0.18	
0.59	1.00	164.00	1.00	0.00	0.00	0.00	
0.07	0.05	36.10	0.22	0.17	0.17	0.03	
				Sum	2.43	0.58	Sq. sum
				Mean error	0.15	0.04	Divided by 16
						0.19	RMSE

Table 4.3 Calculation of ME & RMSE in Electronic Conductivity (EC) image

EC	Normalized	EC (μs)	Normalized lab.	Error	Modulus	Error^2	
0.04	0.00	284.00	0.00	0.00	0.00	0.00	
0.07	0.18	466.00	0.31	0.14	0.14	0.02	
0.08	0.25	566.00	0.49	0.23	0.23	0.05	
0.06	0.14	420.00	0.23	0.09	0.09	0.01	
0.06	0.15	522.00	0.41	0.26	0.26	0.07	
0.06	0.17	350.00	0.11	−0.06	0.06	0.00	
0.06	0.15	520.00	0.41	0.25	0.25	0.06	
0.05	0.10	520.00	0.41	0.31	0.31	0.10	
0.08	0.26	452.00	0.29	0.03	0.03	0.00	
0.08	0.30	410.00	0.22	−0.08	0.08	0.01	
0.06	0.12	400.00	0.20	0.08	0.08	0.01	
0.07	0.23	498.00	0.37	0.14	0.14	0.02	
0.06	0.13	325.00	0.07	−0.06	0.06	0.00	
0.09	0.36	609.00	0.56	0.20	0.20	0.04	
0.20	1.00	865.00	1.00	0.00	0.00	0.00	
0.05	0.11	350.00	0.11	0.01	0.01	0.00	
				Sum	1.94	0.39	Sq. sum
				Mean error	0.23	0.02	Divided by 16
						0.16	RMSE

Table 4.4 Calculation of ME & RMSE in pH image

pH	Normalized	pH	Normalized lab.	Error	Modulus	Error^2	
0.47	0.01	8.15	0.43	0.42	0.42	0.65	
0.66	0.48	9.18	0.91	0.43	0.43	0.65	
0.71	0.62	9.34	0.98	0.36	0.36	0.60	
0.48	0.03	7.78	0.26	0.22	0.22	0.47	
0.48	0.04	7.69	0.21	0.17	0.17	0.42	
0.64	0.44	9.23	0.93	0.49	0.49	0.70	
0.47	0.01	7.65	0.20	0.18	0.18	0.42	
0.47	0.01	7.43	0.09	0.08	0.08	0.29	
0.52	0.14	9.23	0.93	0.79	0.79	0.89	
0.65	0.46	8.94	0.80	0.33	0.33	0.57	
0.47	0.02	8.10	0.40	0.38	0.38	0.62	
0.56	0.24	8.51	0.60	0.36	0.36	0.60	
0.58	0.28	8.95	0.80	0.52	0.52	0.72	
0.70	0.59	9.38	1.00	0.41	0.41	0.64	
0.87	1.00	9.35	0.99	−0.01	0.01	0.12	
0.47	0.00	7.23	0.00	0.00	0.00	0.00	
				Sum	5.17	8.37	Sq. sum
				Mean error	0.61	0.52	Divided by 16
						0.72	RMSE

Table 4.5 Calculation of ME & RMSE in TDS image

TDS	Normalized	TDS (ppt)	Normalized lab.	Error	Modulus	Error^2	
0.13	0.18	319.00	0.28	0.11	0.11	0.01	
0.25	0.43	425.00	0.45	0.02	0.02	0.00	
0.39	0.70	331.00	0.30	−0.40	0.40	0.16	
0.05	0.02	156.00	0.02	0.00	0.00	0.00	
0.04	0.00	202.00	0.10	0.10	0.10	0.01	
0.35	0.62	296.00	0.25	−0.37	0.37	0.14	
0.05	0.02	142.00	0.00	−0.02	0.02	0.00	
0.07	0.06	159.00	0.03	−0.03	0.03	0.00	
0.08	0.09	253.00	0.18	0.08	0.08	0.01	
0.23	0.39	537.00	0.63	0.24	0.24	0.06	
0.05	0.03	276.00	0.21	0.19	0.19	0.03	
0.20	0.33	402.00	0.41	0.09	0.09	0.01	
0.27	0.47	354.00	0.34	−0.13	0.13	0.02	
0.34	0.59	336.00	0.31	−0.28	0.28	0.08	
0.54	1.00	769.00	1.00	0.00	0.00	0.00	
0.06	0.05	240.00	0.16	0.11	0.11	0.01	
				Sum	2.17	0.54	Sq. sum
				Mean error	0.26	0.03	Divided by 16
						0.18	RMSE

Table 4.6 Calculation of ME & RMSE in temperature image

Temperature	Normalized	Temperature (°C)	Normalized temp. lab.	Error	Modulus	Error^2	
0.09	0.18	27.50	0.07	−0.11	0.11	0.01	
0.19	0.64	29.20	0.43	−0.21	0.21	0.04	
0.20	0.70	28.90	0.37	−0.33	0.33	0.11	
0.11	0.29	29.50	0.50	0.21	0.21	0.04	
0.19	0.66	30.50	0.72	0.06	0.06	0.00	
0.16	0.53	30.20	0.65	0.13	0.13	0.02	
0.09	0.22	27.30	0.02	−0.20	0.20	0.04	
0.05	0.00	27.80	0.13	0.13	0.13	0.02	
0.13	0.36	27.20	0.00	−0.36	0.36	0.13	
0.25	0.93	31.30	0.89	−0.04	0.04	0.00	
0.08	0.15	28.20	0.22	0.07	0.07	0.00	
0.15	0.48	30.50	0.72	0.23	0.23	0.05	
0.18	0.61	28.50	0.28	−0.33	0.33	0.11	
0.17	0.56	30.20	0.65	0.10	0.10	0.01	
0.27	1.00	31.80	1.00	0.00	0.00	0.00	
0.14	0.43	29.20	0.43	0.00	0.00	0.00	
				Sum	2.50	0.59	Sq. sum
				Mean error	0.16	0.04	Divided by 16
						0.19	RMSE

In total, five parameters were tested and overall accuracy found at a quite satisfactory level. In this whole analysis, river water was excluded, only closed surface water bodies of fringe area of KMDA were taken into the main focus of interest.

7 Findings

According to Bhagat (2003), the unsatisfactory environmental condition in most rural-urban fringe is mainly because of negligence and non-recognition of these areas, which deserve to be in urban and compensating urban facilities relating to sustainable ways of development. Bhagat said that there is a problem in the definition of urban area in census actually. In this analysis, a large area of villages (according to census) is under fringe as detected. On the other hand, several portions of towns are there in fringe as found. It means somewhere the facilities and utilities are not meeting the optimal needs of a particular area. They are in this manner not giving a significant number of the essential administrations underestimated in the city. They should manage all alone and this outcome in expanding levels of neighborhood imbalances as enormous organizations and open foundations just as the upper salary gathering can introduce privatized fundamental administrations, yet there is a finished nonattendance of these administrations for poor people and littler organizations/workshops. These territories, by and large, need funneled water gracefully and need to get it from neighborhood waterways, ponds, or through cylinder wells. Such water isn't dealt with, and on the off chance that the sources are dirtied, at that point the water gracefully isn't sheltered and is of low quality. As to, private property holders and establishments situated in the fringe region make their own offices with the structure of septic tanks and surface depletes that vacant into nearby streams or nullahs. Be that as it may, shouldn't something be said about strong squanders created in these zones? With no civil administrations, strong squanders lie uncollected alongside of the road, or whenever gathered, are dumped in any low-lying land. These practices are pillaging the nearby scene as well as are a huge wellbeing peril. The fast development of populace in fringe zones in the most recent decade has implied that the volume of strong and fluid squanders has expanded, yet the institutional abilities to deal with them, remain to a great extent missing. In this manner, it is anything but difficult to see the complexity in ecological cleanliness between the peripheral zones and the developed city. Moreover, there is the complexity in other financial markers, all of which, as Kundu et al. (2002: 5046) and Kundu (2003: 364) have appropriately watched, is bringing about the efficient making of a "deteriorated fringe". This begins the disparities in local environmental planning. Here in this analysis, author focused on spatial changes and present status of closed surface water quality of the fringe portion of KMDA, which is found to be a great concern, and it is felt that there should be separate administrative body for fringe to plan uniquely only for fringe areas.

References

Abdelmalik, K. W. (2018). Role of statistical remote sensing for Inland water quality parameters prediction. *The Egyptian Journal of Remote Sensing and Space Science, 21*(2), 193–200.

Banzhaf, E., Grescho, V., & Kindler, A. (2009). Monitoring urban to peri-urban development with integrated remote sensing and GIS information: A Leipzig, Germany case study. *International Journal of Remote Sensing, 30*, 1675–1696.

Bentinck, J. V. (1996, August 4–10). *Brick quarries in Delhi's rural-urban fringe: A model of land degradation in socioeconomic terms*. Paper presented at the 28th International Geographical Congress, The Hague.

Bhagat, R. B. (2003). *Challenges of rural-urban classification for decentralised governance*. Research report of the International Institute of Population Studies, Mumbai. Downloaded from www.iipsindia.org/rp/decegovernance.pdf

Brush, J. E. (1968). Spatial patterns of population in Indian cities. *Geographical Review, 58*, 362–691.

Census. (2011). *Primary census abstracts*. Registrar General of India, Ministry of Home Affairs, Government of India. Available at http://www.censusindia.gov.in/2011census/PCA/pca_highlights/pe_data.html. Accessed on 22 March 2020.

Corry, R. C., & Nassauer, J. I. (2005). Limitations of using landscape pattern indices to evaluate the ecological consequences of alternative plans and designs. *Landscape and Urban Planning, 72*, 265–280.

Desai, A., & Gupta, S. S. (1987). *Problem of changing land use pattern in the rural-urban fringe*. Concept Publishing Company.

Friedberger, M. (2000). The rural-urban fringe in the late twentieth century. *Agricultural History, 74*, 502–514.

Ginsburg, N., Koppel, B., & McGee, T. G. (Eds.). (1991). *The extended metropolis: Settlement transition in Asia*. University of Hawaii Press.

Graves, R. (2016). *Ecotone*. Available online http://www.eoearth.org/view/article/152345. Accessed on 20 March 2020.

Hall, A. S., Kaufman, S. J., & Ricketts, C. T. (2006). Defining urban and rural areas in U.-S. epidemiologic studies. *Journal of Urban Health, 8*, 162–175.

Huang, J., Zhou, Q., & Wu, Z. (2016). Delineating urban fringe area by land cover information entropy—An empirical study of Guangzhou-Foshan metropolitan area, China. *International Journal of Geo-Information, MDPI*; 2016; 5, 59; pp 1–12.

Jindrich, J. (2010). When the urban fringe is not suburban. *Geographical Review, 100*, 35–55.

Kundu, A. (2003). Institutional innovations for infrastructure development in India and the emerging urban scenario. In R. Sengupta & A. K. Sinha (Eds.), *Challenge of sustainable development: The Indian dynamics* (pp. 339–369). Manak Publications.

Kundu, A., Pradhan, B. K., & Subramanian, A. (2002). Dichotomy or continuum: Analysis of impact of urban centres on their periphery. *Economic and Political Weekly, 37*, 5039–5046.

Lesage, P. J., & Charles, S. J. (2008). Using home buyers' revealed preferences to define the urban-rural fringe. *Journal of Geographical Systems, 10*, 1–21.

Louis, H. (1936). Die geographische Gliederung von Gross-Berlin (The geographical structure of Greater Berlin). In H. Louis & W. Panzer (Eds.), *Landerkundliche Forschung: Krebs-Festschrift* (pp. 146–171). Engelhorn. (In German).

Nangia, S. (1976). *Delhi metropolitan region: A study in settlement geography*. K B Publications.

Peterson, D. L., & Parker, T. V. (1998). *Ecological scale theory and applications*. Columbia University Press.

Quattrochi, D. A., & Goodchild, M. F. (1997). *Scale in remote sensing and GIS*. CRC Press.

Ramachandran, R. (1989). *Urbanisation and urban systems in India*. Oxford University Press.

Russwurm, L. (1975). Urban fringe and urban shadow. In *Urban problems: A Canadian reader* (pp. 148–164). Holt, Rinehart and Winston.

Schneider, A., Friedl, M. A., & Potere, D. (2010). Mapping global urban areas using MODIS 500-m data: New methods and datasets based on "urban ecoregions". *Remote Sensing of Environment, 114*, 1733–1746.

Senft, A. R. (2009). *Species diversity patterns at ecotones*. University of North Carolina.

Shafer, E. L., & Brush, R. O. (1977). How to measure preferences for photographs of natural landscapes. *Landscape Planning, 4*, 237–256.

Weeks, J. R. (2010). Defining urban areas. In T. Rashed & C. Jürgens (Eds.), *Remote sensing of urban and suburban areas* (pp. 33–45). Springer.

Wilson, E. H., Hurd, J. D., Civco, D. L., Prisloe, M. P., & Arnold, C. (2004). Development of a geospatial model to quantify describe and map urban growth. *Remote Sensing of Environment, 86*, 275–285.

Wu, J. (2013). Key concepts and research topics in landscape ecology revisited: 30 years after the Allerton Park workshop. *Landscape Ecology, 28*, 1–11.

Xu, H. (2006). Modification of normalised difference water index (NDWI) to enhance open water features in remotely sensed imagery. *International Journal of Remote Sensing, 27*(14), 3025–3033.

Yang, X., Fu, B., & Chen, L. (2013). Remote sensing and geospatial analysis for landscape pattern characterization. In B. Fu & B. Jones (Eds.), *Landscape ecology for sustainable environment and culture* (pp. 205–221). Springer.

Zhang, W., Fang, X., & Zhang, L. (1999). Method to identify the urban-rural fringe by TM images. *Journal of Remote Sensing, 3*, 199–202. (In Chinese).

Chapter 5
Land Use Dynamics and Environmental Cost Estimation in Lower Gondwana Coalfield, West Bengal, India

Amit Sarkar and Soma Bhattacharya

Abstract Raniganj coalfield area has a complex land use land cover degradation scenario due to opencast mining and associated development activities since 1960. Therefore, it is urgent to identify land use dynamics and landscape patterns in order to sustain environmental health and local indigenous inhabitants. This chapter uses land use land cover transformation technique, landscape fragmentation analysis and environmental impacts assessment to identify the degradation scenario of land use land cover due to anthropogenic deeds in the past 50 years. The results denote that there is a significant decrease in the area of forest (19.05% to 14.12%), agricultural land (39.83% to 31.96%), river (2.58% to 0.62%) and water body (0.49% to 0.47%) from 1992 to 2015; much of these landscapes are transformed into fallow land (20.60–29.68%), excavated land (6.90–8.88%) and urban area (2.48–6.18%). On average 30% forest is converted for other land use within only 10 years in Sonepur Bazari, Satgram, Sripur and Khottadih area. Class-level and landscape-level fragmentations of different landscapes are measured using ArcGIS and fragstats software. Patch density of forest (38.236 to 17.259) and water bodies (6.526 to 0.980) has significantly decreased since 1973 to 2015. However, urban area witnessed an acute increment of mean patch size from 1.65 to 2.45 hectare around the city area bordering Asansol, Durgapur, Barakar, Kulti and Raniganj. Total edge of fallow land (9,988,999.8–127,774,115.2 hectare) and excavated land (3,566,547.8–-4,915,028.1 hectare) has also significantly increased. The increase in the Shannon diversity index (1.45–1.61) and Shannon evenness index (0.60–0.67) indicates that the overall landscape became heterogeneous and scattered. To find out the environmental cost in this area, spatial and temporal variations of suspended particulate

A. Sarkar (✉)
Department of Geography, Vivekananda College for Women (University of Calcutta), Barisha, Kolkata, West Bengal, India

Department of Geography, Gobardanga Hindu College (West Bengal State University), Gobardanga, West Bengal, India

S. Bhattacharya
Department of Geography, Vivekananda College for Women (University of Calcutta), Barisha, Kolkata, West Bengal, India

S. Hazra et al. (eds.), *Geographies of the Indian Subcontinent*,
https://doi.org/10.1007/978-3-031-74813-4_5

matter (SPM), respirable particulate matter (RPM), daily noise of air and land degradation are measured on a regular basis. It is found that SPM is 124.9 μgm^3, which is higher than the normal; simultaneously human-made difference in elevation is maximally 154.6 m in the study area. Furthermore, the results of hedonic pricing method indicate that the cost of environmental degradation is 295.49 crore rupees, which is hidden by the mining authority.

1 Introduction

Opencast mining is a surface mining technique for extracting minerals from the earth by their removal from an open pit when deposits of commercially useful minerals are found near the surface, that is, where the overburden (surface materials covering the mineral) is relatively thin (Zerbe Jr., 1988; Coleman, 1980; Mas et al., 2010). Opencast mining is more effective than almost all type of mining because more ore can be extracted rapidly using this technique (Bianche, 2002; Maitima et al., 2009; Foster, 2012). Therefore, opencast mining is gaining popularity decade by decade (Uuemaa et al., 2009; Kitalika et al., 2018; Rutledge, 2014; Ampofo et al., 2016). But it generates noise, dust and adverse particulate matter that damage the adjoining ambient air very badly if necessary steps are not taken to prevent the air degradation in a proper way (Huang et al., 2009; Kuemmerle et al., 2012). Coal extraction by opencast method brings huge profits to the society, but it gradually deteriorates the environment, which is a crucial matter of concern (Bhawana, 2015; Sapena & Ruiz, 2015). The working method of opencast mining is safe; because there is no risk of cave and toxic gas emmissions (Koellner et al., 2013; Awoniran et al., 2014). Besides, opencast activities degrade the stability of ecosystem as it removes vegetation, alter the land use land cover, destroy the soil conditions, disrupt fundamental ecological relationships, reduce biodiversity and enhance air, water, and soil pollution (Haddad et al., 2015; Garg et al., 2013). These negative impacts input serious threats to the entire environmental landscape in and around the mining site (Le Page et al., 2016; Modica et al., 2012). However, opencast mining generates excessive amount of waste material, overburden and quarry which pile up on the surface and occupy relatively large areas of land, reduces land use availability and increases pressure on land supply (Kumar et al., 2014; Mundia & Murayama, 2009; Batistella et al., 2003). Therefore, vast amount of unproductive land is created due to application of poor backfilling techniques just after completion of mining activities (Baldi & Paruelo, 2008; Das & Das, 2016; Dupin et al., 2013). One of the OC projects of this coalfield Sonepur Bazari annually produces 15 million tons over burden at a stripping ratio of 1:4.72 (Sarkar, 2017a, b, c). It clearly denotes vast amount of land use land cover changes, transformation, fragmentation and conversion in this coalfield. Today, opencast mining and associated development activities are recognized as the most significant factors that have caused widespread change in Raniganj coalfield landscape (Sarkar, 2018, 2019a, b, c, d, e, f). Rapid interaction of mining aids like dragline, shovel and dozer with bare surface causes destruction, transformation and intensive loss of various land cover (Pradesh et al., 2005; Feleke, 2003). These anthropogenic changes on land resources can be measured by assessing the land use land cover dynamics and environmental cost analysis. Hence, the researcher tries to explore land use land cover dynamics and environmental impacts in the focal areas of Raniganj coalfield and its surroundings. Raniganj coalfield area denotes a very complex land use land cover degradation scenario because of high concentration of human population, opencast mining and associated development activities from very ancient time (Sarkar & Bhattacharya, 2016). Since 1960, in this coalfield, continuous land excavation activities are being done for producing coal production using opencast techniques (Sarkar, 2019a). Therefore, it is crucial to understand and compute the environmental influence of opencast coal mining on landscape pattern in and around Raniganj coalfield area. Sonepur Bazari is the biggest OC Project of Raniganj coalfield commenced in 1990. On average, Sonepur Bazari opencast project annually produces 3.5 million tons coal and 15 million tons overburden at a stripping ratio of 1:4.72 (Sarkar, 2017b). This huge amount of overburden degrades the environment vastly. To estimate the environmental degradation in this area, spatial and temporal variations of suspended particulate matter (SPM), respirable particulate matter (RPM) and nitrogen oxide (NO_x) of air are measured on a seasonal basis from March 2013 to February 2015. Further, some mathematical algorithms like Inverse Distance Weighted (IDW) and Kriging are applied to detect the overall air pollution condition in and around the coalfield. The primary concern of this study is to estimate the environmental impact due to opencast mining and to provide suitable remedies. Raniganj coalfield lies over the granite plateau ring of Choto Nagpur in the Damodar valley region surrounded by the Durgapur-Asansol industrial belt with a general elevation of 100 m (Goswami & Goswami, 2014). Coal of Permian age is found in some areas of this region (205 to 108 million years old), in Raniganj and Barakar stages of Gondwana age (Hassan et al., 2015). In Raniganj coalfield area, opencast mining activities generate enormous noise, dust, adverse visual impact, degraded land and, not least, long-term environmental damage, which can be harmful to the rural areas even after restoration (Indra et al., 2017; El-hallaq & Habboub, 2015). Therefore, the tension between the environmental cost and the profitability of opencast coal mining is debatable (Lin et al., 2005; Woldai & Taranik, 2001). The hedonic pricing method is applied in Khottadih and Sonepur Bazari, which is one of the biggest opencast projects in Raniganj coalfield. These two regions are situated in a green belt area in the past, where land is used almost entirely for agricultural purposes (Ramachandra et al., 2014). Therefore, the abandonment of these two mines would generate adverse environmental impacts. The primary concern is whether the benefits outweigh the costs, which this chapter attempts to answer via hedonic pricing method. This chapter tries to identify such a comparison for Khottadih and Sonepur Bazari opencast projects, which produces 76.78 million tons coal and 262.43 million tons overburden during the past 19 years (Khorshiddoust, 2009; Ramalho & Ramalho, 2011).

2 Study Area

Raniganj coalfield overlies the granite plateau ring of Choto Nagpur in the Damodar valley region surrounded by the Durgapur-Asansol industrial belt with a general elevation of about 100 m (Sarkar, 2017a). This coalfield falls under Eastern Coalfields Limited (ECL) and is the birth place of coal mining in India. Raniganj coalfield is located within four districts of West Bengal, i.e. Burdwan (71%), Birbhum (9%), Bankura (8%) and Purulia (7%), as shown in Fig. 5.1 (Sarkar & Bhattacharya, 2016). Origin of coal in this area was of Permian age, estimated to be 205 to 108 million years old in Raniganj and Barakar stages of Gondwana age (Sarkar, 2017a, b, c). This coalfield is elliptical in shape (Sarkar, 2019a, b, c, d, e, f). East-west extension is about 75 km and north-south extension is about 35 km. It has an area of about 1530 sq. km and falls within latitudes 23°30′ N to 23°52′ N and longitudes 86°38′ E to 87°23′ E (Sarkar, 2018). Mean elevation is 98.45 m with broad undulation (Sarkar, 2019a, b, c, d, e, f). As on 2018, this coalfield had 17 running OCPs and 21 abandoned OCPs within 11 areas (ECL, 2013; Sarkar, 2019d). Two opencast mines are selected for analysis of environmental cost namely Khottadih and Sonepur Bazari. These two collieries belong to Raniganj coalfield and are the departmental mine of Eastern Coalfields Limited. Khottadih OCP is part of Pandaveswar area, located within latitude 23°23′59″ N to 23°29′30″ N and longitudes 87°04′10″ E to 87°12′25″ E (Ahmed et al., 2013). It is commenced in 1982 and situated at the downfield area of Raniganj coalfield (Sarkar, 2019c). Coal is extracted from R_{v} and R_{vi} seams at a depth of 50 m with stripping ratio of 1:4.33 (Sarkar, 2019b). Total leased area of this project is about 334.5 hectare (including UG project), of which total opencast project area is 6.4 hectare (Sarkar, 2017a, b, c). Backfilling process is completed at the southern part of this OCP. This mine will run for further 5 years (estimate) from 2018 (Punia & Raman, 2006). In 2014 and 2015, 25,000 plants were

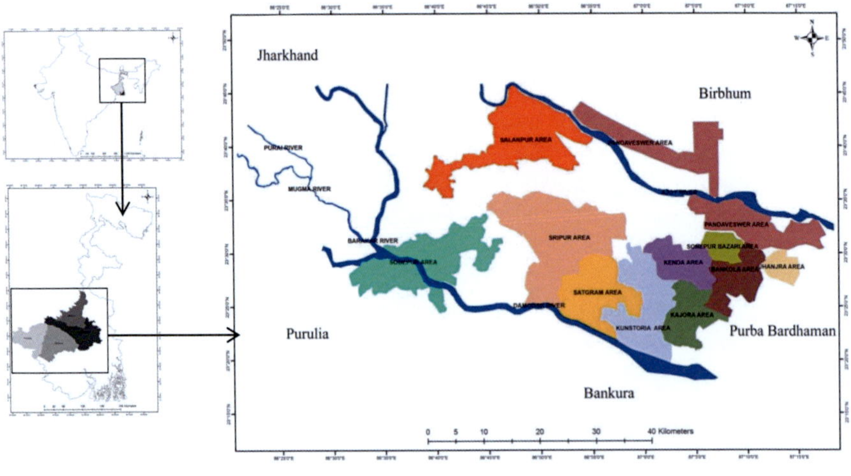

Fig. 5.1 Location map of the study area. (Source: ECL/CMPDI/author)

planted (Katoria et al., 2013). Laterite soil is massively found at the surface of this mine (Guha, 2014). Anthropogenic agents like dumper (34), dozer (7) and shovel (8) keep functioning continuously in this mine. On the other hand, Sonepur Bazari OCP is part of Sonepur Bazari area, located within latitude 23°39′40″ N to 23°42′ 10″ N and longitudes 87°11′0″ E to 87°16′16″ E (Tiwary & Dhar, 2006; Sarkar, 2019a, b, c, d, e, f). This is the biggest OCP of Raniganj coalfield which commenced in 1990 (Sarkar, 2019b). The total working area of this mine is 1575.40 hectare (Sarkar, 2017a, b, c). Coal is extracted from R_{ii} to R_{viii} seams at the depth of 150 m with Stripping ratio of is 1:4.72 (Sarkar, 2019b). The quality of coal is G_4. According to 2014 estimation, the mine has a total reserve of 172.86 million tons, a life of 58 years (Reddy et al., 2016; Pandey & Kumar, 2013). Since the birth, 225,500 plants are planted in this OCP and total planted area is about 92.5 hectare (Choudhury, 2011). Anthropogenic agents like dumper (63), dozer (17), dragline (1) and shovel (3) keep functioning continuously in this mine (Srivastav et al., 2016; Kuchay & Ramachandra, 2016).

3 Database and Methodology

3.1 Remote Sensing Data Sources

The satellite data are gathered from USGS (US Geological Survey) Earth Explorer portal. Five temporal cloud-free satellite images of summer season are used for preparing land use land cover map from 1973 to 2015 (shown in Table 5.1). The visible bands, near infrared bands and middle-infrared band are utilized in this study.

Table 5.1 Details of remote sensing satellite data

Year	Date of acquisition	Path/row	Spatial resolution	Sensor	Projection
1973	18th and 20th March	149/44 and 150/43	60 m	Landsat MSS	World Geological Survey 84/UTM, Zone 45
1992	15th March	139/44	30 m	Landsat 5 (TM)	
2002	19th March	139/44	30 m	Landsat 7 (TM)	
2010	25th March	139/44	30 m	Landsat 7 (TM)	
2015	15th March	139/44	30 m	Landsat 8 (ETM+)	

Further, five individual topographical maps on 1:50,000 scales are used namely 73M/1, 73M/2, 73M/5, 73M/6 and 73I/14 from the Survey of India to build the base layer of these satellite data.

3.2 *Data Processing and Image Classification*

Geometric correction is done using linear rubber sheeting method. Image rectification is also performed using map and topographical sheet at different scale (Yu & Ng, 2007), after which images are resampled by using the bilinear interpolation algorithm method. Image enhancement technique such as histogram equalization, contrast stretching and tail trimming is performed to improve the visual interpretability of the image (Csorba & Szabó, 2010). Colour space transformation functions like RGB (red, green, blue) to IHS (intensity, hue, saturation) and the reverse IHS to RGB function are applied for the year 2010 and 2015 to enhance the image information (Mustard et al., 2004). Finally, supervised image classification technique under parametric decision rule, maximum likelihood algorithm is used (Neill et al., 1988). Post-classification filtering is also applied to enhance the accuracy (Görcs, 2011). Entire tasks are done in ERDAS Imagine 9.2 software. The prepared land use land cover map in Raniganj coalfield area of different spatial map of 1973, 1992, 2002, 2010 and 2015 in ten distinguished categories—forest, agriculture, fallow land, river, river sand, water body, exposure land, mining lagoon, urban area, excavated land are shown in Fig. 5.2.

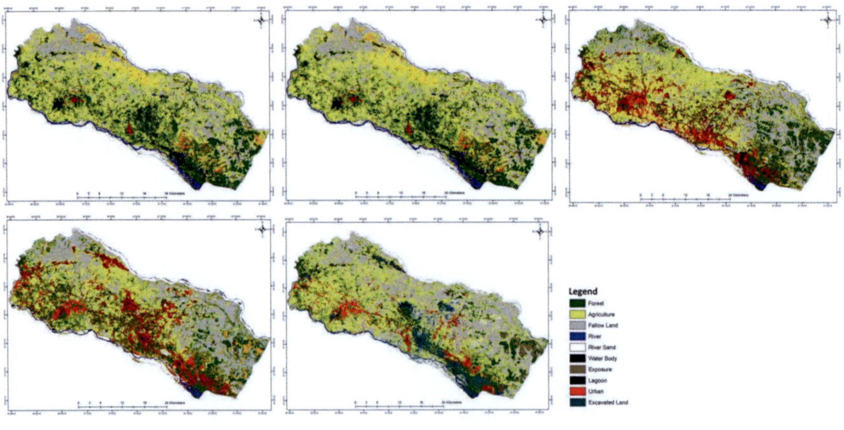

Fig. 5.2 Land use map of 1973, 1992, 2002, 2010 and 2015 (clock wise), Raniganj coalfield. (Source: Landsat, topographical map, author)

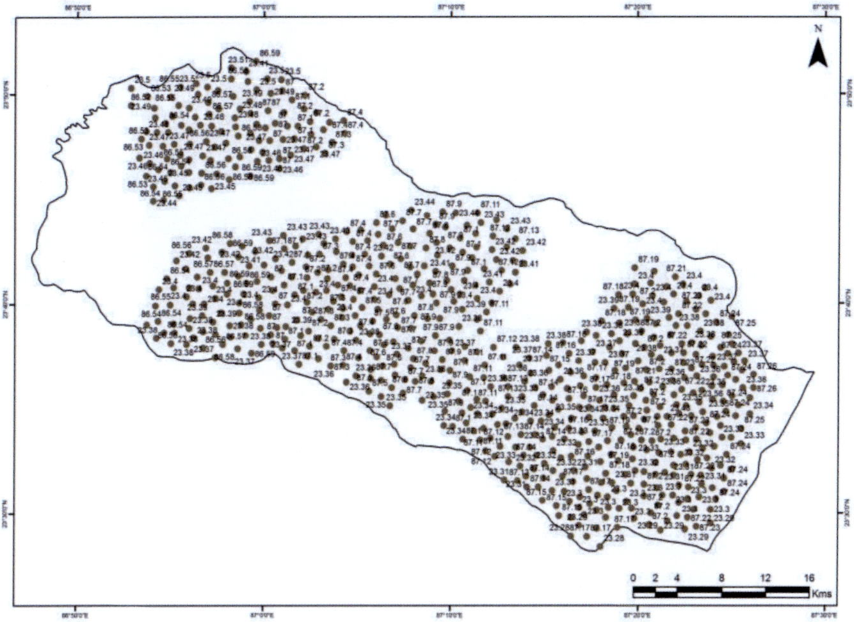

Fig. 5.3 Field validation GPS waypoints. (Source: Primary survey by the author)

3.3 Ground Truth Training Data and Accuracy Assessment

Ground truth verification data is collected through wide range of filed survey during May 2014 and December 2015 with global positioning system (GPS) device (Garmin 72H) representing all land cover classes for in-situ collection of ground truth geographic coordinates and locations; land use and land cover attributes, species information, and other general notes using stratified random sampling method (Alemu et al., 2015; Areendran et al., 2013; Samanta, 2015; Liu et al., 2016). A total of 950 random in situ GPS waypoints are collected to ensure that each class received a minimum of 30 ground truth points (shown in Fig. 5.3).

These randomly generated points are assigned to a land cover class through a combination of expert knowledge, evaluation of spectral signatures, supervised clustering and analysis of topographical map (Stipinovich, 2005; Jia et al., 2018). These GPS waypoints are plotted in the study region according to its geographical coordinate value (Gyawali et al., 2004). Landscape photograph is also shown in Fig. 5.4. In addition to these, the accuracy for each classification is measured by ERDAS Imagine software using a confusion matrix. Hence, stratified random sampling method is applied to furnish the 720 ground truth reference data. These ground truth points overlie the land use land cover map and respective value is obtained by using the following algorithms:

Fig. 5.4 Landscape photograph of forest (Purusattampur area, 23°42′10″ N and 87°16′16″ E), excavated quarry land (Sonepur Bazari area, 23°39′40″ N and 87°11′0″ E), agricultural land (Kendra area, 23°25′42″ N and 86°42′41″ E), urban area (Barakar railway station, 23°54′22″ N and 87°03′34″ E), fallow land and (Mithali area, 23°43′00″ N and 87°03′00″ E), river sand (river bed of Ajay, 23°32′25″ N and 86°20′15″ E), exposure land (Sonepur Bazari area, 23°38′26″ N and 87°12′12″ E), water body (Dalurband area, 23°42′36″ N and 87°01′52″ E), river (Barakar river, 23° 59′12″ N and 87°04′10″ E) and mining lagoon (Poidih abandoned mine, 23°29′30″ N and 87°12′ 25″ E) (clock wise), Raniganj coalfield. (Source: Primary survey by the author)

$$\text{Overall accuracy} = \frac{\text{Total number of classified pixels (diagonal)}}{\text{Total number of reference pixel}} \times 100$$

$$\text{User accuracy} = \frac{\text{Number of correctly classified pixels in each category}}{\text{Total number of classified pixels in that category}} \times 100$$

$$\text{Producver accuracy} = \frac{\text{Number of correctly classified pixels in each category}}{\text{Total number of classified pixels in that category (column total)}} \times 100$$

$$\text{Kappa coefficient } (T) = \frac{(\text{Total sample} \times \text{total corrected sample}) - \Sigma(\text{column total} \times \text{row total})}{\text{Total sample}^2 - \Sigma(\text{column total} \times \text{row total})}$$

Field validation data are not available for 1973, 2002 and 2010; therefore, it is assumed that a similar accuracy is obtained using the same methods from the 2015 image (Brown et al., 2000; Kindu et al., 2013; Kendall & Antrobus, 2013). Hence, 878 ground truth points are accurately classified resulting in an overall accuracy of 90.74% and a Kappa coefficient of 0.92 in 2015 as shown in Table 5.2.

Table 5.2 Confusion matrix for accuracy assessment of landscape map, 2015, Raniganj coalfield

Classified data	Reference data												Users accuracy (%)
	LULC classes	F	AL	FL	R	RS	WB	E	L	U	EL	Total	
	F	**1612**	48	21			4			17		1702	94.7
	AL	20	**1180**	57	5	7	8	5		5	3	1290	91.5
	FL	56	19	**1008**				29		34		1146	88.0
	R	9	20		**523**	9	124		10			695	75.3
	RS	5	2	3		**716**	12				2	740	96.8
	WB	9	17		8		**506**		4			544	93.0
	E	13	35	36		2		**315**		2	5	408	77.2
	L						23		**423**	20		466	90.8
	U	8	25	28	2			6		**1522**		1593	95.5
	EL		8	7	4					3	**682**	702	97.2
	Total	1732	1354	1160	542	734	677	355	437	1603	692	**9286**	
Producer accuracy (%)		93.1	87.1	86.9	96.5	97.5	74.7	88.7	96.8	94.9	98.6		
Overall accuracy: 90.74%													
Kappa coefficient: 0.92													

Source: Computed by the author

3.4 Land Use Land Cover Transformation

Land use land cover transformation is measured to calculate the effective and effected land use between 2 successive years (Baodong et al., 2008; Cao et al., 2007; Lahiri-dutt, 2003; Eggert, 2002; Freudenburg & Madison, 2002). This transformation is done in Arc GIS software following the combined algorithm from spatial analyst tool (local analysis) (Homère et al., 2016; Biali et al., 2014). Typically this transformation activity depicts the converted land uses from individual to individual and individual to other (Kosmas et al., 2013; Hamdy et al., 2017). The change matrix data is collected from attribute table (Tibbett & Fourie, 2007). Land use land cover transformation matrix for individual land use category is computed for the year 1973–1992, 1992–2002, 2002–2010 and 2010–2015 as shown in Tables 5.3, 5.4, 5.5 and 5.6 respectively.

3.5 Landscape Fragmentation

Landscape fragmentation is measured both at class level and landscape level. In individual class level, the author computes composition of landscape, fragmentation process or patterns through temporal pattern (1973–2015). While in landscape level, the author explores fragmentation patterns, diversity, adjacency and isolation (Abdallah, 2016; Wodaje, 2016; Yuksel et al., 2008). These entire tasks are performed using Arc GIS 10.2 and Fragstats 4.2 software. Firstly, each individual land use land cover map is converted from raster to polygon following conversion from raster algorithm (3d analyst tool) using Arc GIS (Vázquez-Quintero et al., 2016; Mukhopadhaya, 2016). After that fragmentation is computed using class analysing tool and landscape analysing tool, respectively, from patch analyst tool (Adedeji et al., 2015; Abdul Rahaman et al., 2017). Land use fragmentation is performed using Fragstats software by grid method followed by input layer and analysis tab (Eastman, 2012; Handayani et al., 2017; Mishra et al., 2014). Patch metrics, class metrics and landscape metrics indices are also computed. Further, area edge, shape, core area, contrast, aggregation and diversity are also computed for different indices. The selected landscape metrics are shown in Table 5.7.

The above-mentioned algorithms are used to compute 14 class-level indices and two landscape-level indices. ArcGIS and Fragstats software are used to measure the landscape indices in Raniganj coalfield from 1973 to b2015. The results are shown in Tables 5.8 and 5.9, respectively.

Table 5.3 Land use land cover change matrix, 1973–1992, Raniganj coalfield

Area (hectare)	Forest	Agricultural land	Fallow land	River	River sand	Water body	Exposure land	Lagoon	Urban	Excavated land
Forest	513,260,600	68,706,900	28,380,700	0	0	595,800	46,746,900	741,600	22,330,700	49,147,200
Agricultural land	53,988,100	797,263,700	119,354,600	0	4,387,500	201,600	30,654,900	527,400	46,169,100	38,226,600
Fallow land	44,679,600	71,918,100	462,366,300	0	23,744,700	137,700	3,593,700	331,200	15,194,800	3,851,100
River	1,572,100	1,645,200	4,828,000	18,764,200	19,677,700	2,157,300	0	1,338,300	462,100	2,040,800
River sand	818,100	550,800	8,413,000	1,807,200	43,362,900	39,600	3600	43,200	526,100	548,100
Water body	237,500	1,803,600	1,547,200	1,280,700	354,900	7,133,000	0	1,580,400	248,100	839,700
Exposure land	12,530,800	10,348,300	2,953,800	0	0	0	68,514,000	0	589,300	12,816,100
Lagoon	149,900	84,600	171,000	0	0	3,453,200	486,000	7,944,900	6900	1,417,100
Urban	0	0	0	0	0	0	0	0	95,044,000	4,186,800
Excavated land	15,033,700	0	7,890,000	0	0	1,845,000	37,253,700	12,400,100	0	140,378,500

Source: Computed by the author

Table 5.4 Land use land cover change matrix, 1992–2002, Raniganj coalfield

Area (hectare)	Forest	Agricultural land	Fallow land	River	River sand	Water body	Exposure land	Lagoon	Urban	Excavated land
Forest	399,739,800	78,228,500	58,999,100	0	0	671,525	25,675,900	425,400	626,000	88,494,700
Agricultural land	56,026,869	911,839,258	106,322,564	0	1,652,352	256,669	51,458,554	0	51,193,257	15,275,258
Fallow land	35,876,800	69,363,000	664,952,526	0	43,762,569	132,589	3,621,547	0	50,787,000	3,783,600
River	301,256	1,134,125	1,850,259	8,764,259	9,696,600	0	0	0	256,987	575,258
River sand	256,254	785,245	5,423,258	1,102,584	44,271,587	326,587	0	0	256,874	325,478
Water body	0	563,258	895,254	492,147	251,300	6,813,725	0	992,568	152,698	526,489
Exposure land	10,545,258	5,572,148	1,052,670	0	0	0	68,654,521	0	0	8,844,654
Lagoon	149,900	84,600	171,000	0	0	3,482,014	352,158	7,582,530	0	2,157,254
Urban	0	0	0	0	0	0	0	0	102,542,548	3,168,259
Excavated land	4,812,547	0	5,242,458	0	0	1,106,574	9,080,500	3,582,648	0	41,256,247

Source: Computed by the author

Table 5.5 Land use land cover change matrix, 2002–2010, Raniganj coalfield

Area (hectare)	Forest	Agricultural land	Fallow land	River	River sand	Water body	Exposure land	Lagoon	Urban	Excavated land
Forest	822,112,000	81,022,000	141,653,800	0	0	771,300	41,784,300	727,200	20,646,900	18,897,300
Agricultural land	71,878,300	636,270,700	83,979,900	0	1,125,257	0	13,129,200	720,000	49,282,200	11,348,100
Fallow land	14,750,100	40,064,200	400,170,800	0	3,551,400	204,300	3,995,100	0	7,287,300	3,395,700
River	259,500	487,400	784,400	5,243,000	4,724,200	0	0	0	183,100	303,300
River sand	1,296,600	2,403,400	9,821,700	4,004,100	75,097,100	113,400	0	0	1,022,600	1,476,000
Water body	0	145,000	586,300	523,900	90,000	8,494,900	0	854,100	125,400	885,400
Exposure land	12,598,500	5,210,400	4,032,000	0	0	0	78,434,700	0	0	4,651,200
Lagoon	158,400	82,500	17,100	0	0	1,766,700	207,000	4,581,300	0	1,476,200
Urban	0	0	125,489	0	0	0	0	0	3.08E+08	3,500,100
Excavated land	5,021,200	0	5,499,400	0	0	1,427,400	8,099,700	6,508,600	0	44,837,100

Source: Computed by the author

Table 5.6 Land use land cover change matrix, 2010–2015, Raniganj coalfield

Area (hectare)	Forest	Agricultural land	Fallow land	River	River sand	Water body	Exposure land	Lagoon	Urban	Excavated land
Forest	425,455,500	137,512,800	35,200,800	0	0	441,000	19,788,300	185,400	19,853,100	17,612,100
Agricultural land	13,045,500	296,056,000	80,879,500	0	0	274,500	3,993,300	226,800	15,214,500	13,413,600
Fallow land	3,590,100	76,887,000	241,563,600	0	14,289,300	92,700	1,087,200	28,800	2,691,000	2,975,400
River	779,200	634,500	2,428,200	4,869,900	5,551,500	0	0	0	0	1,860,000
River sand	624,600	1,793,700	11,207,700	1,478,700	51,355,800	371,700	42,300	202,500	874,800	4,088,400
Water body	258,700	117,900	71,100	616,500	145,800	9,459,800	0	1,224,900	93,600	2,703,600
Exposure land	21,361,000	10,244,400	6,580,800	0	0	0	52,505,500	0	0	3,009,800
Lagoon	536,400	45,900	30,600	522,000	120,600	943,200	0	5,881,000	60,300	1,027,900
Urban	0	0	0	0	0	0	562,500	0	442,317,100	5,699,000
Excavated land	18,263,700	5,308,200	5,469,300	0	0	1,587,600	438,300	2,027,700	0	53,331,300

Source: Computed by the author

Table 5.7 Selected landscape indices for Raniganj coalfield area (the equations are derived from Fragstats 4.2)

Category	Metrics	Equation	Unit	Description
Patch size and density	Number of patches	$NP = n_i$	Number	This indicates number of patches in each class type and the growth of particular patches whether in an aggregated or fragmented growth in the region (Province, 2007). It is always equal to or greater than 1; the value 1 indicates that there is only 1 patch of a particular class, showing aggregated growth (Ram & Kiran, 2013). Further, the significant increase in the number of patches indicates the occurrence of habitat fragmentation (Rabia, 2016)
	Largest patch index	$LPI = \frac{\max\left(a_{ij}\right)_{j-1}^{n}}{A}$	%	It is the single largest patch at the class level which indicates dominance, changes, homogeneity and fragmentation within a landscape (Fadhil, 2009). It denotes 0 when the largest of corresponding patch type is absent, whereas it denotes 100, when the entire landscape consists of a single patch (Mckenna, 2009)
	Patch density	$PD = \frac{n_i}{A}(10{,}000)(100)$	Patch per hectare	It is a measure of fragmentation of a patch type or landscape (Mulkova & Popelková, 2010). Lower patch density denotes a single homogeneous patch, while greater value denotes fragmented growth; patch density increases with an increase in patches (Estoque, 2012)
	Total edge length	$TE = \sum_{k=1}^{M} e_{ik}$	Hectare	This is an absolute measure of the total edge length for a particular patch type. Greater total edge length indicates larger continuous patches and vice versa (Smith & Palmquist, 2001).
	Mean patch size	$MPS = \frac{NP}{TA}$	Hectare^2	It is the average size of patches of a land use class and represents landscape configuration, which is decreased

(continued)

Table 5.7 (continued)

Category	Metrics	Equation	Unit	Description
				with increasing fragmentation (Gundimeda, 2006). Landscape with a smaller mean patch size for the target patch type than another landscape might be considered more fragmented (Pike, 2008)
Shape and edge	Patch size coefficient of variance	$\text{PSCoV} = \frac{\text{PSSD}}{\text{MPS}}$		It is the coefficient of variation of patches (Nita, 2008). It measures the relative variability of the percentage of the mean and not absolute variability (Saranathan et al., 2010)
	Edge density	$\text{ED} = \frac{\Sigma L}{\text{TA}} \times 1000$	m/km^2	It is the ratio of total edge distance to the total area (Jana, 2013). It explores edge length on a per unit area basis that facilitates comparison among landscapes of varying size (Tarboton, 2003). Zero edge density indicates no class of the landscape (Evans, 2016)
	Patch size standard deviation		Hectare	It is the standard deviation of patch areas (Mini & Manjunatha, 2014). It is an absolute measure of mean patch size and difference in patch size among patches (Paul, 2013)
	Mean shape index	$\text{MSI} = \Sigma \frac{\text{PA}}{A}$	m/km^2	It is a measure of average patch shape for a particular patch type or for all patches in the landscape (Chadderton & Elliott, 2011). When all patches in a landscape are square (grids) or circular (polygons), then mean shape index will 1 (Sayemuzzaman & Jha, 2014). It increases with increasing patch shape irregularity (Paul, 2013)
	Area weighted mean shape index	$\text{AWMSI} = \frac{\text{MSI}}{\text{PA}}$		It is the mean patch shape complexity, weighted by patch area (Evans, 2016). Areas weighted mean shape index indicates all patches are circular (Rabia, 2016).

(continued)

Table 5.7 (continued)

Category	Metrics	Equation	Unit	Description
				Increasing value indicates that patches become complex in shape (Ram & Kiran, 2013)
	Mean perimeter area ratio	$\text{MPAR} = \frac{\Sigma\text{PA}}{\text{AR}} \times \text{Num}P$	m/km^2	It is the measure of shape complexity (Smith & Palmquist, 2001). It is the function of sum of each patch perimeter, area ratio and number of patches (Pike, 2008)
Proximity and isolation	The Shannon diversity index	$\text{SHDI} = -\sum_{I=1}^{M} (p_i \cdot \ln p_i)$	m	It is the relative measure of patch diversity of the community ecology and only available at the landscape level which gives an indication of landscape fragmentation and heterogeneity (Gundimeda, 2006). Shannon diversity index indicates there is only one patch in the landscape; higher Shannon diversity index indicates more diverse land use and cover (Mckenna, 2009). Index value will be bigger as the number of patch types or proportional distribution of patch types increases (Nita, 2008).
	Shannon evenness index	$\text{SHD} = -\sum_{i=1}^{M} p_i \cdot \ln p_i / 1 \text{ nm}$	m	It is a measure of patch distribution or abundance and only available at the landscape level (Chadderton & Elliott, 2011). Shannon evenness index indicates low patch distribution (Gundimeda, 2006). Shannon evenness index of 1 indicates the even distribution of patch types (Mckenna, 2009). An even distribution of the area among the patch types results in maximum evenness (Nita, 2008)

Table 5.8 Class-level indices from 1973 to 2015, Raniganj coalfield

Landscape indices	Year	Forest	Agricultural land	Fallow land	River	River sand	Water body	Exposure land	Lagoon	Urban	Excavated land
Number of patches	1973	22,216	33,061	13,080	6408	689	2794	14,388	1917	7658	9441
	1992	41,415	32,735	19,688	5952	821	2319	17,994	2450	19,178	13,804
	2002	44,170	31,855	22,923	1462	3454	2611	24,696	3163	21,226	17,147
	2010	53,172	36,730	25,175	1242	6043	3104	34,197	1873	32,263	26,033
	2015	59,981	37,166	26,290	1056	7868	3090	25,239	2108	32,603	17,147
Patch density	1973	38.236	28.256	19.256	6.526	0.652	2.102	13.654	1.201	5.625	12.201
	1992	32.564	30.215	17.624	5.216	0.846	1.956	18.256	1.950	16.256	15.265
	2002	40.231	26.364	16.102	1.365	6.126	2.620	28.245	2.695	26.358	19.856
	2010	34.125	27.698	12.365	1.102	4.021	2.056	14.652	1.186	26.254	7.658
	2015	17.259	29.652	20.362	0.980	6.856	2.421	3.654	1.265	16.859	14.521
Largest patch index	1973	18.201	9.526	7.269	1.201	0.039	0.524	2.654	0.216	1.624	2.126
	1992	14.256	11.246	6.387	0.952	0.080	0.236	6.912	0.310	4.325	4.987
	2002	20.354	7.246	4.215	0.547	1.210	0.856	8.856	0.895	8.654	7.568
	2010	16.325	7.698	2.321	0.432	0.562	0.123	3.568	0.120	8.452	1.268
	2015	6.248	10.564	7.658	0.215	1.625	0.756	1.003	0.256	5.846	2.689
Total edge	1973	17,341,756.1	18,226,821.49	9,988,992.8	1,787,515.88	583,497.05	630,478	2,744,298.5	415,808.8	1,362,484.1	3,566,547.8
	1992	13,507,148.7	20,489,816.40	9,101,732.4	1,660,097.41	695,237.13	523,167	4,710,312.4	531,391.71	3,411,983.9	5,284,871.4
	2002	15,913,500	17,846,220	9,385,020	709,140	2,029,920	630,000	7,721,820	709,260	7,780,080	8,142,000
	2010	15,252,698.3	13,695,074.47	7,209,901.9	640,936.775	1,377,647.8	532,291	4,667,691.9	417,124.10	8,675,120.1	3,385,493.5
	2015	7,968,584.76	17,937,756.08	12,774,115	383,651.01	2,533,393.8	584,605	1,326,444.8	401,368.93	5,399,480.7	4,915,028.1
Edge density	1973	55.335	58.168	31.826	5.696	1.854	2.000	8.756	1.322	4.344	11.382
	1992	43.106	65.391	29.0473	5.298	2.218	1.669	15.032	1.695	10.889	16.866
	2002	50.3586	56.474	29.699	2.244	6.423	1.993	24.435	2.244	24.620	25.765
	2010	48.267	43.338	22.815	2.028	4.3596	1.684	14.771	1.32	27.4526	10.713
	2015	25.115	56.535	40.261	1.209	7.984	1.842	4.180	1.265	17.017	15.491

Mean patch size	1973	0.5824	0.5983	0.3709	1.4807	0.1653	3.0178	3.6729	2.7691	2.1558	1.2996
	1992	1.3940	0.5270	0.6128	1.4809	0.1654	3.0185	2.6761	2.7693	2.1558	1.2823
	2002	1.6186	0.6485	0.6659	0.7953	0.4067	3.4062	2.5039	2.8712	1.8918	1.1380
	2010	1.8859	0.9492	0.6133	1.2923	0.7513	4.0709	3.5713	1.4432	2.0139	1.8846
	2015	2.7144	0.8285	0.5734	1.4712	0.8869	4.3355	4.8233	1.2557	1.6592	1.8913
Mean shape index	1973	1.296	1.183	1.370	1.356	1.261	1.263	1.252	1.246	1.256	1.272
	1992	1.323	1.332	1.342	1.255	1.415	1.274	1.281	1.270	1.242	1.280
	2002	1.243	1.346	1.317	1.367	1.255	1.221	1.249	1.229	1.243	1.281
	2010	1.317	1.330	1.344	1.376	1.307	1.258	1.297	1.259	1.289	1.299
	2015	1.334	1.362	1.343	1.317	1.291	1.211	1.265	1.21	1.273	1.267
Area weighted mean shape index	1973	8.293	22.181	10.162	8.226	2.501	2.603	1.666	2.558	0.877	3.792
	1992	6.4555	28.661	9.256	7.641	2.982	2.158	2.856	3.271	2.197	5.619
	2002	6.214	26.609	7.368	7.283	8.480	1.597	2.046	2.708	3.653	5.221
	2010	5.321	19.609	10.573	4.404	6.179	1.646	2.474	2.016	3.095	6.654
	2015	9.052	24.411	13.652	3.454	5.099	1.528	5.874	1.638	4.155	9.030
Patch size coefficient of variance	1973	2363.65	8340.468	3657.92	3033.215	482.584	489.277	836.635	752.756	323.052	1301.804
	1992	1841.280	9376.264	3333.118	2817.434	575.333	406.538	1436.460	962.866	809.257	1929.676
	2002	2122.029	5178.708	1569.066	1456.741	3804.184	213.052	431.978	962.265	1195.015	1309.010
	2010	1427.099	2731.143	2560.325	819.279	2217.867	262.774	868.819	453.559	833.837	2025.092
	2015	1132.689	3088.358	3190.552	655.091	1959.538	313.480	2880.019	467.412	1228.510	3381.010
Patch size standard deviation	1973	16.690	140.549	50.484	19.381	28.535	1.578	2.114	2.629	0.674	6.748
	1992	13.162	158.742	46.440	18.965	34.849	1.313	3.635	3.362	1.690	10.040
	2002	10.361	79.847	23.447	18.314	55.842	0.527	1.245	3.351	4.112	7.576
	2010	11.732	32.472	67.877	13.612	51.568	0.687	4.513	1.649	4.071	21.604
	2015	11.138	42.216	55.831	5.938	20.969	0.762	21.266	1.457	5.535	27.076

(continued)

Table 5.8 (continued)

Landscape indices	Year	Forest	Agricultural land	Fallow land	River	River sand	Water body	Exposure land	Lagoon	Urban	Excavated land
Mean perimeter area ratio	1973	1315.82	1344.857	1437.410	1273.875	1126.561	1320.11	1502.461	1389.405	1402.146	1382.716
	1992	1274.110	1287.594	1309.733	1183.072	1081.363	1273.52	1432.278	1359.826	1350.154	1308.824
	2002	1151.218	1073.240	1079.850	1073.070	1116.738	1136.83	1126.807	1146.740	1143.486	1109.619
	2010	1225.643	1183.373	1199.933	1164.910	1286.983	1257.63	1278.145	1267.195	1231.215	1146.517
	2015	1207.298	1212.615	1215.398	1198.112	1260.597	1192.47	1330.564	1278.060	1257.139	1258.928

Source: Computed by the author

Table 5.9 Landscape-level indices from 1973 to 2015, Raniganj coalfield

Year	Shannon diversity index	Shannon evenness index
1973	1.453	0.606
1992	1.524	0.635
2002	1.536	0.640
2010	1.610	0.671
2015	1.612	0.672

Source: Computed by the author

Table 5.10 Details of the monitoring station, Raniganj coalfield

Sl no	Monitoring station	Latitude	Longitude	Elevation (m)
1	Training Centre	23°24′9.36″ N	87°7′22.36″ E	100.32
2	Dispensary	23°26′58.54″ N	87°8′30.25″ E	99.26
3	Kenda Area Office	23°24′16.29″ N	87°8′36.75″ E	104.36
4	CISF Camp/Office	23°25′53.57″ N	87°6′52.36″ E	110.00

Source: Computed by the author

3.6 Evaluation of Environmental Impacts

3.6.1 Pollution Data

Four monitoring stations of Sonepur Bazari OCP are selected in a way that these stations can cover the entire region shown in Table 5.10. Global positioning system (GPS) receiver is used to get the latitude, longitude and elevation of the monitoring stations. Three pollutants of air such as suspended particulate matter (SPM), respirable particulate matter (RPM) and nitrogen oxide (NO_x) are measured in the monitoring stations. Primarily the concentration of air pollutant data is gathered from CMPDI and ECL. After that, the researcher collects the concentration of air pollutant from the field survey with the help of the personnel of ECL.

Suspended particulate matter is measured by volumetric filtration and gravimetric weighing, where sampling of air through a filter medium at a known flow rate for a specified time is performed (Zurayk et al., 2001). Testo 460 pocket-sized RPM meter was used to measure the respirable particulate matter (Goswami, 2013a). Simple/total NO_x industrial emission analyser, model no. E4500 (hand-held industrial combustion gas and emissions analyser), portable, is used to measure nitrogen oxide (Kumar et al., 2015). Seasonal (8 hourly), 2-year combined concentration of suspended particulate matter, respirable particulate matter and nitrogen oxide data are shown in Table 5.11.

The air quality data is imported in Arc GIS 10.2 software where inverse distance weighted (IDW) a method for multivariate interpolation, a process of assigning values to unknown points by using values from usually scattered set of known points is applied (El Jazouli et al., 2017; Eckert et al., 2015).

Table 5.11 Seasonal (8 hourly) 2 years combined concentration of SPM, RPM (μg/m^3) and NO_x (ppm) from March 2015 to February 2017

Name of station	Summer (March to June)			Monson (July to September)			Post-monsoon (October to November)			Winter (December to February)		
	SPM	RPM	NO_x	SPM	RPM	NO_x	SPM	RPM	NO_x	SPM	RPM	NO_x
Training Centre	276	156	27	204	108	18	220	112	18	234	142	24
Dispensary	202	110	23	120	82	9	124	78	12	170	98	18
Kenda Area Office	304	128	29	212	106	19	230	82	11	270	110	20
CISF Camp/Office	296	153	26	116	74	10	216	109	19	234	130	23
Average	269.5	136.75	26.25	163	92.5	14	197.5	95.25	15	227	120	21.25
Standard deviation	46.52	21.81	2.5	52.08	17.07	5.22	49.35	17.72	4.08	41.62	19.73	2.75
% CV	17.26	15.95	9.52	31.95	18.46	37.34	24.98	18.61	27.22	18.34	16.44	12.95

3.6.2 Wage Studies and Value of Health Risks

This method uses information on risk premium to value people willingness to pay to avoid hazards using the following algorithm:

$$P = f(s1.s2.s3 \ldots si; n1.n2.n3 \ldots nj; e1.e2.e3 \ldots ej$$

Where, $s1.$ $s2.$ $s3 \ldots si$ are the structural variables; $n1.$ $n2.$ $n3 \ldots nj$ are the neighbourhood variables; $e1.$ $e2.$ $e3 \ldots ej$ are environmental variables. Hence, the hedonic price method of environmental evaluation uses surrogate markets for placing a value on environmental quality (Higginbottom & Symeonakis, 2014). Typically, this technique uses the effects of the project as a proxy for its environmental costs (Masoudi et al., 2018). Indeed, the hedonic price method is a revealed preference method of evaluation, which interprets market goods as a bundle of characteristics (Street et al., 2010; Higginbottom & Symeonakis, 2014). This method includes the structural features like type or age of house and whether a house is with concrete roof; neighbourhood features like existence of amenities, local transportation and distance from OCP, the accessibility of the location; environmental features like quality of air, availability of vegetation and status of land use usually delineated by location (Masoudi et al., 2018; Rashid et al., 2011; Cena et al., 2010).

4 Discussion and Findings

4.1 Land Use Dynamics

Massive temporal changes are observed in all categories of land use from 1973 to 2015. It certainly indicates that there is subsequent decrease and increase in land uses. Areal extent of different land use land cover is represented through pie diagram and is shown in Fig. 5.5. Percentages of various land use are also shown in Fig. 5.6 and rate of land use change is shown in Fig. 5.7.

Native forest in Raniganj coalfield area is classified into three sub-categories on the basis of crown density: dense forest whose crown density is above 40% and appears in dark red colour on FCC; open forest whose crown density is about 10–40% and appears light red to pinkish colour on FCC with smooth to medium texture; and degraded forest (Goswami, 2013a, b).

Dalbergia sisso and *Butea monosperma* are the most dominant plants species; *Lantana camara* and *Eupatorium odoratum* have been found to be most frequent shrubs in this area (Goswami, 2013a, b). Besides, major herbaceous species are *Saccharum spontaneum, Eragrostis tenella, Dichanthium annulatum, Cynodon dactylon* and *Desmodium triflorum* (Goswami, 2012). Continuous change in land use land cover has a serious threat on these vegetation community. In the beginning, the mining activities are carried out mostly in dense forest areas of West Burdwan

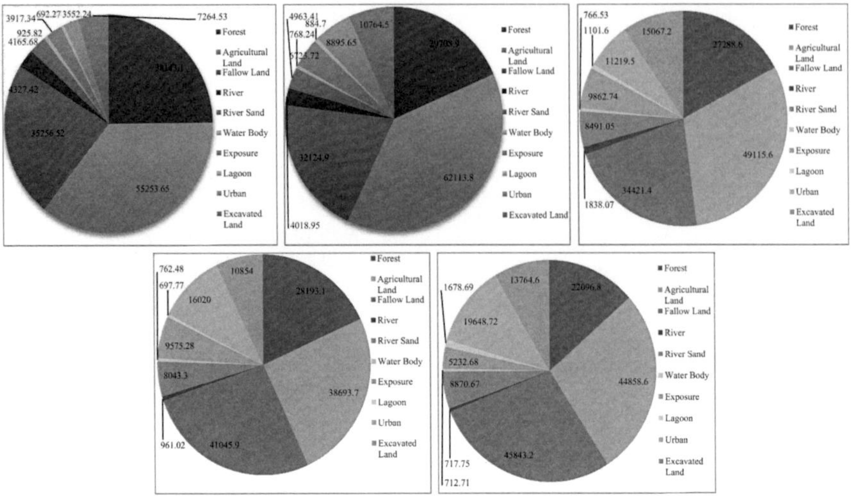

Fig. 5.5 Temporal change of land use in 1973, 1992, 2002, 2010 and 2015 (clock wise), Raniganj coalfield. (Source: Computed by the author)

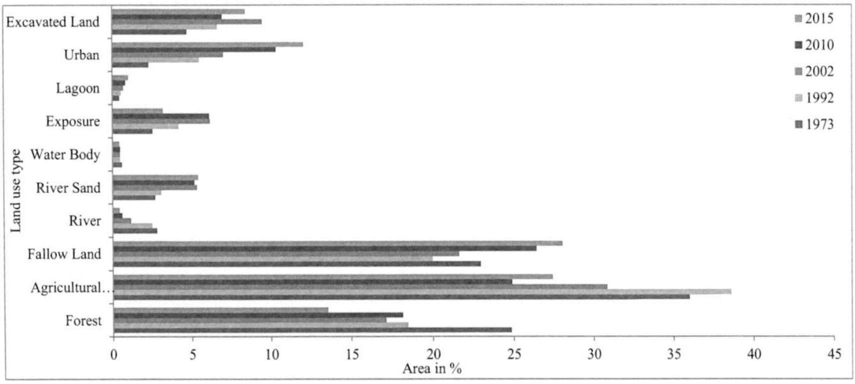

Fig. 5.6 Percentage of land uses, 1973–2015, Raniganj coalfield. (Source: Computed by the author)

district (Bandopadhyay, 2014). Therefore, the dense forest is converted into open forest or degraded forest area. Further, the decrease in the area of dense forests is also attributed to the removal of trees to initiate development of mining infrastructure (Sarkar, 2018, 2019a, b, c, d, e, f). However, due to the location of open forest in the peripheral area of dense forest, they are largely affected by local people intervention for requirement of fuel wood (Srivastav et al., 2016). Indeed the degraded forest is located primarily in the peripheral zones of the open and dense forests or around the human settlement (Pandey & Kumar, 2013). It is also noticed that complete destruction of forests occurs in the vicinity of coal mining region like Khottadih, Sonepur Bazari, Dabour, Pandaveswar and Purusottampur area. In addition to these,

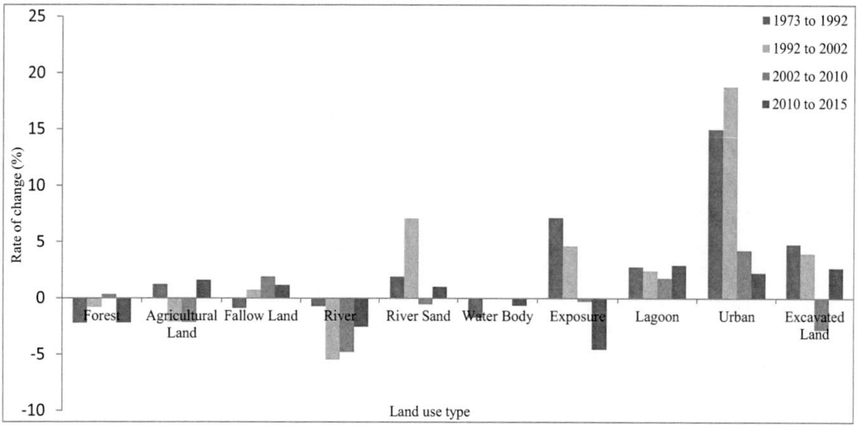

Fig. 5.7 Rate of land use change in 1973–2015, Raniganj coalfield. (Source: Computed by the author)

overburden from coal mine dumped in unmined areas creates mine spoils that later affect the surrounding vegetation. It is observed that almost all the coal areas currently being allocated for mining are located in forest land (Goswami, 2012). Forest showed a massive decline in their areal extent from 38,143.1 hectare in 1973 to 22,096.8 hectare in 2015. However, forest cover was 25.12% in 1973 but it decreased to 13.52% in 2015. Highly affected forest cover is mostly converted into fallow, agricultural and quarry land. Hence, the rate of decrease in forest is found to rise from −2% during 1973 and 1992 to −2.16% during 2010 and 2015. The results of percent change analysis show that a huge deforestation has occurred in the vicinity of coal mining areas. Former forest area is converted into quarry, mining overburden, sparse forest, fallow land and settlement in Purusattampur, Dalmiya, Kenda and Shibpur abandoned, mine which is located in the eastern and south western parts of the study area. However, blank forest indicates complete destruction of forests basically located around the active mining areas of Sonepur Bazari, Dalmiya, Mahalaxmi. Loss of plant vegetation also occurs due to haphazard spread out of waste material, poor environmental conditions, agricultural expansion into forest land and clearing of tree for fuel wood requirement by local inhabitants (Abdul Rahaman et al., 2017). Agriculture is the prime economic activity in Raniganj coalfield area; thus, the local people intend to meet their subsistence requirements through expansion of agriculture (Ahmed et al., 2013). In this area, the agricultural land can be classified into two classes—cropland and plantation. Cropped area appears in bright red to red in colour with varying shape and size in a contiguous to non-contiguous pattern with regular to sub-regular outline shape on FCC (Sarkar, 2017a). The main crop is rice which is monsoon season crop. Plantation appears in dark red to red tone of different sizes with regular and sharp edges indicating the presence of a fence around it on FCC (Goswami, 2013a, b). Infrastructure of mining and development of residential area cause loss of agricultural land. Agricultural land holds the highest percentage area of 55,253.65 hectare in 1973 and 44,858.6 hectare in 2015. Agricultural land is fairly distributed in the study area with the highest concentration around north, central, eastern and southern parts—Kalna, Jamalpur, Saktigarh, Sathgachiya, Bohar. Agricultural land dominates among all land uses, but the area has declined constantly from 55,253.65 hectare in 1973 to 44,858.6 hectare in 2015. Primarily from 1973 to 1992, agricultural land has increased from 36.21% to 38.58% at a rate of 1% due to conversion of forest land into it. But after 1992, agricultural land has continuously decreased from 38.58% in 1992 to 24.89% in 2010 at a rate of −2.12%. In the last 5 years, agricultural land exhibited a very small increase from 24.89% in 2010 to 27.45% in 2015 with a rate of 1.59%. Therefore, it can be said that agricultural land retains its own land and is moderately affected. During the ground truth survey, it was seen that in Satgram, Pandaveswar, Kunstoria, Kajora, Bankola, Jhanjra and Sonepur Bazari area, most of the forest and agricultural lands are lost due to opencast mining activities including land excavation, construction of new houses and buildings, and mining quarry located in the south, east and western part of the coalfield. In some region, the agricultural land is increased.

Fallow land appears light brown to light yellow with medium to smooth texture having non-contiguous pattern with irregular outline on FCC (Goswami, 2013a, b). The fallow land is increased due to loss of crop land which changed to fallow land as a result of the rapid increase in mining activities. Fallow land exhibits both positive and negative change over the years.

Indeed, fallow land is concentrated around the south-eastern and south-western parts of the area with very small patches around the northern part of the coalfield. Fallow land is the second largest land use in this coalfield, which has continuously increased from 35,256.52 hectare in 1973 to 45,843.2 hectare in 2015. During opencast mining activities, virgin land is to be extracted to meet the coal (Sarkar, 2019b, c). The virgin land can be either forest or agricultural land, which after the completion of mining operation turns into fallow land. For instance, there is a vast transformation of forest and agricultural land into fallow land in the south east and north western part of the study area, i.e. at Purusattampur, Sripur and Dabor. During 1973–2002, fallow land decreased from 23.24% to 21.62% with a rate of −1%. Later on, fallow land increased rapidly from 21.62% in 2002 to 28.05% in 2015 with a rate of 1.92%. A vast forest and quarry were converted into fallow land. A huge amount of mining waste and overburden dumped on the surrounding fertile land destructed the quality of soil and productive capacity turning them into fallow land. Damodar River in South and Ajay River in North constitute the main stream draining the study area. However, Raniganj coalfield is located on the bank of Damodar River. Further, there are several rivers like Barakar and Tumni in this coalfield. The river Damodar that flows through the southern parts of the study area following WNW to ESE direction is the only perennial river in this area. The areal percentage of river is almost at the stage of elimination, which had come down from 4327.42 hectare in 1973 to 717.75 hectare in 2015. Percentage of river land is continuously declined from 3% in 1973 to 0.43% in 2015. Most of the river land has converted into river sand and fallow land at a rate of −5.42% from 1992 to 2002, −4.77 from 2002 to 2010 and −2.53 from 2010 to 2015. River sand land use drastically increasded due to ceasation of river, rising population pressure, agricultural activity on river bed. Few segments of Damodar and Ajay rivers turned into river sand in recent time. River sand has increased from 4165.68 hectare in 1973 to 8870.67 hectare in 2015. River sand was 3.12% in 1973 but it increased to 5.42% in 2015. The rate of change of river sand was 7.10% in 1992–2002, but it reduced to 1.02% in 2010–2015 and transformed into crop area, river water and moist fallow area. Water bodies appear blue to dark blue or cyan colour depending on the depth of water on FCC (Goswami, 2013a, b). The areal cover of water bodies was minimum and also was decreasing at a rapid rate due to opencast mining and urbanization processes from 925.82 hectare in 1973 to 712.71 hectare in 2015. Water body was 1.02% in 1973 and reduced to 0.43% in 2015 at a rate of −2% from 1973 to 1992. After that, the rate of change has been settled about 0.52%. The area of water body recorded a negative change although very minimal in nature. Whereas land use class under rivers indicated no significant change during the entire period. This indicates that coal mining and allied activities don't have much influence on perennial river system. Exposure land is the vegeted land which is formed directly due to opencast mining activities. The overburden is dumped around the mine area which looks like small hillocks. These hills are 60 m in height and have a standard of 30̊ slope. At the time of completion of mining project, the scarp is stabilizes with native vegetation. Therefore, breakdown of this features is uneconomic. Exposure land has increased from 3917.34 hectare in 1973 to 5232.68 hectare in 2015 respectively. Exposure land was 3% in 1992 and 6.19% in 2002; but areal cover has reduced to 3.20% in 2015. The rate of change was 7% from 1973 to 1992 and −4.53% from 2010 to 2015. Mining lagoon is elongated water bodies measuring up to 1 km in length and is located within the mining site. However, it is the outcome of opencast mining. Initially mining lagoon area is renowned for mining activity. The mining lagoon located at Poidih, Khottadih and Dabour area is created at the time of digging of land to extract both coal and overburden. When extraction and production of coal are exhausted, mined areas are left abandoned. In due course of time, these areas are covered mostly with grasses and water. Mining lagoon is that waterlogged abandoned opencast land. Indeed, this mining lagoon is formed due to accumulation of rain water or ground water in the depression areas formed due to mining. Mining lagoon increased from 692.27 hectare in 1973 to 1678.69 hectare in 2015. Mining lagoon has increased almost 0.2% in 1973 to 1.03% in 2015. The rate of change was 3.14% from 1973 to 1992 and 2.93% from 2010 to 2015. Urban land use appears in cyan to light grey tone and as coarse texture, scattered pattern and irregular outline on FCC (Goswami, 2013a, b). Indeed, the study area was urbanized prior to independence. Durgapur and Asansol are two mighty cities in the study area. However, Raniganj coalfield has both industrial land use and rural settlement. These two cities have coal-based industry. Therefore, this area is urbanizing at a rapid rate. Urban areas show positive change in all the time periods. Urban land has increasded from 3552.24 hectare in 1973 to 19,648.72 hetare in 2015. South, central and western parts of this area are intensively urbanizing. These area include Barakar, Dishergarh, Sanctoria, Panjabi, etc. Urban land was 2.14% in 1973 but it increased to 12.02% in 2015. The increase in the rate of urbanization is primarily due to conversion of fallow land into urban land and construction of new building. The devastating rate of urbanization was 15% from 1973 to 1992 and 18.80% from 1992 to 2002. After 2002, the rate of urbanization has reduced to 2.27% from 2010 to 2015. Further, the rapid increase of urban area is caused due to industrial buildings, coal mining allied industrial settlement, development of industrial sector which requires residential colonies, school, community halls, etc.

Hence, major industrial settlements are located in the southern and south eastern part of the study area in the vicinity of Damodar River and around coal mining areas. However, thermal power station and Durgapur Steel Plant limited are the main industrial belt along with a number of small industries like sponge factory, glass factory, cement factory, coal washeries, etc. (Mangai et al., 2010). Excavated land appears black in colour with medium to smooth texture having linear to curvilinear pattern and irregular shape on FCC (Goswami & Goswami, 2014). Overburden dump appears white to light blue in colour with coarse to medium texture having contiguous pattern and irregular outer shape on FCC (Saranathan et al., 2010). In Raniganj coalfield area, an active mining area denotes huge surface excavation of coal and overburden, whereas abandoned mining area indicates old mines where mining operation is ended. Mining quarry land increased from 7264.53 hectare in 1973 to 13,764.6 hectare in 2015. Excavated land was 5.02% in 1973, and it increased to 9.47% in 1992 but later it reduced to 8.42% in 2015. The rate of change was 5% from 1973 to 1992 and 2.68% from 2010 to 2015. Dense forest, agricultural land and water body of Sodepur, Mahalaxmi, Slanpur and Sonepur Bazari are converted to excavated land or quarry or mining lagoon located in the western, central and southern part of the study area. It is immensely interesting to note that excavated area witnessed 4213.2 hectare reduction during the period of 2002–2010 as areas of abundant mines were largely converted into water bodies, mining lagoon, fallow land or forest within coal mine areas. For example, during 2002–2010, Sheebpur, Gangaramchak, Palasthali, Poidih, Dalmiya, Chapapur and Kapasara opencast mines are abandoned. The area of these mine is about 3823.2 hectare that is almost equal to the reduced quarry land. This reduction in Raniganj mining area indicates shortening of coal reserves and decline in the coal production. However, numbers of coal mines are now abundant and filled with water. Hence, the spread of opencast coal mining activity leads to destruction of forest cover within the mine area and conversion of surrounding agricultural lands into fallow land (Jana, 2013; Pike, 2008). There was an increase of 10,586.48 hectare fallow land, 4704.32 hectare river sand, 16,095.76 hectare urban land and 6499.47 hectare excavated land from 1973 to 2015. It clearly denotes that there is gradual decrease of dense or open forest, agricultural land, water body and river in the study area. In due course of time, considerable segment of forest and agricultural land is converted into open cast quarry, construction, mining lagoon, road and fallow land. Further, rapid growth of population and urbanization triggers the land use land cover changes. These factors are being accelerated due to open cast mining activities in Raniganj coalfield area. As a result there was a loss of 16,046.2 hectare forest, 10,394.4 hectare agricultural land, 3609.25 hectare river land and 213.11 hectare water body from 1973 to 2015. But there was an increase of 10,586.48 hectare fallow land, 4704.32 hectare river sand, 16,095.76 hectare urban land and 6499.47 hectare excavated land from

1973 to 2015. However, the result of percentage of land shows that agricultural land occupies highest percentage of land followed by fallow and forest. On the other hand, water body occupies lowest percentage of land followed by river and mining lagoon. In 1973, the forest land use covered an area of 25.23%, but in 2015, it decreased to 13.52%.The area covered by agricultural land also decreased from 36.14% in 1973 to 27.44% in 2015. Henceforth, the river land use decreased from 3.02% in 1973 to 0.43% in 2015. It is worth mentioning that fallow land has increased by 10.23% in the past 50 years. The result shows that the area under the forest has turned into fallow, agricultural and urban lands. It is also evident that urban land use increased by 2.32% in 1973 to 12.02% in 2015.

4.2 *Effective Analysis*

Forest conversion is shown in Fig. 5.8. The southern part of this coalfield like Satgram, Kajora and Sripur experienced maximum forest conversion. Results indicate that considerably large proportion of forest gets transformed into agricultural land, that is, 68,706,900 hectare (9.41%) from 1973 to 1992 and 137,512,800 (20.96%) hectare from 2010 to 2015. In 2002–2010, 141,653,800 (12.56%) hectare forest was converted into fallow land; in 1992–2002, 88,494,700 hectare (11.98%) forest was converted into excavated land. A maximum of 822,112,000 hectare (72.91%) of forest land was retained from 2002 to 2010 and a minimum of

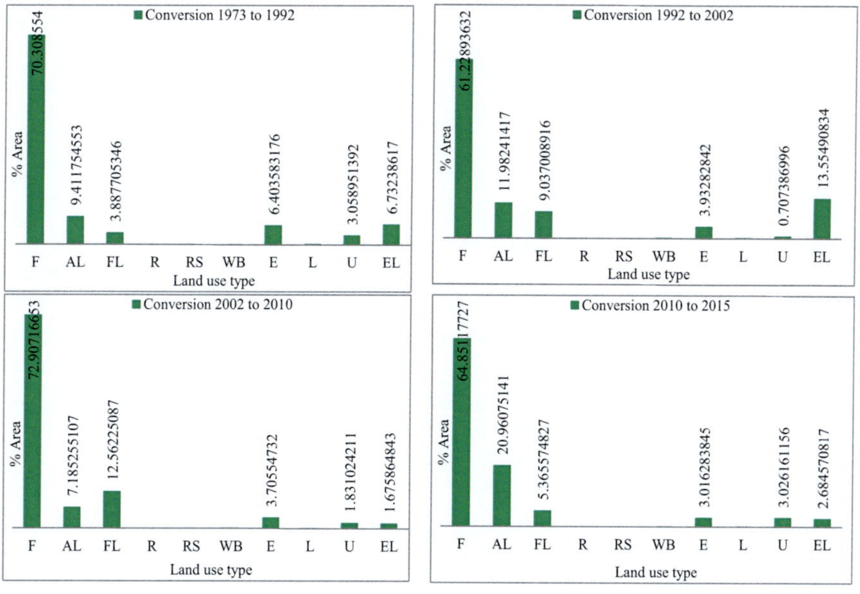

Fig. 5.8 Forest conversion, 1973–2015, Raniganj coalfield. (Source: Computed by the author)

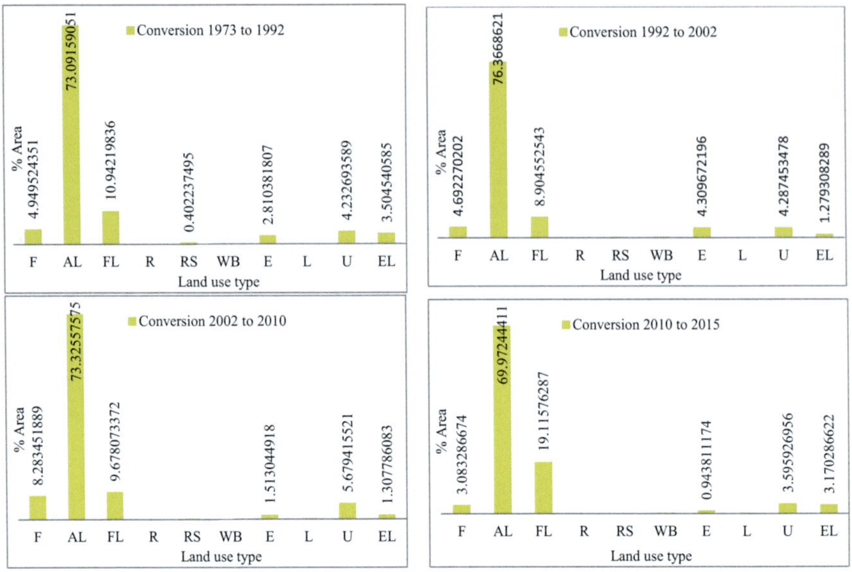

Fig. 5.9 Agricultural land conversion, 1973–2015, Raniganj coalfield. (Source: Computed by the author)

399,739,800 hectare (61.23%) from 1992 to 2002. Forest loss to agriculture and fallow is due to conversion of natural forest, both dense and open, into croplands and increasing industrial activities.

Conversion of agricultural land is shown in Fig. 5.9. Result indicates that maximum agricultural land is converted into fallow land throughout the time period. However, agricultural abandonment appears to be lowest in the study area. Agricultural conversion happened at a rate of 119,354,600 hectare (10.25%) from 1973 to 1992, 106,322,564 hectare (8.90%) from 1992 to 2002, 83,979,900 (9.68%) hectare from 2002 to 2010 and 80,879,500 (19.11%) hectare from 2010 to 2015. A maximum of 911,839,258 hectare (76.36%) of agricultural land was retained in 1992–2002 and a minimum of 296,056,000 (69.97%) hectare in 2010–2015 in north-western area such as Sodepur, Slanpur and Pandaveswar area of Raniganj coalfield. The major transition is from the agricultural land use to the fallow land use which denotes abandonment. The transition of fallow land is shown in Fig. 5.10. The result indicates that maximum fallow land is converted into agricultural land throughout the time period. Fallow land conversion happened at a rate of 71,918,100 hectare (11.49%) in 1973–1992, 69,363,000 hectare (7.95%) in 1992–2002, 40,064,200 (8.46%) hectare in 2002–2010 and 76,887,000 (22.46%) hectare in 2010–2015 around Sonepur Bazari, Bankola and Kenda area of Raniganj coalfield. Fallow land retains its own land, i.e. a maximum of 400,170,800 hectare (84.52%) in 1992–2002 and a minimum of 241,563,600 hectare (70.38%) in 2010–2015. At the same time, fallow land is converted into forest and river sand

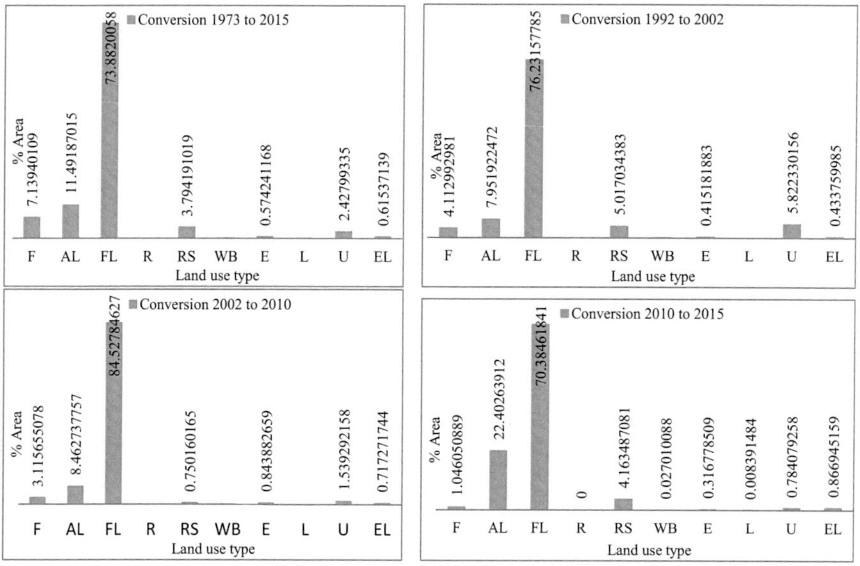

Fig. 5.10 Fallow land conversion, 1973–2015, Raniganj coalfield. (Source: Computed by the author)

in some places like Sheebpur and Purushottampur. However, most of the area under fallow land is obtained from either forest or agricultural land use.

Figure 5.11 shows the conversion of river. It is evident that river is converted into river sand at Ajay River near Pandaveswar and at Damodar River near Sodepur and Kunstoria at a rate of 19,677,700 (37.49%) hectare in 1973–1992, 9,696,600 (42.54%) hectare in 1992–2002, 4,724,200 (39.42%) hectare in 2002–2010 and 5,551,500 (34.43%) hectare in 2010–2015. River retains its own land, i.e. a maximum of 5,243,000 (43.76%) hectare in 2002–2010 and a minimum of 4,869,900 (30.20%) hectare in 2010–2015. About 62.51% river is converted into other classes in every transition period. Indeed river is the largest inter-categorical transitional land use. Conversion of river sand is shown in Fig. 5.12. Result denotes that large portion of river sand is converted into fallow land near Slanpur and Pandaveswar at a rate of 8,413,000 hectare (14.99%) in 1973–1992, 5,423,258 (10.28%) hectare in 1992–2002, 9,821,700 (10.31%) hectare in 2002–2010 and 11,207,700 (15.56%) hectare in 2010–2015. In 1973–1992, 1,803,600 hectare (12%) river sand is converted into agricultural land, 616,500 (4%) hectare into river in 2010–2015. River sand retains its own land, i.e. a maximum of 44,271,587 (83.93%) hectare in 1992–2002 and a minimum of 51,355,800 (71.28%) hectare in 2010–2015 at Slanpur and Satgram. A large portion of water body is converted into fallow at a rate of 895,254 (8.37%) hectare in 1992–2002 and 586,300 (5%) hectare in 2002–2010 near Sonepur Bazari and Kenda area as shown in Fig. 5.13. Water body retains its own land, maximally 8,494,900 hectare (72.57%) in 2002–2010 and minimally 7,133,000 hectare (47.37%) in 1973–1992. Maximum exposure is

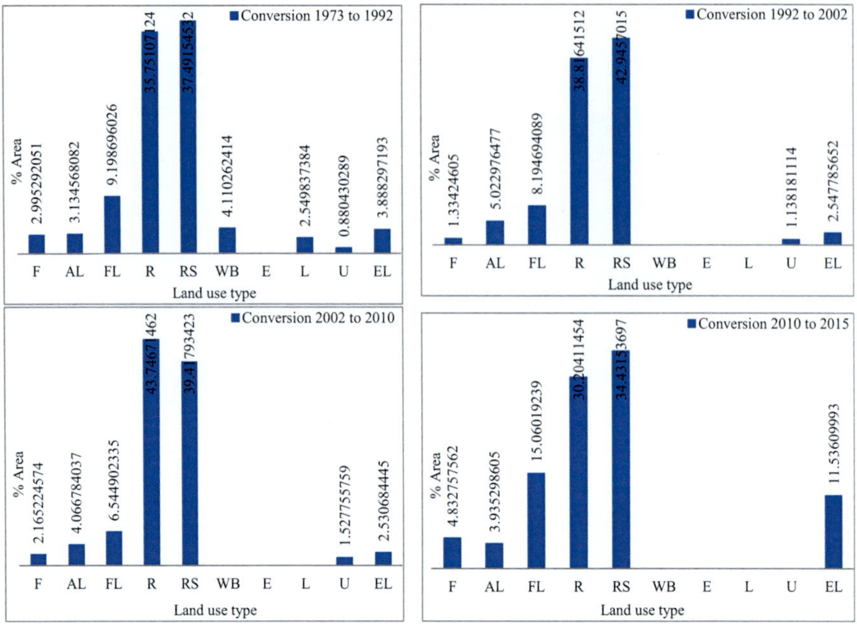

Fig. 5.11 River conversion, 1973–2015, Raniganj coalfield. (Source: Computed by the author)

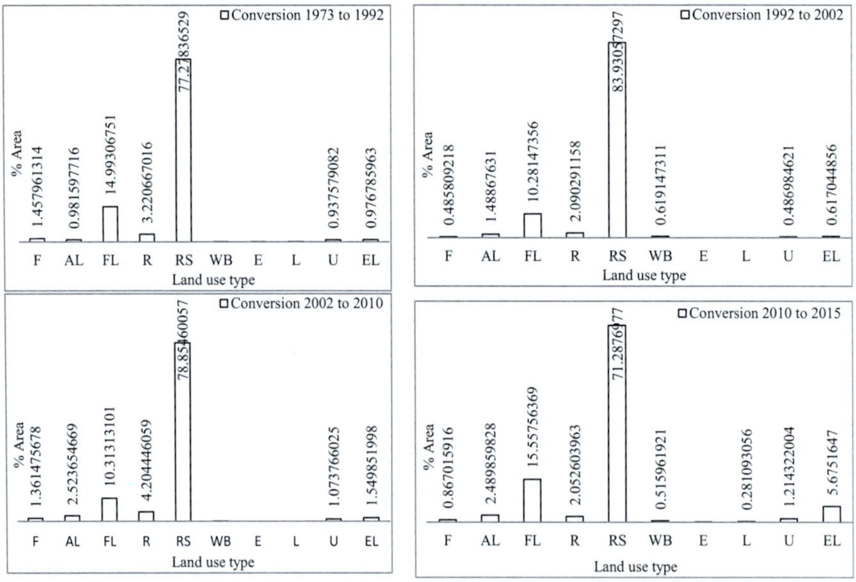

Fig. 5.12 River sand conversion, 1973–2015, Raniganj coalfield. (Source: Computed by the author)

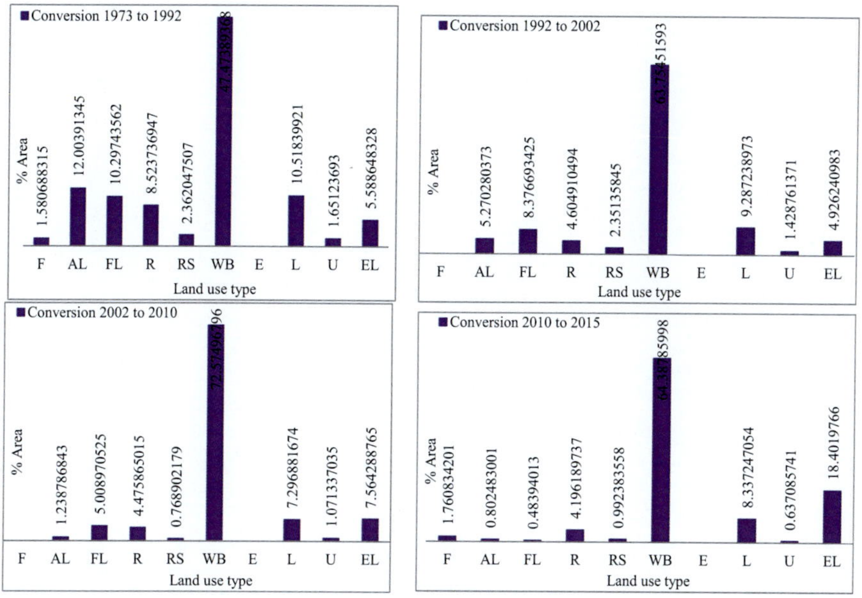

Fig. 5.13 Water body transitions, 1973–2015, Raniganj coalfield. (Source: Computed by the author)

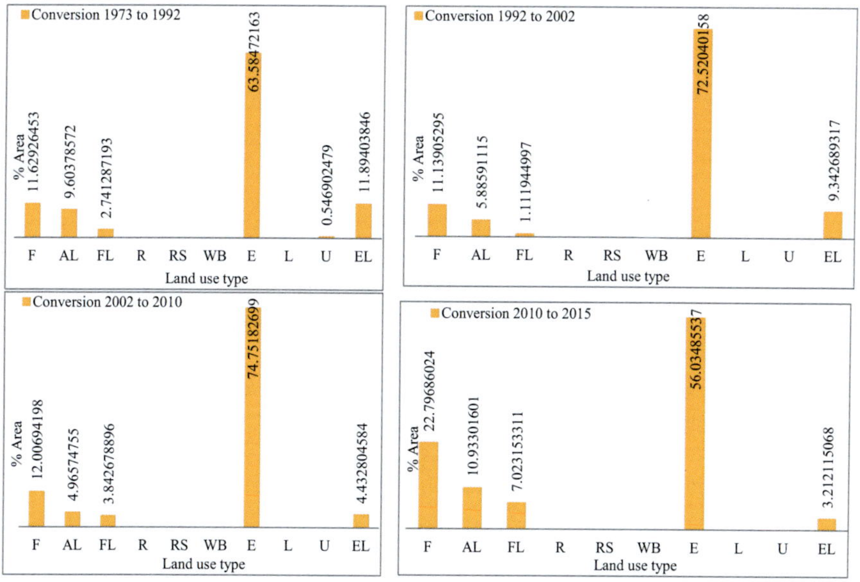

Fig. 5.14 Exposure land conversion, 1973–2015, Raniganj coalfield. (Source: Computed by the author)

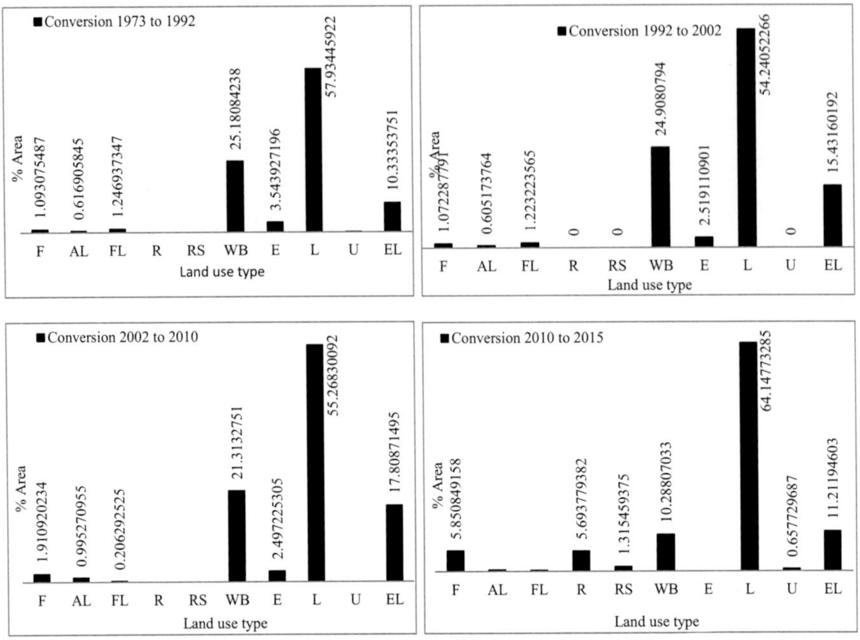

Fig. 5.15 Mining lagoon Conversion 1973–2015, Raniganj coalfield. (Source: Computed by the author)

converted into forest area throughout the study phrase shown in Fig. 5.14. Conversion occurred at a rate of 12,530,800 (11.62%) hectare in 1973–1992, 10,545,258 (11.14%) hectare in 1992–2002, 12,598,500 (12.01%) hectare in 2002–2010 and 21,361,000 (22.80%) hectare in 2010–2015 near Jhanjra, Bankola and Sonepur Bazari area. Exposure land retains its own area, maximally 78,434,700 (74.75%) hectare in 2002–2010 and minimally 52,505,500 (56.03%) hectare in 2010–2015 near Kenda area. Maximum mining lagoon is converted into water body throughout the study phase (Fig. 5.15) at a rate of 3,453,200 (25.18%) hectare in 1973–1992, 3,482,014 (24.91%) hectare in 1992–2002, 1,766,700 (21.31%) hectare in 2002–2010 and 943,200 (10.29%) hectare in 2010–2015 around Sripur, Kenda and Slanpur area. Mining lagoon retains its own land, maximally 5,881,000 (64.17%) hectare in 2010–2015 and minimally 7,582,530 (54.24%) hectare in 1992–2002 near Satgram.

Conversion of urban land is shown in Fig. 5.16. Result denotes that maximum urban land is converted into fallow land during the last 50 years. Conversion occurred at a rate of 4,186,800 (4.21%) hectare in 1973–1992, 3,168,259 (2.99%) hectare in 1992–2002, 3,500,100 (1.12%) hectare in 2002–2010 and 5,699,000 (1.27%) hectare in 2010–2015 around Bankola and Kajora area.

Urban land retains its own land, maximally 307,685,700 (98.83%) hectare in 2002–2010 and minimally 95,044,000 (95.78%) hectare in 1973–1992 near Kunstoria, Sripur, Satgram and Sodepur area. Conversion of excavated land is shown in Fig. 5.17. Result indicates that maximum quarry is converted into exposure

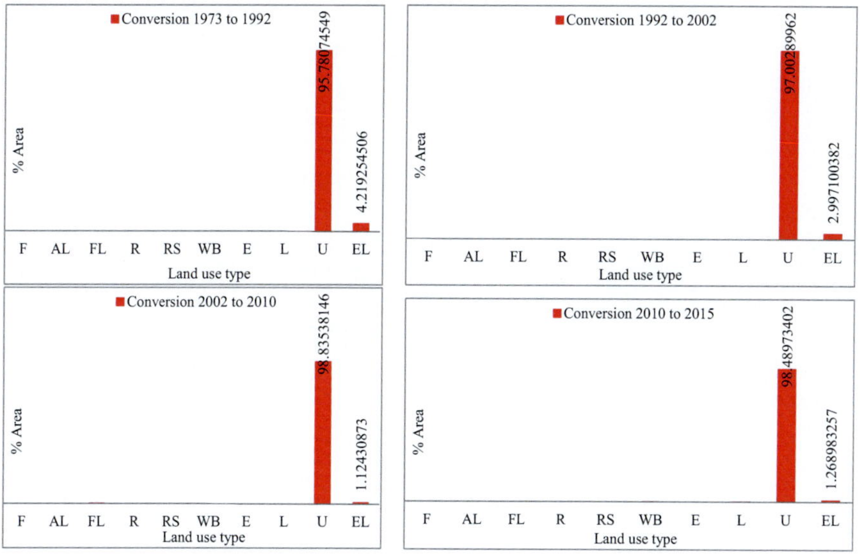

Fig. 5.16 Urban land conversion, 1973–2015, Raniganj coalfield. (Source: Computed by the author)

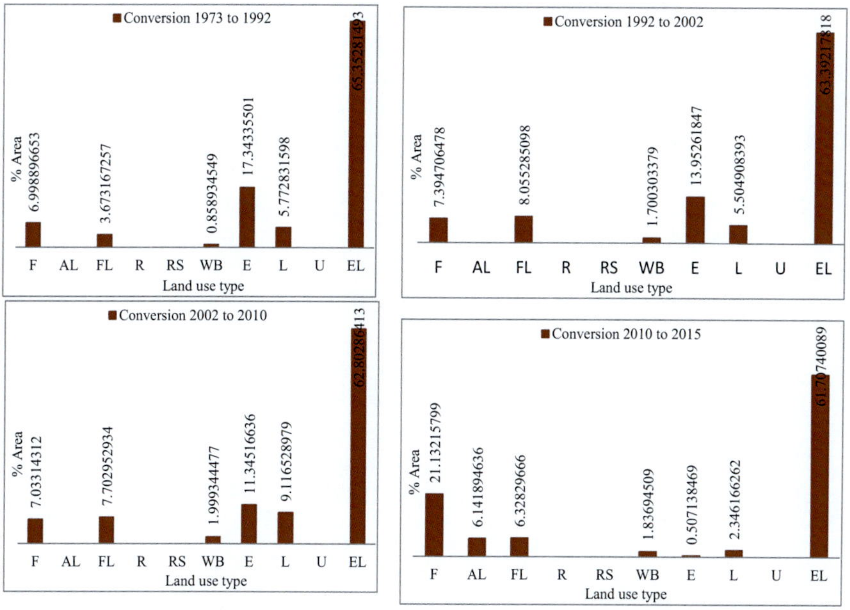

Fig. 5.17 Excavated land conversion 1973–2015, Raniganj coalfield. (Source: Computed by the author)

land at a rate of 37,253,700 (17.34%) hectare in 1973–1992, 9,080,500 (13.95%) hectare in 1992–2002, 8,099,700 (11.34%) hectare in 2002–2010 around Jhanjra, Bankola, Pandaveswar and Slanpur area. However, 18,263,700 (21.13%) hectare excavated land was converted into forest in 2010–2015. Excavated land retains its own land, maximally 140,378,500 (65.35%) hectare in 1973–1992 and minimally 53,331,300 hectare (61.71%) in 2010–2015. Indeed, excavated land is converted into exposure land, fallow land and forest due to efforts in greening the mined out area by putting plantation on overburden dumps. These extensive levels of land use conversion in Raniganj coalfield area occurred due to extension of mining areas, development of infrastructure and residential complexes of mining industry and thermal power plants. This effective land use analysis indicates that forest, water body and river are acutely affected land use. Contrary, fallow land, agricultural land, river sand and urban built up area is acutely effective land use in Raniganj coalfield.

4.3 Landscape Fragmentation

Vast statistical information of different land use is measured to quantify the fragmentation of the landscapes in Raniganj coalfield area. Class area and percentage of land of different land use have already been discussed earlier. Class area sharply denotes that there is a major temporal change in forest followed by agricultural and fallow lands. However, there was also reduction in forest, agriculture, water body and river in 1973–2015. On the other hand, area of fallow land, river sand, exposure land, lagoon, urban land and excavated land increased in the abovementioned time period in northern and eastern parts of this coalfield namely Purushottompur, Sonepur and Pandaveswar area. Result shows that temporal changes in number of patches increased (Fig. 5.18). Numbers of forest patches increased from 22,216 in 1973 to 59,981 in 2015. However, many forest patches were isolated and converted into agricultural, fallow, quarry or urban land in recent times near north-western area, i.e. Sodepur, Slanpur and Pandaveswar. Number of patches for agriculture increased from 33,061 to 36,730 during 1973 to 2015, which is due to conversion of

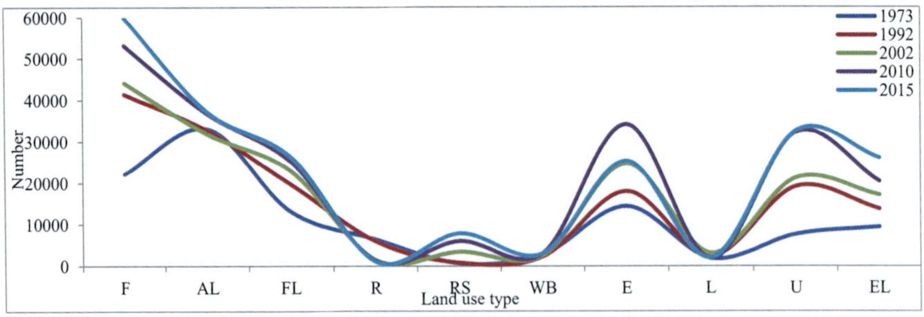

Fig. 5.18 Number of patches, 1973–2015, Raniganj coalfield. (Source: Computed by the author)

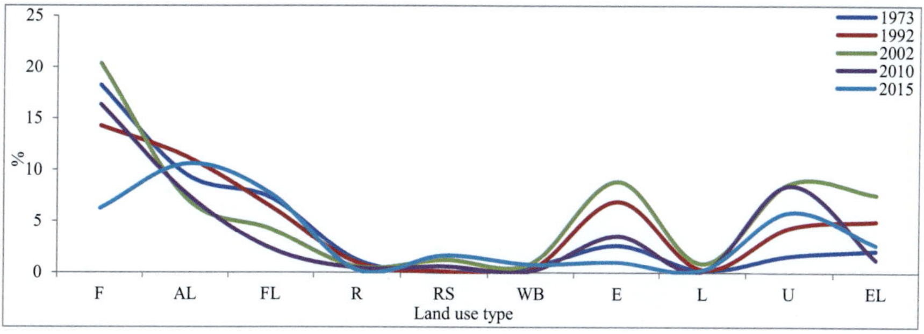

Fig. 5.19 Largest patch index, 1973–2015, Raniganj coalfield. (Source: Computed by the author)

farmland into grassland. Further, the number of fallow patches also increased from 25,175 to 26,290, river sand patches increased from 689 to 7868 and urban patches increased from 7658 to 212,226 in 1973 to 2015 respectively. Agricultural patches increased slightly because of aggregation of smaller fields into larger ones. Hence, urban and excavated patches increased due to growing urbanization and mining activities in Asansol and Raniganj area. Only river patches decreased from 6408 to 1056 in 1973 to 2015. Indeed, acute increase in the number of patches denotes fragmentation in the Raniganj coalfield area.

Figure 5.19 shows class-level largest patch index in Raniganj coalfield area. Forest is the largest patch in each class as a percentage of the total landscape in the study area. In 1973, largest patch index for forest was about 18.201% of total area and then it decreased to 6.248% of total area in 2015. In 1973, the patches of forest were large and continuous. However, in 2015, the patches of forest were scattered compared to 1992, 2002 and 2010. It is the sign of more prominent fragmentation process. The largest patch index of urban land was 1.624% of total area in 1973 and it increased to 5.846% of total area in 2015. It denotes that urban land is increasingly becoming a dominant land use within this landscape. Urban land expansion is primarily at the southern part of the study area bordering Asansol, Durgapur, Barakar, Kulti, Raniganj city. However, largest patch index for agricultural land and river sand increased by about 1%. But largest path index for fallow land decreased from 7.269% in 1973 to 2.321% in 2010.

Interestingly, fallow land increased to 7.658% in 2015. In 2010, largest patch index of all land uses decreased. However, in 2015, the largest patch index for agricultural land, river sand, water body, mining lagoon and excavated land slightly increased and the entire landscape becomes more complex with comparison of simple shape in 1973. Decrease in the largest patch size and increase in patches of distinctive land use land cover classes clearly denote that patches were getting clumped and to form a single patch. Therefore, it can be said that landscape is becoming fragmented at an acute rate.

Figure 5.20 shows that patch density for forest, agricultural land, river and exposure land significantly decreased with time. The main decrease was observed in forest, 38.236 to 17.259, and river, 6.526 to 0.980, since 1973–2015. However,

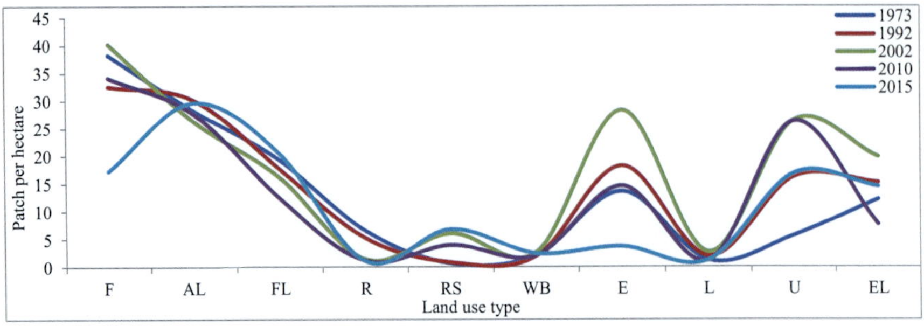

Fig. 5.20 Patch density, 1973–2015, Raniganj coalfield. (Source: Computed by the author)

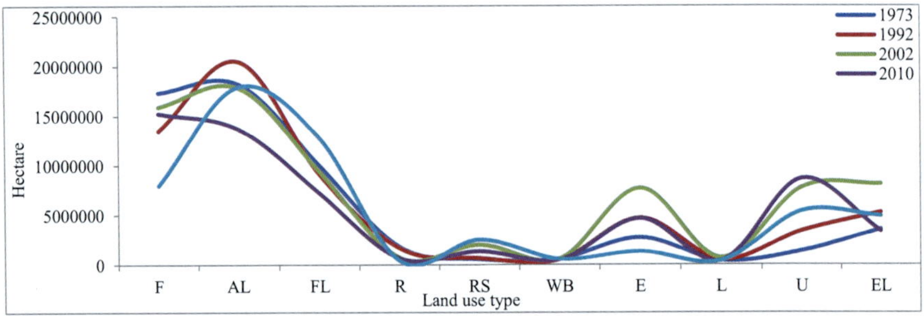

Fig. 5.21 Total edge, 1973–2015, Raniganj coalfield. (Source: Computed by the author)

there is a little increase in patch density for agricultural land, 28.256–29.652, and fallow land, 19.256–20.262. Further, patch density drastically accelerated for river sand, 0.652–6.856; urban, 5.626–16.859; and excavated land, 12.201–14.521.

Higher patch density denotes low fragmentation within these land uses (Punia & Raman, 2006). The result of patch density and number of patches is almost similar. Figure 5.21 shows temporal changes in total edge. The result indicates that there is a significant increase in total edge for fallow land, river sand, and urban and excavated lands. Fallow land increased from 9,988,999.8 to 127,774,115.2 hectare, whereas urban land increased from 1,362,484.1 to 5,399,480.7 hectare during 1973–2015 near Kunstoria, Sripur, Satgram and Sodepur areas. However, agricultural land, exposure land and mining lagoon do not exhibit much difference in total edge. During the study period, the edges of forest, river and exposure land decreased drastically. The edges of forest decreased from 17,341,756.1 to 7,968,584 hectare, river from 1,787,515.8 to 383,651.1 hectare and exposure land from decreased 2,744,298.5 hectare to 13,264,448.6 hectare respectively in 1973–2015.

The mean patch size is computed at class level as shown in Fig. 5.22. According to the result, that area covered with vegetation showed a decrease in mean patch size from 2.71 to 0.51 hectare in 1973 to 2015.

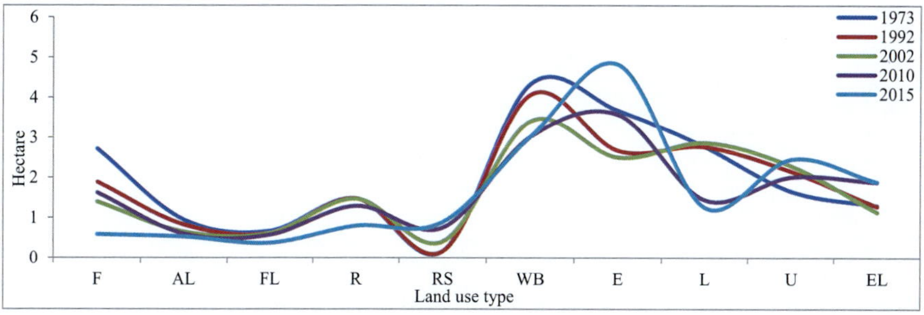

Fig. 5.22 Mean patch size, 1973–2015, Raniganj coalfield. (Source: Computed by the author)

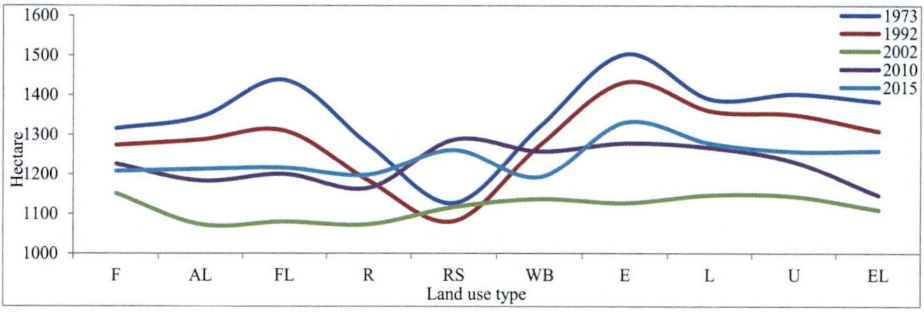

Fig. 5.23 Mean perimeter area ratio, 1973–2015, Raniganj coalfield. (Source: Computed by the author)

On the other hand, urban area witnessed an acute increase from 1.65 to 2.45 hectare in 1973 to 2015. However, forest agricultural land, river, water body and mining lagoon are undergoing conversion and fragmentation into quarry, fallow and urban area near Jhanjra, Bankola, Kenda and Sonepur Bazari. Beside, subsequent decrease in mean patch size in major land use classes which are associated with increased fragmentation (Rutledge, 2014). Therefore, it can be said that mean patch size is likely the most crucial indicator of fragmentation (Garg et al., 2013), and the process of fragmentation has increased in Raniganj coalfield area. Figure 5.23 shows the mean perimeter area ratio. Except river sand, the mean perimeter area ratio declined for all land use categories. It denotes that whole landscape has become more complicated and fragmented in the present day. Indeed, fallow land exhibits an acute decrease from 1437.41 in 1973 to 1215.39 in 2015. But there is an increase of mean perimeter area ratio for river sand from 1126.56 in 1973 to 1260.59 in 2017. Therefore, it can be said that river sand will dominate in Raniganj coalfield area in terms of mean perimeter area ratio. Figure 5.24 shows temporal change of edge density. There is an intensive decrease of edge density for forest, agriculture, river, water body and exposure land use. Forest decreased from 55.33 to 25.11 m/hectare, river decreased from 5.69 to 1.20 m/hectare and exposure land declined from 8.75 to 4.18 m/hectare respectively during 1973–2015. This result indicates that fragmentation in these land uses is accelerated due to opencast mining and human

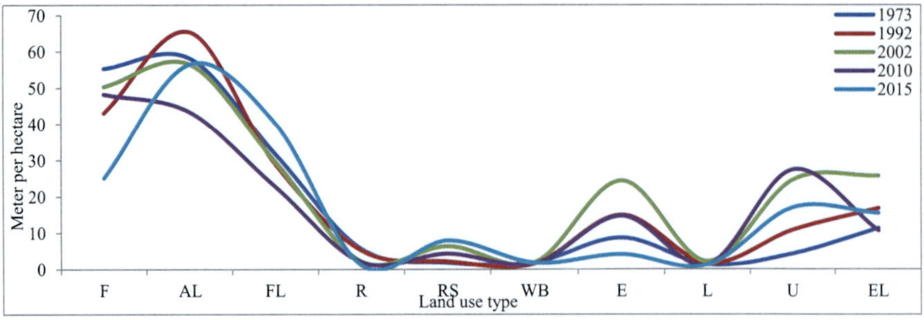

Fig. 5.24 Edge density, 1973–2015, Raniganj coalfield. (Source: Computed by the author)

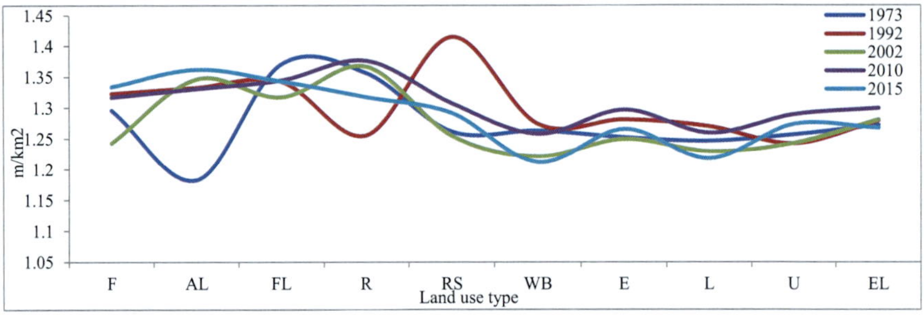

Fig. 5.25 Mean shape index, 1973–2015, Raniganj coalfield. (Source: Computed by the author)

pressure on existing land. However, some land use indicates huge increase in edge density such as fallow 31.82–40.26 m/hectare, river sand 1.85–7.98 m/hectare, urban land 4.34–17.01 m/hectare, respectively, during 1973–2015. In real circumstances, it happens due to rapid urbanization process and industrial development in and around Raniganj coalfield area.

Mean shape index is shown in Fig. 5.25. Result indicates that there is little change in mean shape index for all land uses. Forest increased from 1.29 to 1.33 m/km^2, agricultural land from 1.18 to 1.36 m/km^2, river sand from 1.26 to 1.29 m/km^2 and urban land from 1.25 to 1.27 m/km^2 in 1973 to 2015. Increase in mean shape index in these landscapes indicates patch shape irregularity and fragmentation process. However, fallow, river and mining lagoon exhibit a decrease from 1.37 to 1.34 m/km^2, from 1.35 to 1.31 m/km^2 and from 1.24 to 1.21 m/km^2, respectively, in 1973 to 2015. Decrease in mean shape index in these landscapes indicates less fragmented and less irregular patch. Figure 5.26 represents area weighted mean shape index. Result indicates that forest patches are converted into numerous small patches and isolated in recent times. The area weighted mean shape index denotes small change for forest from 8.29 in 1973 to 9.05 in 2015. The result indicates an increasing trend in shape complexity and fragmentation of forest patches. The area weighted mean shape index for agricultural land, fallow, river sand, excavated land and urban land increased. Such changes denote an expansion of these land uses around urban and coalfield areas of Raniganj coalfield. However, agricultural land

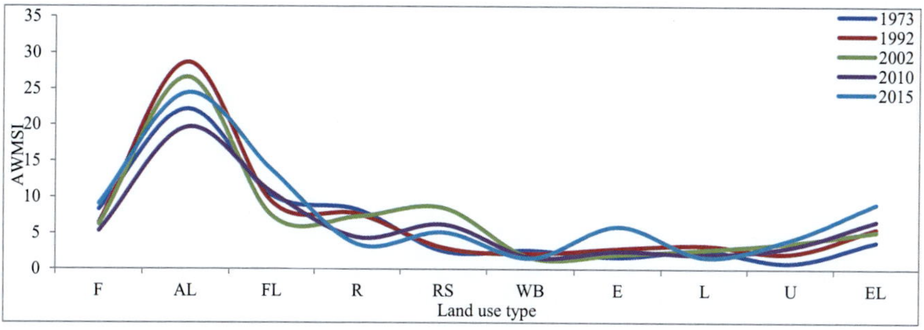

Fig. 5.26 Area weighted mean shape index, 1973–2015, Raniganj coalfield. (Source: Computed by the author)

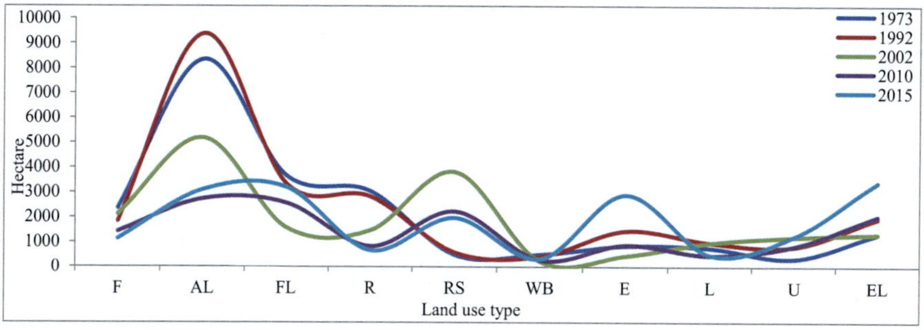

Fig. 5.27 Patch size standard deviation, 1973–2015, Raniganj coalfield. (Source: Computed by the author)

increased from 22.18 to 24.41 and urban land from 0.87 to 4.15, respectively, in 1973 to 2015. It indicates that inter-patch connectivity among these land uses decreased. This also indicates an increase in shape complexity. Area weighted mean shape index for mining lagoon, water body and river declined from 2.55 to 1.63, 2.60 to 1.52 and 8.22 to 3.45, respectively, in 1973 to 2015. This result indicates that these landscapes tend to be circular in shape and decrease in shape complexity.

Figure 5.27 shows the standard deviation of patch size. The result shows that the standard deviation of patch size decreased at a lower rate for forest, river, river sand, water body and mining lagoon. On the other hand, fallow exhibits a huge decrease in patch size standard deviation from 140.54 hectare in 1973 to 42.21 hectare in 2015. This indicates fragmentation growth in these landscapes. However, there is a huge increase in patch size standard deviation for urban and excavated lands from 0.67 to 5.53 hectare and 2.11 to 21.26 hectare, respectively, in the above-mentioned study period. It indicates that these landscapes become dominant in Raniganj coalfield area. Patch size standard deviation and patch size coefficient of variation can be misleading with regard to landscape structure in the absence of information on the number of patches or patch density (Areendran et al., 2013; Samanta, 2015). Hence, patch size coefficient of variation is measured due to some drawback of patch

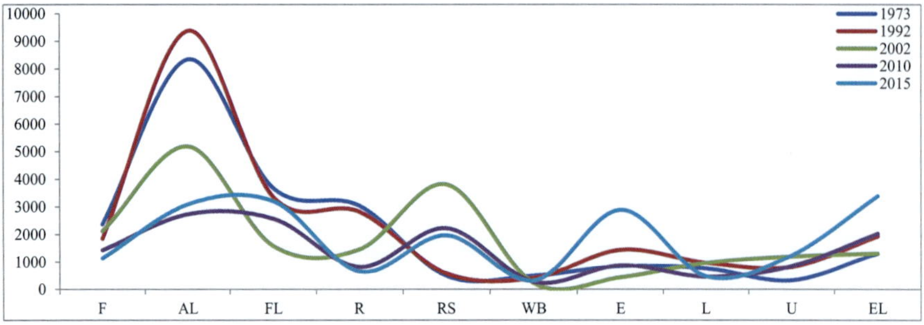

Fig. 5.28 Patch size coefficient of variation, 1973–2015, Raniganj coalfield. (Source: Computed by the author)

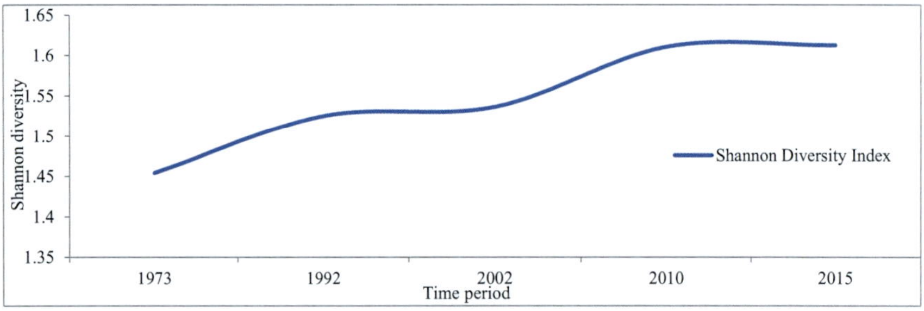

Fig. 5.29 Shannon diversity index, 1973–2015, Raniganj coalfield. (Source: Computed by the author)

size standard deviation and shown in Fig. 5.28. Patch size coefficient of variation is acutely decreased for forest from 2363.65 to 1132.68, agricultural land from 8340.46 to 3088.35, and river from 3033.21 to 655.09. However, water body and mining lagoon decreased at a lower rate. Further, there is a sharp increase in patch size coefficient of variation for excavated land, urban land and river sand. Henceforth, urban area dominantly increased from 323.05 to 1228.25 during the study period. Two landscape-level fragmentations are also measured for the study area. Firstly, Shannon diversity index is measured which is shown in Fig. 5.29.

The increase in the Shannon diversity index from 1.45 in 1973 to 1.61 in 2015 denotes that the landscape heterogeneity and diversity have increased in Raniganj coalfield. It also indicates proportional distribution of different patch in the landscape; therefore, landscape becomes evenly distributed and fragmented. Further, Fig. 5.30 shows Shannon evenness index. The result indicates that it increased from 0.60 in 1973 to 0.67 in 2015. The increase in the Shannon evenness index shows that the distribution of landscape patches is more even in 2015 than it was in 1973. However, clear patterns of spatial change are also found, which means landscape became heterogeneous and scattered. In the real sense, the Shannon diversity index and Shannon evenness index are likely skewed due to many small patches such as in water body, excavated land, exposure land and mining lagoon (Liu et al., 2016).

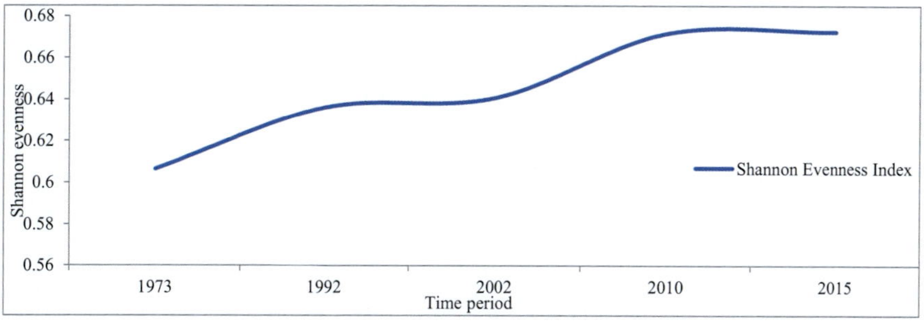

Fig. 5.30 Shannon evenness index, 1973–2015, Raniganj coalfield. (Source: Computed by the author)

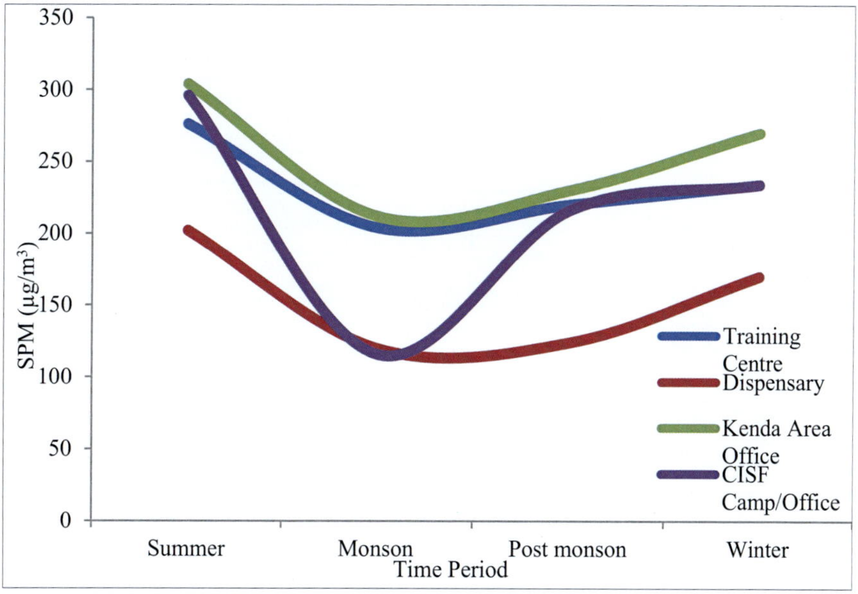

Fig. 5.31 Seasonal SPM concentration (µg/m^3)—period March 2015 to February 2017. (Source: Computed by the author)

4.4 Evaluation of the Environmental Impacts

The concentration of SPM, RPM and NO_x is computed by combining the value of 2-year data as shown in Figs. 5.31, 5.32, and 5.33 respectively. The continuous 2-year data is divided into four distinctive seasons—summer (March to June), monsoon (July to September), post-monsoon (October to November) and winter (December to February). Correlation among different pollutants taking two years combined annual average SPM is observed maximally at Kenda Area Office is

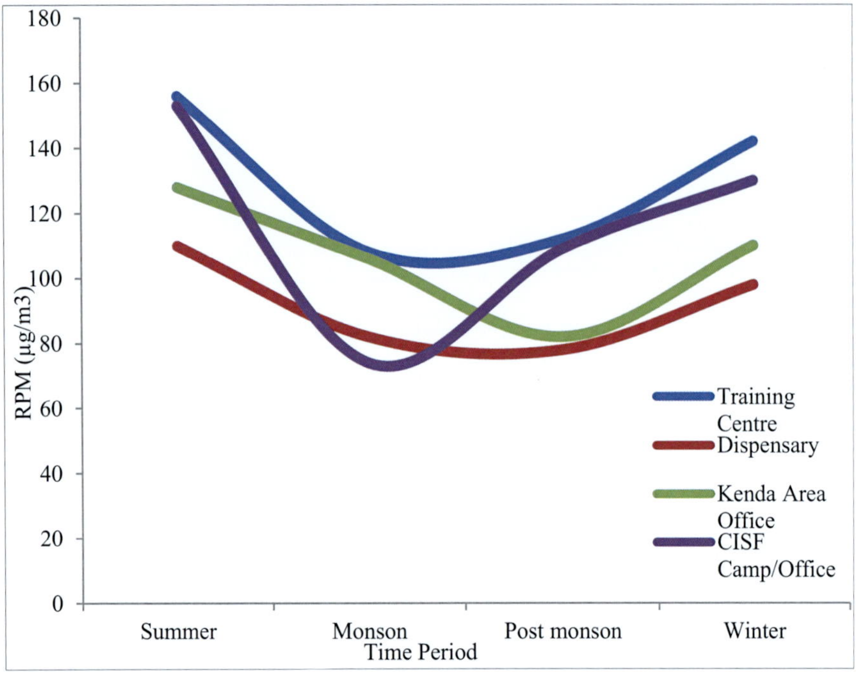

Fig. 5.32 Seasonal RPM concentration (µg/m^3)—period March 2015 to February 2017. (Source: Computed by the author)

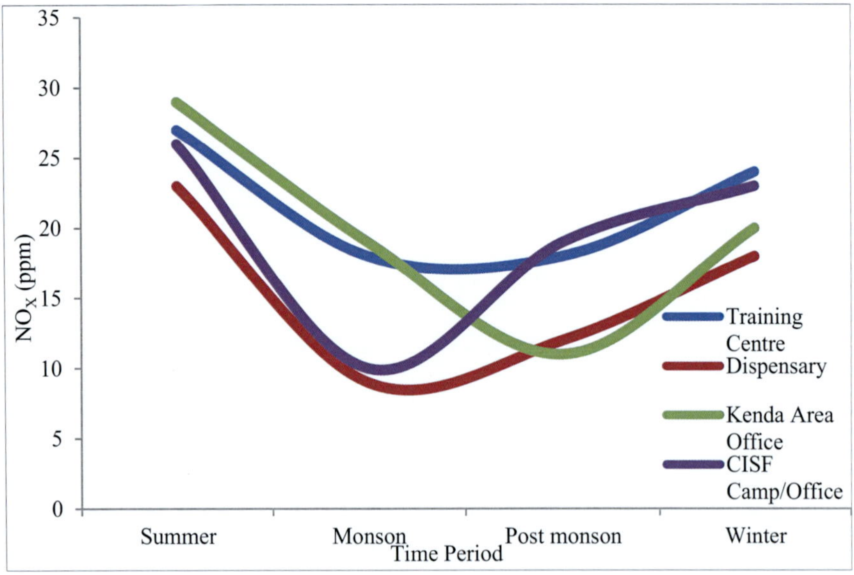

Fig. 5.33 Seasonal NO$_x$ concentration (ppm)—period March 2015 to February 2017. (Source: Computed by the author)

304 μg/m^3 (SD 46.52, % CV 17.26), RPM is observed maximally at Training Centre 156 μg/m^3 (SD 21.81, % CV 15.95) and NO_x is observed maximally at Kenda Area Office is 29 ppm (SD 2.5, % CV 9.52). Further, the combined annual average SPM is observed minimally at CISF Camp/Office (116 μg/m^3; SD 52.08, % CV 31.95), RPM is observed minimally at Dispensary (78 μg/m^3; SD 17.72, % CV 18.61) and NO_x is observed minimally at Dispensary (9 ppm; SD 5.22, % CV 24.98). During the monitoring period, SPM is drastically low in monsoon season, that is in July to September. But rest of the year this area exhibits greater SPM concentration. Almost entire precipitation occurred in monsoon period in the study area; that's why SPM concentration is very low in this particular time period. However, SPM is measured higher at Kenda area office (304 μg/m^3). Further, minimum SPM is found at CISF Camp/Office (116 μgm^3). The trends clearly denote that higher concentration of SPM is recorded in summer; but lower concentration of SPM is recorded in monsoon period. The average seasonal SPM concentrations are above the critical level in the study area according to National Ambient Air Quality Standards (UK).

This result clearly depicts the poor quality of air in the study area due to enormous suspended particles in the air as a result of excavation, digging, and transportation of mining vehicles and lack of air pollution prevention measures. Indeed, the higher suspended particulate matter is found during summer, whereas moderate suspended particulate matter and lower suspended particulate matter are found during post-monsoon/winter and monsoon period respectively. The trend (Fig. 5.32) shows that during the monitoring period, RPM is drastically high in hot summer period (March to June). RPM is maximally observed at Training Centre (156 μg/m^3). The hot summer wind causes high concentration of SPM contributed to this higher value of RPM. RPM is minimally observed at CISF Camp/Office (74 μgm^3). The trends sharply denote that maximum RPM is recorded in summer, whereas minimum RPM is recorded in monsoon. RPM concentration is very low in the air during the monsoon period due to excessive rainfall. According to National Ambient Air Quality Standards, Central Pollution Control Board of India (New Delhi), the average seasonal RPM concentrations are above the critical level in the study area. Therefore, it can be said that the quality of air is not good enough in the study area due to the presence of enormous respirable particles in the air as a result of excavation, digging, and transportation of mining vehicles and lack of air pollution prevention measures. However, maximum, moderate and minimum concentrations of respirable particulate matter are same as suspended particulate matter as mentioned earlier. The trends (Fig. 5.33) show that during the monitoring period NO_x is drastically low in monsoon period (July to September). During this survey, maximum NO_x is recorded at Training Centre (27 ppm).

Minimum NO_x is found at Dispensary (9 ppm). This trend clearly shows that maximum NO_x is recorded in summer; on the other hand, minimum NO_x is recorded in monsoon period. Indeed, this happened due to precipitation effect. According to National Ambient Air Quality Standards, Central Pollution Control Board of India (New Delhi), the quality of air is poor in the study area because of the critical NO_x concentration level. Enormous NO_x is present in the air due to transportation of mining vehicles and lack of air pollution prevention measures. However, maximum,

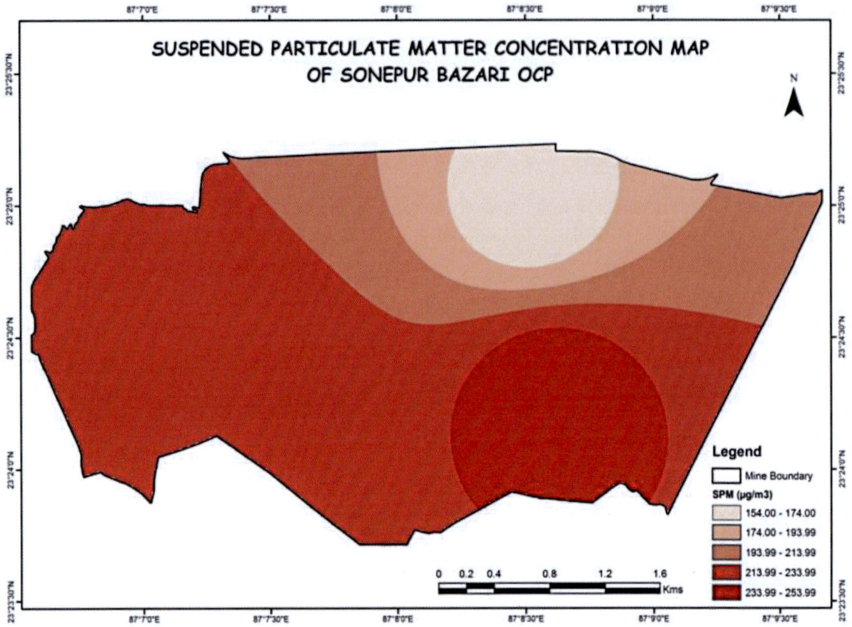

Fig. 5.34 Suspended particulate matter (SPM) concentration map of Sonepur Bazari mine. (Source: Computed by the author)

moderate and minimum concentration trends of oxide of nitrogen is same as suspended particulate matter and respirable particulate matter during the monitoring period as mentioned earlier.

Figure 5.34 shows the suspended particulate matter concentration map of Sonepur Bazari mine. This pollution map is prepared to show the maximally, moderately and minimally polluted area. Result indicates that dark coloured area is highly polluted in terms of suspended particulate matter concentration of air that is 233.99–253.99 μgm^3 in and adjoining area of Kenda area office. This area is excessively polluted due to greater opencast mining activities. However, light coloured area depicts minimum suspended particulate matter concentration of air, that is 154.00–174.00 μgm^3, in dispensary and its adjoining area. This area is least polluted because it is a residential area for the mining workers and engineers. Further, moderate coloured area is moderately polluted in terms of suspended particulate matter concentration. Further, it can be divided into two sub classes—high moderately polluted area (213.99–233.99 μgm^3) and low moderately polluted area (174.00–213.99 μgm^3).

CISF camp office and its adjoining area are the high moderately polluted areas; in contrast, Kenda area office and its adjoining area is the low moderately polluted area. Hence, the observed value is successfully validated with the monitoring value. The result strongly indicates that Sonepur Bazari mine generates huge amount of suspended particulate matter and the entire region of the mine is moderate to highly polluted with suspended particulate matter.

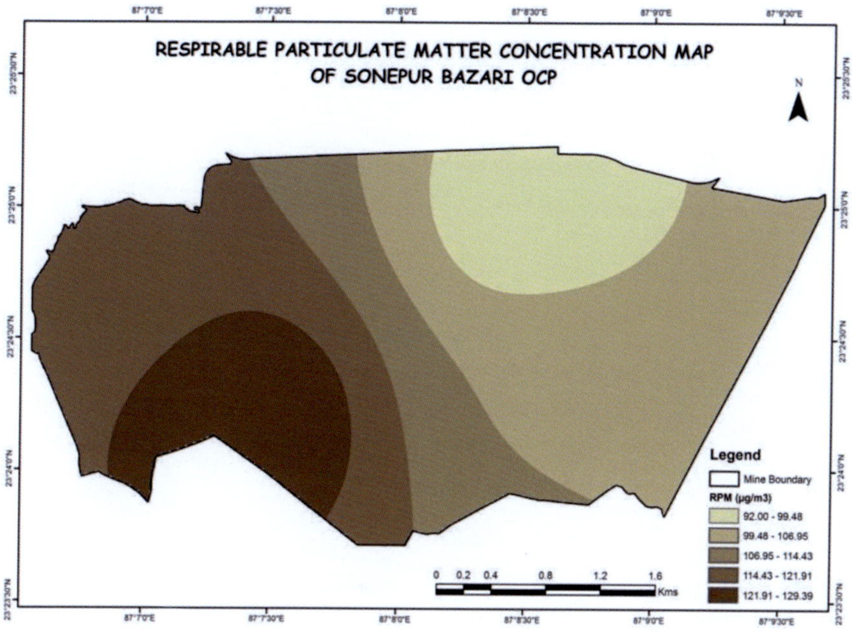

Fig. 5.35 Respirable particulate matter (RPM) concentration map of Sonepur Bazari mine. (Source: Computed by the author)

Figure 5.35 shows the respirable particulate matter concentration map of Sonepur Bazari mine. The study area is divided into five distinctive polluted classes in terms of RPM concentration. Results indicate that light coloured area is minimally polluted (92.00–99.48 μgm^3); the rate of pollution indispensary and its adjoining residential area is relatively low. Moderate coloured area depicts moderately polluted area in terms of respirable particulate matter concentration. Further, it can be divided into two sub classes—high moderately polluted area (114.43–121.91 μgm^3) and low moderately polluted area (99.48–114.43 μgm^3). CISF office camp and its adjoining is high moderately polluted area; on the other hand, Kenda area office and its adjoining is low moderately polluted area. However, dark coloured area is highly polluted in terms of respirable particulate matter concentration of air, that is 121.91–129.39 μgm^3, in and adjoining area of training centre. This area is immensely polluted due to high opencast mining activities. The result strongly indicates that Sonepur Bazari OCP generates huge amount of respirable particulate matter and the entire region of the mine is moderate to highly polluted with respirable particulate matter.

Figure 5.36 shows the oxide of nitrogen concentration map of Sonepur Bazari mine. Highly polluted area is represented with dark colour (20.49–21.74 ppm) in and adjoining area of training centre; due to rapid opencast coal mining activities. Light coloured area indicates the minimal nitrogen oxide concentration of air (15.50–16.75 ppm) because it is dispensary and its adjoining area is the residential

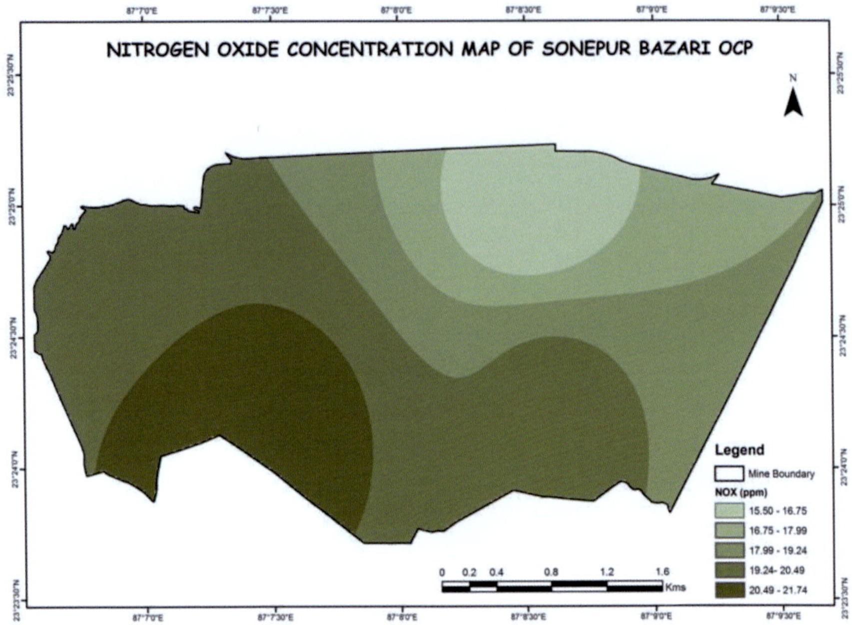

Fig. 5.36 Nitrogen oxide (NO_x) concentration map of Sonepur Bazari mine. (Source: Computed by the author)

area for the mining workers and engineers. However, moderate colour indicates moderately polluted areas in terms of nitrogen oxide concentration. Further, it can be divided into two sub classes—high moderately polluted area (19.24–20.49 ppm) and low moderately polluted area (16.75–19.24 ppm). CISF camp office and its adjacent area are the high moderately polluted area; in contrast, Kenda area office and its adjoining area are the low moderately polluted area. The result strongly indicates that Sonepur Bazari OCP generates huge amount of nitrogen oxide and the entire region of the mine is moderate to highly polluted with nitrogen oxide.

To measure the environmental costs of Khottadih and Sonepur Bazari mines for the production of opencast coal, the researcher examined the spatial-temporal concentration of suspended particulate matter in Pandaveswar, Khottadih and Sonepur Bazari area. Figure 5.37 shows the concentration of suspended particulate matter of Khottadih, Sonepur Bazari and Pandaveswar. During the study period (September–October, 2016), SPM concentration is low at Pandaveswar area.

But the concentration of SPM is higher in Khottadih and Sonepur Bazari area. Figure 5.38 is a Theissan polygon map, where the three regions indicate different pollution levels. Highly polluted area is represented with dark colour in terms of SPM concentration of air that is 244.5 μgm^3 in and adjoining area of Sonepur Bazari OC project. Khottadih OC project and its adjoining area are moderately polluted and average SPM concentration is 237.45 μgm^3. These two areas are maximally polluted

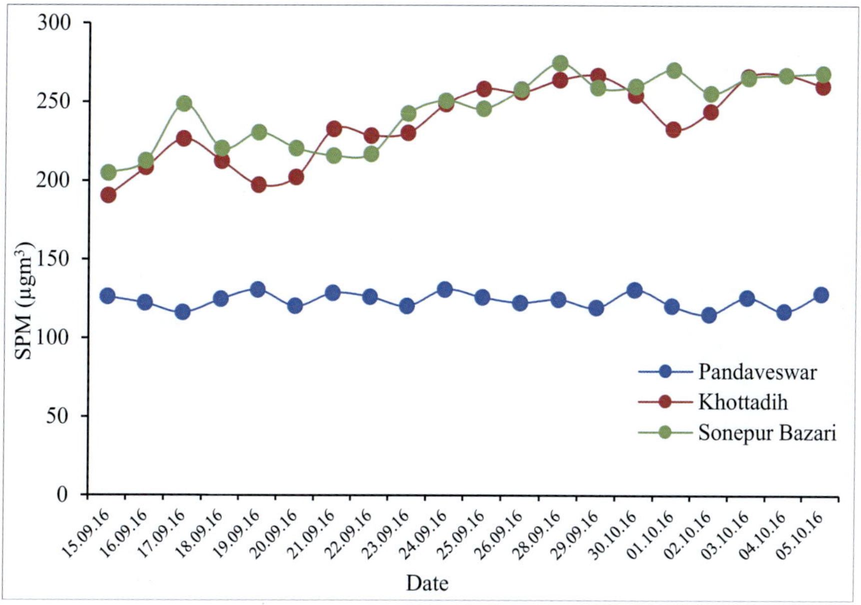

Fig. 5.37 Suspended particulate matter (SPM) concentration of Khottadih, Sonepur Bazari and Pandaveswar. (Source: Computed by the author)

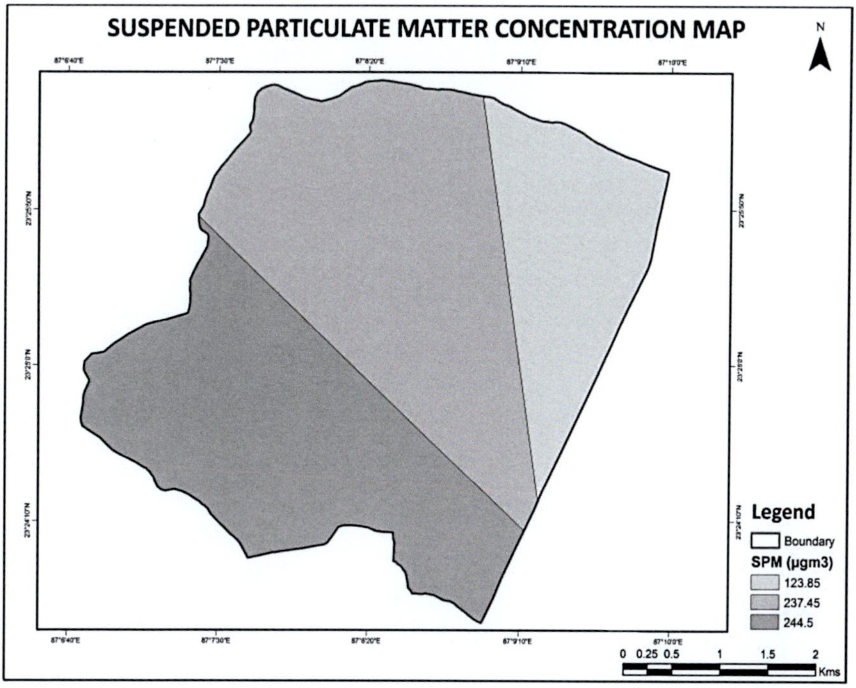

Fig. 5.38 Suspended particulate matter (SPM) concentration map of Khottadih, Sonepur Bazari and Pandaveswar. (Source: Computed by the author)

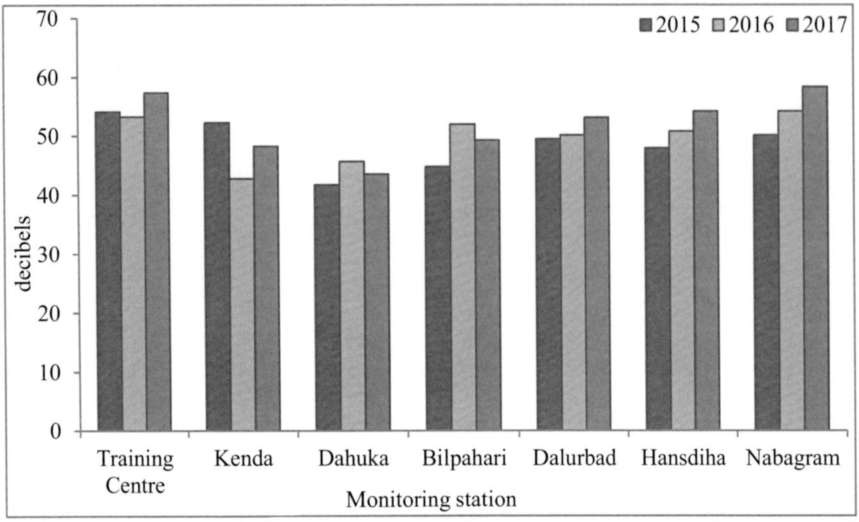

Fig. 5.39 Daily noise concentration of seven monitoring stations. (Source: Computed by the author)

due to intensive opencast mining activities. In Pandaveswar area, the SPM concentration of air is optimum, that is 123.85 μgm^3, because it is a residential area. However, highly polluted, moderately polluted and less polluted area is 0.000315, 0.000679 and 0.000625 respectively.

Figure 5.39 shows the annual mean spatial and temporal variations of daily noise in seven different monitoring stations in three different years. The result denotes that the noise is under desirable rate. But the crucial thing is that the villagers suffer from above-mentioned noise generated from mining activities throughout the day. In the year of 2015, noise is recorded highest in Training Centre (57.4 dB) and lowest in Dahuka (43.6 dB). Further, in 2017, highest noise is observed in Nabagram (54.2 dB) and lowest in Kenda (42.8 dB). However, on 23 August, 2016, noise is recorded highest in Training Centre (54.2 dB) and lowest in Dahuka (41.8 dB). To find out the environmental cost due to opencast mining, hedonic pricing method is applied where land value is used as a proxy of environmental degradation. For this purposes, air pollution and land degradation are taken into account; because these two environmental degradations have direct impact on the land value. Hence, the researcher concentrates on the possible impact on land quality in the surrounding villages of the mine. Alongside, an estimate of the possible impacts is made through field survey during June 2016 and August 2017. Four property values such as—20,000, 60,000, 100,000 and 200,000—are considered in the questionnaire and interview schedule to measure the changes of land value from 1980 to 2017. However, few lands belong to Eastern Coalfields Limited; therefore, the researcher gets plenty of data from ECL too. It is argued that opencast mining in Khottadih and Sonepur Bazari areas have been a miscellaneous effect on house and land prices, the health of environment is getting poor in the area during last 37 years. But in some areas like Pandaveswar, it is found that land price increases due to development purposes associated with mining. However, it should be remembered that the Khottadih and Sonepur Bazari areas are very inaccessible from beginning; therefore, the rate of average fall in land prices due to air pollution is low compared to other developed countries. Table 5.12 shows the average rise and fall in house prices for each of these three areas. Result shows that overall land prices fall by 15–40%. In Sonepur Bazari area, the land value is decreased maximally worth Rs. 120,000 (fall is 33.33%). Further, in Khottadih area, land value is most severely affected; highest decreased land value i.e. worth Rs. 100,000 (fall is 40%). But in Pandaveswar area, land value rises; the maximum increase in land value is Rs. 300,000 (rise is 66.67%).

Figure 5.40 (prepared with hedonic pricing method) shows the map of affected areas due to mining activities. Result denotes that villagers of Sonepur Bazari and Khottadih areas are most severely affected because of their close proximity to the mining site shown on the map with saddle brown colour. Indeed, Bengal Para village in Sonepur Bazari and Sukhbazar village in Khottadih are the most severely affected villages (price of land is Rs. 60,000–100,000 per cotta). Kenda, Dalurbad, Nabagram, Dahuka, etc., are moderately affected areas and is shown in dark goldenrod and pale goldenrod colour on the map (price of land is Rs. 100,000–200,000 per cotta). However, the value of land in Pandaveswar and its adjoining area is higher due to the existence of railway station, large market and other facilities provided by ECL, as shown on map in steel blue colour (price of land is Rs. 400,000 per cotta). Further, Bilpahari, Hansdiha, Kumarkhala, etc., areas are the low- to moderate-price valued land due to unhealthy environmental condition (price of land is Rs. 60,000–300,000 per cotta).

To measure people's marginal willingness to pay for healthy environment, the researcher uses pollution level and price of land to get the trend of willingness in the study area. Firstly, suspended particulate matters (SPM), daily noise and land degradation are sorted into higher to lower value; after that the values are added together to get one value. Level of pollution is marked on x axis as it is an independent variable (0 indicates no pollution; 300 indicates highest pollution) as shown in Fig. 5.41. The price of land is sorted from greater to smaller value and is marked on y axis as it is a dependent variable.

Seven different stations are surveyed with questionnaire and interview schedule. Trend line (Fig. 5.41) denotes that people are willing to pay higher price for pollution-free environment and not so much interested to pay for polluted environment. This trend line also shows that the value of land is decreased with an increased pollution. Nevertheless, properties across a range of values are confirmed by the strong negative correlation between the pollution level and property values. However, this result is also consistent with the expectation that money valuations of environmental impacts should be an increasing function of individual income (Neill et al., 1988; Alemu et al., 2015). To measure the overall profit and environmental degradation of this project, it is necessary to compute the investment, trade and environmental cost caused by mining activities. Data regarding investment and benefit of the projects is gathered from ECL. But the environmental cost was very difficult to gather (Gyawali et al., 2004; Choudhury, 2011) as no adequate procedure

Table 5.12 Average rise and fall of land prices in Sonepur Bazari, Khottadih and Pandaveswar area

	Sonepur Bazari			Pandaveswar			Khottadih		
Property value (rupees)	1980	2017	%	1980	2017	%	1980	2017	%
20,000	20,000	17,000	15	15,000	20,000	25	20,000	15,000	25
60,000	40,000	50,000	25	50,000	80,000	37.5	50,000	50,000	25
100,000	120,000	80,000	33.33	80,000	150,000	46.67	100,000	60,000	40
200,000	200,000	150,000	25	300,000	500,000	66.67	150,000	100,000	33.33

Source: Computed by the author

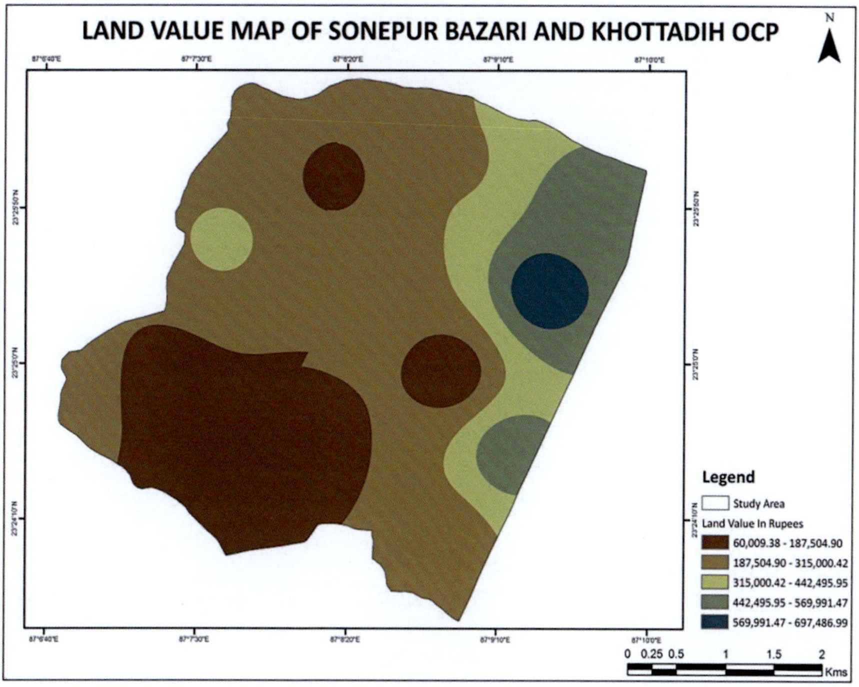

Fig. 5.40 Land value map in Sonepur Bazari, Khottadih and Pandaveswar area. (Source: Computed by the author)

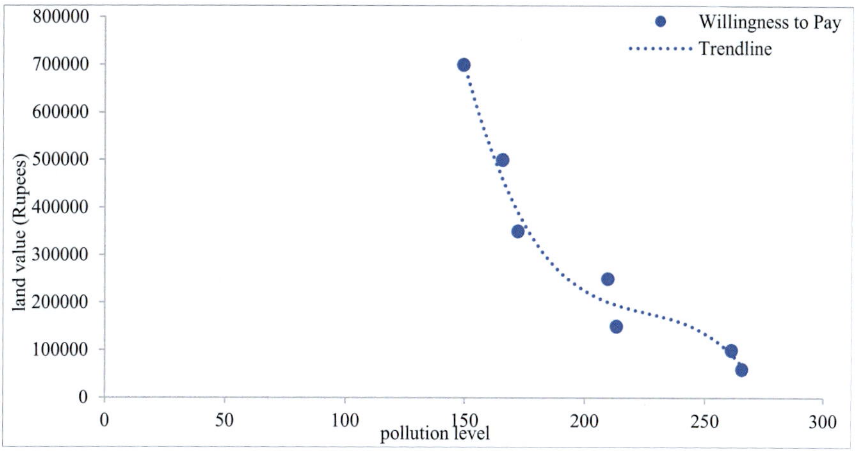

Fig. 5.41 People's willingness to pay for healthy environment in Sonepur Bazari, Khottadih and Pandaveswar area. (Source: Computed by the author)

Table 5.13 Comparisons of cost and benefit in Sonepur Bazari, Khottadih and Pandaveswar area

Balance sheet 1 (crore)		Balance sheet 2 (crore)	
Profit	14,696.86	Profit	14,696.86
Cost	697.36	Cost	1092.848
Net benefit	13,999.5	Net benefit	13,604.012

Source: Computed by the author

by which environmental cost can be measured was available (Kuchay & Ramachandra, 2016). Therefore, price of land as a proxy for environmental degradation was used. It is appropriate to get the price of environment degradation with hedonic pricing method. The environmental cost is computed by summing all prices of land shown in Table 5.13 and considering the hedonic method, which have been decreasing since the commencement of the project due to pollution. By multiplying the decreased price of land during 26 years with area, the author gets the hidden or environmental cost.

Table 5.13 shows a comparison of these costs with the benefits from coal production. Balance sheet 1 compares the profit of Khottadih and Sonepur Bazari sites with cost, in which total cost during this period is 711.37 crore and profit is 14,696.86 crore. Therefore, the net benefit from these opencast sites is 13,999.5 crore. Net economic benefits are gained from these sites, which excludes the environmental costs. On the other hand, balance sheet 2 compares the environmental cost with the benefit provided by the opencast sites, in which total cost is increased to 1092.848 crore due to the addition of environmental or hidden cost. Total calculated environmental cost is 395.488 crore. Therefore, the net benefit is 13,604.012 crore. This means that once the environmental cost of opencast mining is taken into account, the net benefit of the project is decreased. This decreased amount is the cost of environment. However, it can be said that the production of coal by using opencast technique is environmentally costly in Khottadih and Sonepur Bazari mine.

5 Conclusion

In Raniganj coalfield area, an increase of 10,586.48 hectare fallow land, 4704.32 hectare river sand, 16,095.76 hectare urban land and 6499.47 hectare excavated land is observed in the past five decades. A gradual decrease of dense or open forest, agricultural land, water body and river in due course of time is also observed. Further, considerable proportion of forest and agricultural land is converted into opencast quarry, construction, mining lagoon, road and fallow land. Opencast mining activity, rapid growth of population and urbanization trigger the land use land cover changes (Sarkar, 2017a, b, c). However, 16,046.2 hectare forest, 10,394.4 hectare agricultural land, 3609.25 hectare river land and 213.11 hectare water body are also lost. Whereas 10,586.48 hectare fallow land, 4704.32 hectare river sand, 16,095.76 hectare urban land and 6499.47 hectare excavated land increased. Therefore, it can be said that a strong relationship is there between loss in different land use land cover and growth of opencast mining. It can also be argued that forested area is most prone to future transformation. It is found that agricultural expansion, illegal small-scale mining, large-scale opencast mining, land abandonment, deforestation and rapid urbanization predominately occur in and around the periphery of Raniganj mining area, which will make the landscape more complex and fragmented in near future. Study also indicates that forest land, water body and river are badly affected land use in this coalfield. These land uses are more affected in past 40 years because of around 38% change area. Hence, forest and river are most prone to future transformation. Fallow, agriculture, river sand land use are moderately used because of around 75% retention in the last 40 years. Urban land is extensively effectively used in Raniganj coalfield area because of about 93% retention area in the past 40 years. Therefore, it can be concluded that agricultural and urban lands will dominate this area in near future. Gradual change in land use land cover pattern which is associated with various running opencast mining projects and developmental projects in Raniganj coalfield area is bringing a chronic challenge to the landscape. Land use land cover analysis reveals an increase in the non-forest area increases human-made land use. If the rate of land use land cover transformation happens with the present rate, it definitely will lead to deforestation, destruction of fertile soil and more mine abandoned land in Raniganj mining area. Sonepur Bazari, Khottadih, Bomjemehri and Dalurband, Shibpur, Pandaveswar, Damaliya and Purusattampur area are the mostly affected areas in terms of loss of forest cover in this mining site. In order to minimize the impact of mining deeds on landscape, scientific approach during and after mining activities has to be taken, like proper backfilling, separate storage of top soil, afforestation and monitoring soil erosion from dumped overburden. In recent time, opencast mining and associated development activities are recognized as the most significant factor for widespread fragmentation and transformation of the landscape in Raniganj coalfield area. After assessing the result, it can be argued that forest patches have converted into numerous small patches and isolated in recent times. Further, urban, fallow and excavated land will dominate because of their increasing patch size. This analysis also denotes that forest, water body and river are acutely fragmented land use because these land uses are more affected in the past 40 years because of declining class area, percentage of land, patch density and largest patch size. Whereas fallow, agricultural land and river sand are moderately fragmented land use because of moderate patch density, largest patch index and percentage of land. Urban land is effectively used because of inclining class area, patch density and percentage of land in the past 40 years. However, this huge level of land use fragmentation in Raniganj coalfield area occurred due to extension of mining areas, development of infrastructure and residential complexes of mining industry and thermal power plants (Mishra et al., 2014). In the Indian coal industry, the dependence on opencast mining has increased rapidly during the last three decades due to mechanization/modernization of mining operation, low gestation period, higher recovery of coal and more amenable to modern technologies than underground mines, ensuring speed and economy in implementation (Das & Das, 2016). The environmental cost which is a proxy of land value is measured using Hedonic Pricing method. And this environmental cost is higher in khottadih and Sonepur Bazari as the land value is decreased in this area due to opencast mining. This environmental cost denotes that opencast mining is not perfectly desirable in Khottadih and Sonpur Bazari areas because of high environmental costs. Therefore, opencast coal mining in this area may be more costly to produce than underground mining; however, the researcher did not explore the UG mining operations. The researcher only excludes the air pollution from the cost; thus, it can be argued that UG mining operation could be a better option than opencast technique in order to prevent pollution along with sustainable site development. In some cases the environmental cost can not describe the actual scenario of land degradation due to some inaccessible issue of environemntal degradation menu. Nevertheless, if UG mining is not possible due to unfavourable geological condition, the authority will have to go for opencast mining. Then several techniques should be employed to arrest air pollution and land degradation like developed countries. These techniques are applied to prevent the air pollution must be suppressed by sprinkling water on the main haul roads and other roads of the mine where vehicles play, wet drilling and provision of dust collector in drilling machine, water sprinkling at the various points of the CHP where coal is handled, etc. Literally, it is not impossible to check land degradation in opencast technique. But the authority can reduce it to some extent by using proper backfilling techniques and plantation of herbs and trees. However, the people of Khottadih and Sonepur Bazari area have adjusted with the polluted environment; they often argued that colliery is the source of life here. They are entirely depended on mining operations; therefore, pollution does not play any crucial role in their life. Further, ECL provides enormous facilities to them.

Acknowledgments The authors thank the Eastern Coalfields Limited (ECL) and Coal Mining Planning and Development Institute (CMPDI) for providing the facilities to manufacture the composite.

References

Abdallah, S. A. (2016). *Use of geospatial techniques in modelling land degradation case study of Kenya's Lake Victoria Basin.*

Abdul Rahaman, S., Aruchamy, S., Balasubramani, K., & Jegankumar, R. (2017). Land use/land cover changes in semi-arid mountain landscape in Southern India: A geoinformatics based Markov chain approach. *The International Archives of the Photogrammetry Remote Sensing and Spatial Information Sciences, XLII*, 6–9. https://doi.org/10.5194/isprs-archives-XLII-1-W1-231-2017

Adedeji, O. H., Tope-Ajayi, O. O., & Abegunde, O. L. (2015). Assessing and predicting changes in the status of Gambari forest reserve, Nigeria using remote sensing and GIS techniques. *Journal of Geographic Information System, 7*(3), 301–318.

Ahmed, B., Ahmed, R., & Zhu, X. (2013). Evaluation of model validation techniques in land cover dynamics. *ISPRS International Journal of Geo-Information, 2*, 577–597. https://doi.org/10.3390/ijgi2030577

Alemu, B., Garedew, E., Eshetu, Z., & Kassa, H. (2015). Land use and land cover changes and associated driving forces in North Western Lowlands of Ethiopia. *International Research Journal of Agricultural Science and Soil Science, 5*(1), 28–44.

Ampofo, S., Sackey, I., & Ampadu, B. (2016). Landscape changes and fragmentation analysis in a Guinea Savannah ecosystem: Case study of Talensi and Nabdam districts of the Upper East region, Ghana. *Journal of Geography and Geology, 8*(1), 41–54. https://doi.org/10.5539/jgg.v8n1p41

Areendran, G., Rao, P., Raj, K., Mazumdar, S., & Puri, K. (2013). Land use/land cover change dynamics analysis in mining areas of Singrauli district in Madhya Pradesh, India. *Tropical Ecology, 54*(2), 239–250.

Awoniran, D. R., Adewole, M. B., Adegboyega, S. A., & Anifowose, A. Y. B. (2014). Assessment of environmental responses to land use/land cover dynamics in the Lower Ogun River Basin, Southwestern Nigeria. *African Journal of Environmental Science and Technology, 8*, 152–165. https://doi.org/10.5897/AJEST2013.1607

Baldi, G., & Paruelo, J. M. (2008). Land-use and land cover dynamics in South American temperate grasslands. *Ecology and Society, 13*(2), 6.

Bandopadhyay, S. (2014). Crucial factors influencing environmental degradation: A case study on Raniganj coal mine. *International Journal of Engineering Development and Research, 3*(1), 70–73.

Baodong, M. A., Lixin, W., & Shanjun, L. (2008). Remote sensing detection for subsidence-resulted water body and solid-waste dump in coal mine: Yanzhou being a case. *The International Archives of the Photogrammetry, Remote Sensing and Spatial Information Sciences, 37*, 269–272.

Batistella, M., Robeson, S., & Moran, E. F. (2003). Settlement design, forest fragmentation, and landscape change in Rondônia, Amazônia. *Photogrammetric Engineering & Remote Sensing, 69*(7), 805–812.

Bhawana, K. C. (2015). *Land use and land cover change in relation to internal migration and human settlement in the middle mountains of Nepal.*

Biali, G., Patriche, C. V., & Pavel, V. L. (2014). Application of GIS techniques for the quantification of land degradation caused by water erosion. *Environmental Engineering and Management Journal, 13*(10), 2665–2673.

Bianche, V. B. (2002). *Automatic land use/land cover classification system with rules based both on objects attributes and landscape indicators.*

Brown, D. G., Duh, J., & Drzyzga, S. A. (2000). Estimating error in an analysis of forest fragmentation change using North American landscape characterization (NALC) data. *Remote Sensing of Environment, 71*, 106–117.

Cao, L., Zhang, Y., He, J., Liu, G., Yue, H., Wang, R., & Ge, L. (2007). Coal mine land subsidence monitoring by using spaceborne InSAR data-A case study in Fengfeng, Hebei Province, China. *The International Archives of the Photogrammetry, Remote Sensing and Spatial Information Sciences, 37*(495), 255–262.

Cena, B., Aliu, M., & Musliu, T. (2010). Measurements of emission of gases SO2, NOx, CO and CO2 from the burning process in the furnaces of power plant "Kosova B". *Journal of International Environmental Application and Science, 5*(2), 172–174.

Chadderton, C., & Elliott, E. (2011, March). *A guide to assessing the health and wellbeing impacts of opencast mining.*

Choudhury, S. (2011). *Change detection analysis of Talcher coalfield using Remote Sensing and GIS.*

Coleman, J. L. (1980). Efficiency, utility, and wealth maximization. *Hofstra Law Review, 8*(3), 3.

Csorba, P., & Szabó, S. (2010). The application of landscape indices in landscape ecology. In *Perspectives on nature conservation – Patterns, pressures and prospects*. IntechOpen.

Das, G., & Das, R. (2016). "Land use/land cover status relating the coal fire of Jharia coal field" – An analytical case study by RS-GIS techniques. *International Journal of Science and Research (IJSR), 5*(9), 1–11.

Dupin, L., Nkono, C., Burlet, C., Muhashi, F., & Vanbrabant, Y. (2013). Land cover fragmentation using multi-temporal remote sensing on major mine sites in southern Katanga (Democratic Republic of Congo). *Advances in Remote Sensing, 2*(2), 127–139.

Eastman, J. R. (2012, January). *IDRISI Selva chapter one: IDRISI Selva introduction.*

Eckert, S., Hüsler, F., Liniger, H., & Hodel, E. (2015). Trend analysis of Modis NDVI time series for detecting land degradation and regeneration in Mongolia. *Journal of Arid Environments, 113*, 16–28. https://doi.org/10.1016/j.jaridenv.2014.09.001

ECL. (2013). *Annual report and accounts*. ECL.

Eggert, R. G. (2002). *Mining and economic sustainability: National economies and local communities* (MMSD Working Paper No. 19). International Institute for Environment and Development.

El Jazouli, A., Barakat, A., Ghafiri, A., El Moutaki, S., & Ettaqy, A. (2017). Soil erosion modeled with USLE, GIS, and remote sensing: A case study of Ikkour watershed in Middle Atlas (Morocco). *Geoscience Letters, 4*, 25. https://doi.org/10.1186/s40562-017-0091-6

El-hallaq, M. A., & Habboub, M. O. (2015). Using Cellular Automata-Markov analysis and multi criteria evaluation for predicting the shape of the Dead Sea. *Advances in Remote Sensing, 4*, 83–95.

Estoque, R. C. (2012). *How to use IDRISI Selva.*

Evans, M. (2016). *Landscape systems*, 1–11.

Fadhil, A. M. (2009). Land degradation detection using geo-information technology for some sites in Iraq. *Journal of Al Nahrain University-Science, 12*(3), 94–108.

Feleke, A. K. (2003). *Land use and land cover in relation to Chromolaena odorata distribution*. ITC.

Foster, D. R. (2012). *Integrated studies of the drivers, dynamics and consequences of landscape change in New England.*

Freudenburg, W. R., & Madison, W. (2002). Mining the data: Analyzing the economic implications of mining for nonmetropolitan regions. *Sociological Inquiry, 72*(4), 549–575.

Garg, M., Joshie, A., & Choudhary, S. (2013). Land use land cover classification in Jamuna-Kotma coal field region, Anuppur District MP. *IOSR Journal of Agriculture and Veterinary Science, 6*(2), 38–43.

Görcs, N. L. (2011). Predictive modelling of surface subsidence above an underground coal mine at Máza-Váralja-South (Northeast Mecsek, Hungary). *Hungarian Geographical Bulletin, 60*(4), 343–356.

Goswami, S. (2012). *Impact of coal mining on environment: A study of Raniganj and Jharia coal field in India*, 1.

Goswami, S. (2013a). Environment management in mining areas (A study of Raniganj and Jharia coal field in India). *Global Journal of Human Social Science, 13*(7), 1–11.

Goswami, S. (2013b). Challenges of environmental management in Indian coal mining sector. *Universal Journal of Environmental Research and Technology, 3*(6), 616–629.

Goswami, S., & Goswami, R. (2014). Coal mining vis-à-vis agriculture in India: A question of sustainability. *International Research Journal of Geology and Mining (IRJGM), 4*(6), 154–162.

Guha, D. (2014). A case study on the effects of coal mining in the environment particularly in relation to Soil, Water and Air causing a Socio-economic Hazard in Asansol-Raniganj Area, India. *International Research Journal of Social Sciences, 3*(8), 39–42. Retrieved from www.isca.in, www.isca.me

Gundimeda, H. (2006). *Hedonic price method – A concept note.*

Gyawali, B., Fraser, R., & Wang, Y. (2004). *Land cover and socio-economic characteristics in the eight counties of Alabama: A spatial analysis*, 1–13.

Haddad, N. M., Brudvig, L. A., Clobert, J., Davies, K. F., Gonzalez, A., Holt, R. D., et al. (2015). Habitat fragmentation and its lasting impact on Earth's ecosystems. *Science Advances, 1*, 1–9.

Hamdy, O., Zhao, S., Salheen, M. A., & Eid, Y. Y. (2017). Analyses the driving forces for urban growth by using IDRISI®Selva models Abouelreesh – Aswan as a case study. *International*

Journal of Engineering and Technology, 9(3), 226–232. https://doi.org/10.7763/IJET.2017.V9.975

Handayani, L. D. W., Tejaningrum, M. A., & Damrah, F. (2017). Modelling of land use change in Indramayu District, West Java Province. *IOP Conference Series: Earth and Environmental Science, 54*, 012021. https://doi.org/10.1088/1742-6596/755/1/011001

Hassan, S., Mahmud, S., Islam, U., & Rahman, M. T. (2015). Integration of remote sensing and GIS to assess vulnerability of environmental degradation in North-Western Bangladesh. *Journal of Geographic Information System, 7*, 494–505.

Higginbottom, T. P., & Symeonakis, E. (2014). Assessing land degradation and desertification using vegetation index data: Current frameworks and future directions. *Remote Sensing, 6*(10), 9552–9575. https://doi.org/10.3390/rs6109552

Homère, A., Mfondoum, N., Etouna, J., Nongsi, B. K., & Mvogo, F. A. (2016). Assessment of land degradation status and its impact in arid and semi-arid areas by correlating spectral and principal component analysis neo-bands. *International Journal of Advanced Remote Sensing and GIS, 5*(2), 1539–1560.

Huang, J., Tu, Z., & Lin, J. (2009). Land-use dynamics and landscape pattern change in a coastal gulf region, southeast China. *International Journal of Sustainable Development & World Ecology, 16*, 61–66. https://doi.org/10.1080/13504500902771891

Indra, T. L., Harmantyo, D., Rukmana, I. P., & Nadia, Z. (2017). Simulation of land use change and effect on potential deforestation using Markov chain – Cellular Automata. *AIP Conference Proceedings, 1862*, 30177. https://doi.org/10.1063/1.4991281

Jana, S. (2013). Riverine geomorphology, sedimentology and sub-surface water potentiality – A case study along the section of Dwarkeswar River, Onda, Bankura, West Bengal. *International Journal of Geology, Earth & Environmental Sciences, 3*(3), 30–39.

Jia, N., Jiyuan, L. I. U., Wenhui, K., Xinliang, X. U., & Shuwen, Z. (2018). Spatiotemporal patterns and characteristics of land-use change in China during 2010–2015. *Journal of Geographical Sciences, 28*(2017), 547–562.

Katoria, D., Sehgal, D., & Kumar, S. (2013). Environment impact assessment of coal mining. *International Journal of Environmental Engineering and Management, 4*(3), 245–250.

Kendall, L., & Antrobus, R. (2013). *Proposed Msebe opencast anthracite mine, Kwazulu-final report of the prepared of terms and of the regulation R543 of 2010 environmental act, environmental impact assessment regulations*. Prepared for Mbila Resources (Pty) Limited.

Khorshiddoust, A. (2009). The application of hedonic pricing method in estimating the relationship between services levels, the pollution, and house prices in selected areas of Tabriz. *Archive of SID, 35*(51), 25–27.

Kindu, M., Schneider, T., Teketay, D., & Knoke, T. (2013). Land use/land cover change analysis using object-based classification approach in Munessa-Shashemene landscape of the Ethiopian highlands. *Remote Sensing, 5*, 2411–2435. https://doi.org/10.3390/rs5052411

Kitalika, A. J., Komakech, H., & Njau, K. N. (2018). Land-use and land cover changes on the slopes of Mount Meru-Tanzania. *Current World Environment, 13*, 331–352. https://doi.org/10.12944/Cwe.13.3.07

Koellner, T., De Baan, L., & Beck, T. (2013). Principles for life cycle inventories of land use on a global scale. *International Journal of Life Cycle Assessment, 18*, 1203–1215. https://doi.org/10.1007/s11367-012-0392-0

Kosmas, C., Ritsema, C., Salvati, L., Acikalin, S., Alcala, M., Fassouli, V., et al. (2013). Evaluation and selection of indicators for land degradation and desertification monitoring: Types of degradation, causes, and implications for management. *Environmental Management, 54*, 971–982. https://doi.org/10.1007/s00267-013-0110-0

Kuchay, S. A., & Ramachandra, T. V. (2016). Land use land cover change analysis of Uttara Kannada. *Imperial Journal of Interdisciplinary Research, 2*(4), 460–471.

Kuemmerle, T., Hickler, T., Olofsson, J., Schurgers, G., & Radeloff, V. C. (2012). Reconstructing range dynamics and range fragmentation of European bison for the last 8000 years. *Diversity and Distributions, 18*, 47–59. https://doi.org/10.1111/j.1472-4642.2011.00849.x

Kumar, R., Nandy, S., Agarwal, R., & Kushwaha, S. P. S. (2014). Forest cover dynamics analysis and prediction modeling using logistic regression model. *Ecological Indicators, 45*, 444–455. https://doi.org/10.1016/j.ecolind.2014.05.003

Kumar, K. S., Valasala, N. V. A. S. S., Subrahmanyam, J. V. V., & Mallampati, M. (2015). Prediction of future land use land cover changes of Vijayawada city using remote sensing and GIS. *International Journal of Innovative Research in Advanced Engineering, 2*(3), 91–97.

Lahiri-Dutt, K. (2003). Informal coal mining in Eastern India: Evidence from the Raniganj Coalbelt. *Natural Resources Forum, 27*, 68–77.

Le Page, Y., West, T. O., Link, R., & Patel, P. (2016). Downscaling land use and land cover from the Global Change Assessment Model for coupling with Earth system models. *Geoscientific Model Development, 9*, 3055–3069. https://doi.org/10.5194/gmd-9-3055-2016

Lin, W., Chou, W., Lin, C., Huang, P., & Tsai, J. (2005). Automated suitable drainage network extraction from digital elevation models in Taiwan's upstream watersheds. *Hydrological Processes, 20*, 289–306. https://doi.org/10.1002/hyp.5911

Liu, L., Cao, W., Shao, Q., Huang, L., & He, T. (2016). Characteristics of land use/cover and macroscopic ecological changes in the headwaters of the Yangtze River and of the Yellow River over the past 30 years. *Sustainability, 8*(3), 237. https://doi.org/10.3390/su8030237

Maitima, J. M., Mugatha, S. M., Reid, R. S., Gachimbi, L. N., Majule, A., Lyaruu, H., et al. (2009). The linkages between land use change, land degradation and biodiversity across East Africa. *African Journal of Environmental Science and Technology, 3*(10), 310–325.

Mangai, U. G., Samanta, S., & Das, S. (2010). A hierarchical multi-classifier framework for landform segmentation using multi-spectral satellite images – A case study over the Indian subcontinent. In *Proceedings of the 2010 fourth Pacific-Rim symposium on image and video technology* (pp. 306–313). IEEE Computer Society. https://doi.org/10.1109/Psivt.2010.58

Mas, J.-F., Vega, A. P., & Clarke, K. C. (2010). Assessing simulated land use/cover maps using similarity and fragmentation indices. In *ASPRS 2010 annual conference*. ASPRS.

Masoudi, M., Jokar, P., & Pradhan, B. (2018). A new approach for land degradation and desertification assessment using geospatial techniques. *Natural Hazards and Earth System Sciences, 18*, 1133–1140.

Mckenna, G. (2009). Techniques for creating mining landforms with natural appearance. In *Proceedings of tailings and mine waste '09*. The University of Alberta Geotechnical Centre.

Mini, K., & Manjunatha, B. M. (2014). Estimation of suspended particulate matter (SPM), respirable particulate matter (RPM) and ^{232}Thorium concentration variations in ambient air due to beach placer mining in Manavalakurichi, South West Coast of Tamilnadu, India. *International Journal of Innovative Research in Science, Engineering and Technology, 3*(6), 13793–13801.

Mishra, V. N., Rai, P. K., & Mohan, K. (2014). Prediction of land use changes based on land change modeler (LCM) using remote sensing: A case study of Muzaffarpur (Bihar), India. *Journal of the Geographical Institute "Jovan Cvijic", SASA, 64*(1), 111–127. https://doi.org/10.2298/IJGI1401111M

Modica, G., Vizzari, M., Pollino, M., Fichera, C. R., Zoccali, P., Di Fazio, S., & Development, S. E. (2012). Spatio-temporal analysis of the urban – Rural gradient structure: An application in a Mediterranean mountainous landscape (Serra San Bruno, Italy). *Earth System Dynamics, 3*, 263–279. https://doi.org/10.5194/esd-3-263-2012

Morgado, P., Toger, M., Abrantes, P., & Fiegel, J. (2012). A bottom up approach to modeling habitat connectivity dynamics through networks analysis. In *Sustainable development – Authoritative and leading edge content for environmental management*. IntechOpen.

Mukhopadhaya, S. (2016). Land use and land cover change modelling using CA-Markov case study: Deforestation analysis of Doon Valley. *Journal of Agroecology and Natural Resource Management, 3*(1), 1–5.

Mulkova, M., & Popelková, R. (2010). The aerial photos to detect changes in the landscape affected by black coal deep mining. In *ISPRS TC VII symposium – 100 years ISPRS* (Vol. XXXVIII, pp. 401–406). IAPRS.

Mundia, C. N., & Murayama, Y. (2009). Analysis of land use/cover changes and animal population dynamics in a wildlife sanctuary in East Africa. *Remote Sensing, 1*, 952–970. https://doi.org/10.3390/rs1040952

Mustard, J. F., Fisher, T. O. M., & Moran, E. (2004). Land-use and land-cover change pathways and impacts. In *Land change science. Remote sensing and digital image processing* (pp. 411–429). Springer.

Neill, R. V. O., Krummel, J. R., Gardner, R. H., Sugihara, G., Jackson, B., Deangelis, D. L., et al. (1988). Indices of landscape pattern. *Landscape Ecology, 1*(3), 153–162.

Nita, U. M. J. (2008). The scenic value of abandoned mining areas in Poland. *Landscape & Environment, 2*, 132–142.

Pandey, A. C., & Kumar, A. (2013). Evaluating impact of coal mining activity on landuse/landcover using temporal satellite images in South Karanpura coalfields and environs, Jharkhand State, India. *International Journal of Advanced Remote Sensing and GIS, 2*(1), 183–197.

Paul, S. S. (2013). *Analysis of land use and land cover change in Kiskatinaw River Watershed: A remote sensing, GIS & modeling approach.*

Pike, A. C. (2008). *Assessment of terrain attribute models for the identification of erosion prone areas suitable for the establishment of grassed waterways in an agricultural field setting in the out Bluegrass region of Kentucky.*

Pradesh, H., Ramachandra, T. V., Kumar, U., & Joshi, N. V. (2005). *Landscape dynamics in Western Himalaya – Mandhala.*

Province, J. (2007). *The impact of disposal and treatment of coal mining wastes on environment and farmland.*

Punia, M., & Raman, R. (2006). *Land use dynamics and landscape fragmentation in higher Himalaya.*

Rabia, A. H. (2016, January). *GIS spatial modeling for land degradation assessment in Tigray, Ethiopia.*

Ram, A. M., & Kiran, B. R. (2013). A brief review on the status of coal fire in India. *International Journal of Environmental Biology, 3*(3), 133–142.

Ramachandra, T. V., Bharath, H. A., Sowmyashree, M. V., Transportation, S., & Planning, U. (2014). Urban footprint of Mumbai – The commercial capital of India. *Journal of Urban and Regional Analysis, 6*, 71–94.

Ramalho, E. A., & Ramalho, J. J. S. (2011, September). *Hedonic functions, hedonic methods, estimation methods and Dutot and Jevons house price indexes: Are there any links?*

Rashid, M., Lone, M. A., & Romshoo, S. A. (2011). Geospatial tools for assessing land degradation in Budgam district, Kashmir Himalaya, India. *Journal of Earth System Science, 120*(3), 423–433.

Reddy, C. S., Satish, K. V., Pasha, S. V., Jha, C. S., & Dadhwal, V. K. (2016). Assessment and monitoring of deforestation and land-use changes (1976–2014) in Andaman and Nicobar Islands, India using remote sensing and GIS. *Current Science, 111*(9), 1492–1499.

Rutledge, D. T. (2014). *Landscape indices as measures of the effects of fragmentation: Can pattern reflect process?* (January 2003).

Samanta, P. (2015). Impact assessment and changes analysis of land use/land cover due to open cast coal mining activity: A case study of Raniganj coal field area. *International Journal of IT, Engineering and Applied Sciences Research, 4*(5), 17–27.

Sapena, M., & Ruiz, L. A. (2015). Analysis of urban development by means of multi-temporal fragmentation metrics from LULC data. *The International Archives of the Photogrammetry, Remote Sensing and Spatial Information Sciences, XL*, 11–15. https://doi.org/10.5194/isprsarchives-XL-7-W3-1411-2015

Saranathan, E., Loveson, V., & Victor Rajamanickam, G. (2010). *Land suitability study in land degraded area due to mining in Dhanbad district, Jharkhand.*

Sarkar, A. (2017a). A remote sensing approach to define landscape dynamics in Raniganj coalfield area through trend line analysis. *Asian Journal of Science and Technology, 8*(9), 5546–5555.

Sarkar, A. (2017b). Geoinformatics based estimation of suspended particulate matter, respirable particulate matter and nitrogen dioxide concentration in ambient air of Sonepur Bazari OCP, India. *Asian Journal of Research in Social Sciences and Humanities, 7*(5), 230. https://doi.org/10.5958/2249-7315.2017.00311.2

Sarkar, A. (2017c). Mine closure plan and policy for open cast coal mine of Raniganj coalfield, India. *Asian Journal of Research in Social Sciences and Humanities, 7*(7), 35–47. https://doi.org/10.5958/2249-7315.2017.00363.X

Sarkar, A. (2018). Accuracy assessment and analysis of land use land cover change using geoinformatics technique in Raniganj coalfield area, India. *International Journal of Environmental Sciences and Natural Resources, 11*(1), 25–34. https://doi.org/10.19080/Ijesnr.2018.10.555805

Sarkar, A. (2019a). A geoinformatics approach to assess land use dynamic and landscape fragmentation due to opencast coal mining in Raniganj coalfield, India. *International Journal of Research in Advent Technology, 7*(1), 307–309.

Sarkar, A. (2019b). Accuracy assessment and analysis of land use land cover change using geoinformatics technique in. *International Journal of Research and Analytical Reviews, 6*(1), 997–1008.

Sarkar, A. (2019c). Effective and effected land use assessment through land use transformation and conversion using geoinformatics techniques in Raniganj coalfield, India. *Universal Review (Scientific Information and Technological Board of Sadhana), 10*(2), 25–45.

Sarkar, A. (2019d). Ground truthing and prediction of land use land cover using Camarkov model in Raniganj coalfield, India. *International Journal of Basic and Applied Research, 9*(3), 33–52.

Sarkar, A. (2019e). Land use dynamics and landscape patterns assessment using geoinformatics techniques in Raniganj coalfield, India. *Research Journal of Science and Technology, 11*(1), 27–37. https://doi.org/10.5958/2349-2988.2019.00004.4

Sarkar, A. (2019f). Land use land cover dynamics and prediction analysis using cellular automata Markov model in Raniganj coalfield area, West Bengal, India. *The Research Journal of Social Sciences, 10*(3), 4–19.

Sarkar, A., & Bhattacharya, S. (2016). GIS based assessment and evaluation of the environmental impacts of opencast coal mining in Raniganj coalfield. *IOSR Journal of Environmental Science, Toxicology and Food Technology (IOSR-JESTFT), 10*(12), 45–58. https://doi.org/10.9790/2402-1012024558

Sayemuzzaman, M., & Jha, M. K. (2014). Modeling of future land cover land use change in north carolina using Markov chain and cellular automata model. *American Journal of Engineering and Applied Sciences, 7*(3), 295–306. https://doi.org/10.3844/ajeassp.2014.295.306

Smith, V. K., & Palmquist, R. B. (2001). The use of hedonic property value techniques for policy and litigation. In *The International Yearbook of Environmental and Resource Economics 2002/2003* (pp. 115–164). Edward Elgar Publishing.

Srivastav, S., Roy, S. K., & Krishnamurthy, P. S. (2016). An analysis of land use and land cover dynamics and causative drivers in a thickly populated Yamuna river basin of India. *Applied Ecology and Environmental Research, 14*(3), 773–792.

Stipinovich, B. A. (2005, May). *Change in land cover and water abstraction: Modelling runoff effects in the Bot River catchment.*

Street, S., Eh, E., Tel, B. D., & London, C. S. (2010, August). *Landscape character assessment: Part A*. Prepared for Northumberland County Council by Land Use Consultants.

Tarboton, D. (2003). *Terrain analysis using digital elevation models in hydrology*. Presented at 23rd ESRI International Users Conference, 1–13.

Tibbett, M., & Fourie, A. B. (2007). Post-mining landforms—Engineering a biological system. In *Proceedings of the second international seminar on mine closure 2007* (pp. 3–12). Australian Centre for Geomechanics.

Tiwary, R. K., & Dhar, B. B. (2006). *Environmental pollution from coal mining activities in Damodar river basin, India*. IMWA.

Uuemaa, E., Antrop, M., & Marja, R. (2009). Landscape metrics and indices: An overview of their use in landscape research. *Living Reviews in Landscape Research, 3*, 1–28.

Vázquez-quintero, G., Solís-moreno, R., & Pompa-garcía, M. (2016). Detection and projection of forest changes by using the Markov chain model and cellular automata. *Sustainability, 8*, 1–13. https://doi.org/10.3390/su8030236

Wodaje, S. T. (2016, August). *Land degradation vulnerability assessment using GIS and remote sensing in Beshilo River Basin, Ethiopia.*

Woldai, T., & Taranik, J. (2001). Environmental impact assessment of mining related dewatering activities using integrated InSAR, optical remote sensing and pumping data. *The International Archives of the Photogrammetry, Remote Sensing and Spatial Information Sciences, 37*, 191–198.

Yu, X. J., & Ng, C. N. (2007). Spatial and temporal dynamics of urban sprawl along two urban–rural transects: A case study of Guangzhou, China. *Landscape and Urban Planning, 79*, 96–109. https://doi.org/10.1016/j.landurbplan.2006.03.008

Yuksel, A., Gundogan, R., & Akay, A. E. (2008). Using the remote sensing and GIS technology for erosion risk mapping of Kartalkaya Dam Watershed in Kahramanmaras, Turkey. *Sensors, 8*, 4851–4865. https://doi.org/10.3390/s8084851

Zerbe, R. O., Jr. (1988). *I wish to thank Stewart Jay, David Layton for useful comments*. This paper relies on my recent book, (Zerbe 2001). 1. (Zerbe 2001), 1–40.

Zurayk, R., Hamadeh, S., & Chehab, A. G. (2001). A GIS-based methodology for soil degradation evaluation. In *Sustaining the global farm* (pp. 1082–1089). Purdue University/USSA-ARS National Soil Erosion Research Laboratory.

Chapter 6
Study on the Synergetic Relationship Between Tribes and Their Socioecological Environments—A Study in Jangalmahal of West Bengal, India

Subir Kumar Moyra and Sukla Hazra

Abstract India is home to the world's largest tribal population (8.63%), a large number of whom live in concentrated areas on rugged forest terrain. These Indigenous peoples used to live in a customary socioecological space for their survival. Through major and minor forest resource collection, tribes living along forest buffers have developed sustainable socioecological relationships with their environments. Traditionally, Indigenous people enjoy not only various benefits from living in forest environments but also escalating initiatives to improve the sustainability of those environments. Such interconnections with the environment secure their role as the most influential components of maintaining their ecosystems. Human–ecosystem interconnections happen within complex systems. This study examines the interspace between tribal livability and the socioecological environments in a forest ecosystem. A cross-sectional survey was conducted on Indigenous peoples in Jangalmahal district in West Bengal to monitor ecosystem services and tribal livability. Four ecosystem services in forests are interlinked with multidimensional aspects of tribal livability via the society, environment, livelihood, and food (SELF) method. This study's results revealed that tribal livability significantly harmonized with environmental quality. Exploring such important connections not only helps to empower tribes but also emerges as a feasible way to improve the management of ecosystems.

1 Introduction

Interspaces between forested and settled areas have been recognized by researchers as the most important for examining the livability of Indigenous communities. The most common term used in ecology to refer to such transitional spaces is *ecotones* (Clements, 1905). Traditionally, various Indigenous communities in India prefer to

S. K. Moyra (✉) · S. Hazra
East Calcutta Girls' College, Lake Town, Kolkata, West Bengal, India

S. Hazra et al. (eds.), *Geographies of the Indian Subcontinent*,
https://doi.org/10.1007/978-3-031-74813-4_6

live in forest buffers (Stumpff, 2006.). They not only collect various resources from the forest but also play a vital role in conserving the forest ecosystem. Much has been learned about the functions of forest ecosystems, but no research has yet presented their connections to human communities or all the other components of living in forest ecosystems. India has many socioecological hotspots where tribe–forest relationships can be viewed to assess their multidimensional effects on human wellbeing (Cruz-Garcia et al., 2017).

One biologically rich area featuring human social integration is the Jangalmahal region on the eastern fringe of the Chotanagpur plateau (Hunter, 1897), which boasts tropical wood greenery on an undulated topography dominated by Sal trees (*Shorea robusta*). The area has practical significance to tribe–forest interactions and has a long legacy of Indigenous people's struggling to reclaim their rights to land, water, and the forest itself. Amid the jungle ecosystem, various tribal peoples live in harmony with nature and sustain its available resources. Due to rapid population growth, a majority of the forest buffer could soon become overcrowded due to human pressures. The overexploitation of resources often leads to the degradation of natural ecosystems, which threatens the wellbeing of marginal communities living in or along the periphery of forests (Ninan & Kontoleon, 2016; Verma et al., 2017).

To assess the present status of a given ecosystem, ecologists measure the structure and function of the ecosystem. This study attempts to determine the ecosystem services status in the selected study area. Ecosystem valuation is a well-known method for quantifying the goods and services of an ecosystem. In 2003, the United Nations Millennium Ecosystem Assessment program developed a first-of-its-kind framework for assessing ecosystem services (MA, 2003, 2005). This study assessed the entire study area according to four services, namely provisions, regulations, cultural services, and support services, by using the Rapid Assessment of Wetland Ecosystem Services (RAWES) approach (Ramsar Convention, 2018).

Evaluations of wellbeing at the small scale are by far the most challenging unless they correlate strongly with livability (Chazal, 2010). The livability of socioeconomic and environmental systems is integral to improving quality of life for humans (McCann, 2004; Kaal, 2011). Although the term *livability* is most commonly used in the context of urban areas (Badland et al., 2014; Herrman & Lewis, 2015), it has also been gaining importance in rural areas because evaluations of livability take a holistic approach. Within a community-wellbeing framework, four pillars of livability (Everard, 2020) are assessed in this study according to the society, environment, livelihood, and food (SELF) model (Bijl, 2011).

Various Indigenous communities have passed down several skills for adapting to local socioecological environments to improve their livability (Sanford, 2013). Because of the unique cultures of the Indigenous communities living in the rural areas of India, they face more livability challenges as their forest resources drastically deplete over time. This chapter aims primarily to relate the ecosystem services of forests to Indigenous livability instead of to forests' human development capability. Through a livability approach, a more vibrant ecosystem can be formed, one that will generate enough surpluses to maintain the quality of life of Indigenous communities. In this chapter, a framework that integrates ecosystem services with the livability of Indigenous peoples is proposed.

2 Study Area

In this study, the Jangalmahal region of West Bengal includes Purulia, Bankura, West Medinipur, and Jhargram district (Fig. 6.1). The historical Jungle Mahal estate is also included in the present-day Sinbhum district of Jharkhand and the Balasore district of Orissa. Geographically, the study area lies between 21°50′N to 23°50′N latitude and 85°50′E to 88°55′E longitude, covering an area of 1363.26 km^2. Several east-flowing rivers drain into the area, among which Subarnarekha, Damodar, and Silai are most important. The climate of the area is of the tropical monsoonal type, with a short winter and a humid summer. It receives an annual average rainfall of 1583.961 mm.

The study area is home to a large number of people from different ethnic groups, including Schedule Tribe communities like the Santal, Bhumij, Onrao, Munda, Lodha, etc. Due to adverse conditions, agricultural productivity is not high in most parts of the area. People used to earn a living by collecting natural resources and through wage labor. Various types of animal husbandry are among their significant employment opportunities. Many people engage in the cottage industry, where they manufacture domestic products for sale. Moreover, the area is a rich source of natural flora and fauna, including forest resources such as timber for construction, firewood, and nontimber products.

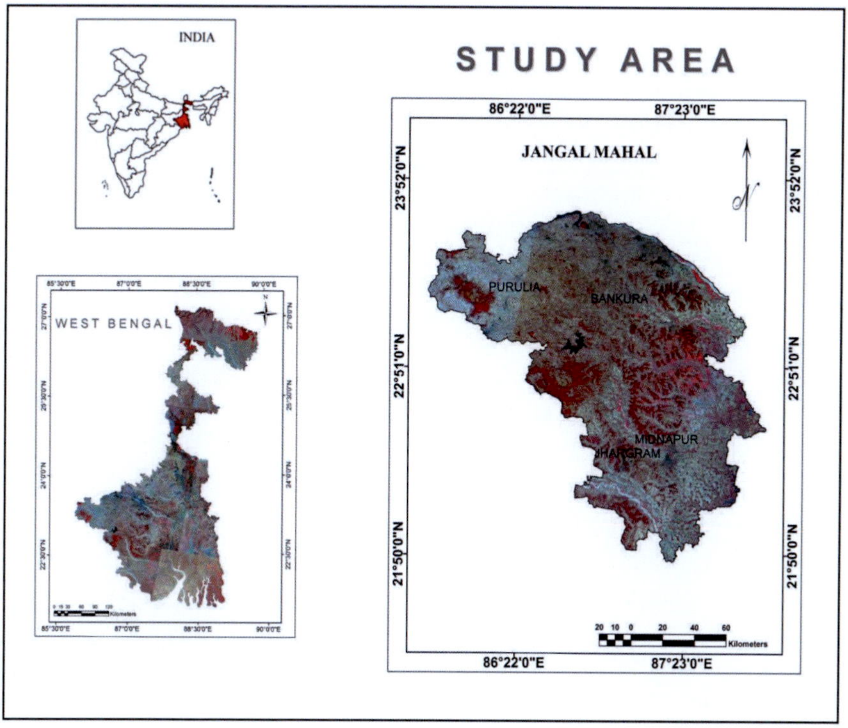

Fig. 6.1 Location of the study area

3 Understanding Capability and Livability Approaches

In the context of ecosystem service management for the forest buffers where Indigenous peoples live, two aspects are vital: (1) increasing human development capability via the sustainable use of available resources and (2) improving livability for Indigenous peoples by ensuring that their basic needs are being met, namely food, livelihood, education, culture, and environmental sustainability. The majority of researchers have relied on the capability approach, instead of the livability approach, to assess ecosystem services (Notess et al., 2018; Sangha et al., 2018, 2019a, b). According to Sen (1993 and 1999), capability comprises the primary components of advancing human wellbeing. Here, *capability* is defined as having certain skills and knowledge for the collection of food and other essentials. Few researchers have claimed that improving capability is essential to seizing a job opportunity (Polishchuk & Rauschmayer, 2012; Sangha & Russell-Smith, 2017).

Studies conducted in India on the interdependencies between a tribe and their ecosystem (Everard et al., 2019a) have focused only on resource use. For natural resource management, the capability approach includes on the high-intensity acquisition of resources and seldom addresses the social and environmental aspects of resource use (Sangha & Russell-Smith, 2017). Even though Indigenous peoples have relevant knowledge and skills in resource use, they are often barred from using certain resources and services, which reduces their capabilities. Despite the rich biological diversity of resources in the forests that they live in and their cultural heritage of using those resources, most tribal populations are now increasingly dependent on state-provided income and food provision.

The livability of forest-dwelling tribal communities depends largely on acquiring services from their ecosystems (Soderqvist et al., 2005; Burger et al., 2008). The surpluses of immature ecosystems remain very high, but most of the forest-dwelling tribal communities in India now live in mature ecosystems, which have very little or no surpluses. Due to such poor surpluses, most of these tribal communities have a poor standard of living and low quality of life. In spite of this, generation after generation of various communities still lead a life in forest ecosystems. In addition, such forest-dwelling tribal communities are essential components of the forest ecosystem in that each maintains a strong socioecological presence there (Li et al., 2009). To determine how best to restore degraded ecosystems and manage natural resources, a comprehensive study on tribal livability is essential.

4 Materials and Methods

The central objective of the research design is to analyze the relationship between the services of an ecosystem and the livability of tribal communities in that ecosystem. Taking a quantitative approach would not allow the incorporation of both Indigenous livability and the environment. This study is based on a combined method that uses qualitative research within a quantitative framework. In the quantitative part, relevant variables are checked through component analysis based on collected data, whereas correlational research is carried out to highlight links between effective ecosystem service indicators and four pillars of livability.

4.1 Study Design

Most of the activity of forest-dwelling Indigenous people takes place in the buffers of forests, also called ecotones. The resources of the socioecological spaces where tribal communities in Jangalmahal live have been extracted according to a geographic information system–based (GIS-based) satellite-image-processing technique (Bhattarjee & Siddique, 2016). A subset of normalized difference vegetation index (NDVI) values is used for the selection of cores, and buffer tools are used to establish buffer spaces.

The socioecological environment in this study is evaluated on the basis of the range of services that it provides (Everard, 2020). The RAWES approach addresses four ecosystem service categories, namely provisions, regulations, cultural services, and support services, as defined by the Millennium Ecosystem Assessment (2005). Temporal changes in major vegetative areas have also been monitored over the past several years to determine the degradation of ecosystem services and forests' health status. Various indicators of livability have been studied by addressing four pillars of livability: society, the environment, livelihood, and food (Balasubramaniana & Sangha, 2021).

4.2 Data Management

The secondary data in this study came from digital images, topographical sheets, administrative maps, and the published reports of various government agencies—like the Census of India, the Agriculture Development Office, and the Ministry of the Environment and Forests, among others. To process the study's satellite data, Arc GIS software and various Internet-based GIS applications were used.

Its primary data were acquired through questionnaires, surveys, semistructured interviews, and observations. For the selection sample, a stratified household-sampling technique was applied. Each Indigenous community that dominates a given district of Jangalmahal was selected, as were the Nayagram from Jhargram district, the Bundwan from Purulia district, the Keshiary from West Medinipur district, and the Ranibundh from Bankura district. In total, 60 tribal households were categorized into blocks according to one of three forest conditions—afforested area (28), degraded areas (16), or no change areas as marked by GIS-based image processing (16)—and were randomly surveyed. From each of the selected blocks, 15 samples were collected, for a total sample size of 60 households. Household surveys were organized according to where forest cores were taken: away from or within buffer areas.

A survey was conducted during the year 2020 to obtain data and information on ecosystem services and livability. Data were collected via door-to-door surveys conducted on the basis of structured interviews and observations. Only those households that earned a living from forest-derived or -related products were studied. Conversations were conducted in local languages. During the survey, female respondents were given priority over their male counterparts. All feedback was transferred to spreadsheets for further processing.

5 Exploration of Forest Resources

The mapping of ecological spaces can be carried out by evaluating the biophysical characteristics of environmental components such as vegetation productivity, animal concentration, and species diversity. This mapping technique is easy to perform thanks to remote-sensing data and GIS images.

5.1 Mapping Forest Areas in Jangalmahal

In this study, the normalized difference vegetation index (NDVI) was used because it has a diversity of digital spatiotemporal data. We used Landsat satellite imagery covering the entire study area of Jangalmahal during the period 1991–2021. Because the NDVI evaluates the productivity of an ecosystem, it is used to delineate ecological spaces (Luck, 2007). NDVI values were calculated by using the following formula (Rouse et al., 1974):

$$\mathrm{NDVI} = (\mathrm{pNIR} - \mathrm{pRED})/(\mathrm{pNIR} + \mathrm{pRED}). \tag{6.1}$$

where pNIR and pRED denote the reflectance value of the near-infrared band and that of the red band, respectively.

The NDVI - 2021 range for each class, divided according to their quartile deviation, is shown in Fig. 6.2. The map shows two classes of forests: first, concentrations of dense forest around five hubs within the Jangalmahal administrative area and, second, less-dense forest areas surrounding the concentrated forest areas. According to their respective NDVI values, 16.45% of the entire Jangalmahal region was occupied by the first class—dense forest areas (0.187–0.44)—and 32.97% by the second class (0.116–0.165)—areas with sparse vegetation growth.

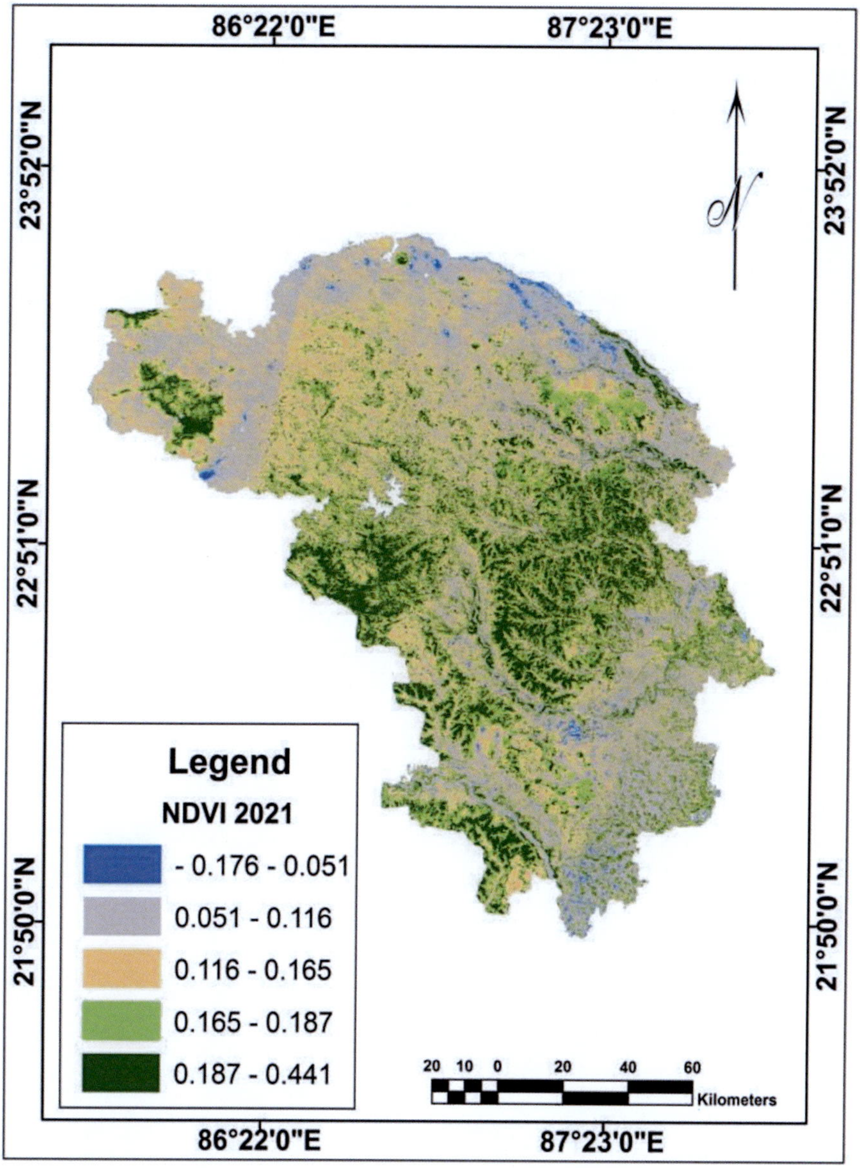

Fig. 6.2 Landsat-image-based NDVI map of Jangalmahal area (2021)

5.2 *Forest Dynamics in Jangalmahal*

Which forest dynamics have changed in this part of the country have remained elusive according to previous records. In the early colonial period, large swathes of forests were cleared in India to expand areas for agriculture (Guha, 1983). The first site of forest reservation was established in a teak forest in Malabar in 1806 to meet the demands of shipbuilding. During the time of railway expansion in India (1870–1910), Indian forests were severely depleted. The mass destruction of forests was authorized by Dalhousie, the governor of British India, who set up the Forest Department with the help of German foresters in 1864. India ratified its first Forest Act in 1927, in which reservation and protection were implemented for forest areas; the act was later amended in 1974 and 2002.

Due to the static nature of secondary data on forest areas, the situation on the ground is very hard to assess. In this study, we tracked changes to forest' areal extents by using NDVI values on the dense forest class over a 30-year period (1991–2021). The total area of dense forest in Jangalmahal was 17.29% in 1991. Due to human interference, however, it decreased to 16.45% in 2021.

Degradation has been comparatively higher in the eastern and southern parts (than in other parts) of Jangalmahal over the past 3 decades (Fig. 6.3). However, the extents of major forest areas increased from 1991 to 2021, due in large part to afforestation initiatives by government agencies. In the contemporary world, forest clearing is commonly used to provide space for agriculture and infrastructural needs. Although various socioecological causes are relevant in the degradation of forest areas, most of this degradation is due to political instability, adverse climatic conditions, a lack of other economic opportunities, etc.

If we consider the trends of deforestation and afforestation in Jangalmahal over the past 3 decades, the scenario seems encouraging (Table 6.1). According to one

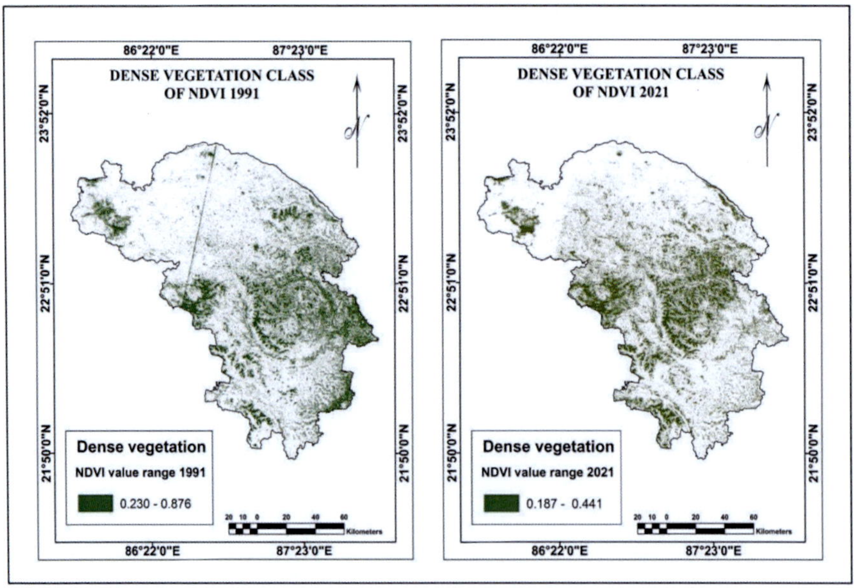

Fig. 6.3 NDVI-based forest-density map of Jangalmahal for 1991 and 2021

Table 6.1 Numerical expressions of ecosystem service values

Temporal span	Deforestation (%)	Afforestation (%)	Balance (%)
1991–2001	8.25	7.84	−0.41
2001–2011	10.10	8.35	−1.75
2011–2021	8.97	10.51	+1.54

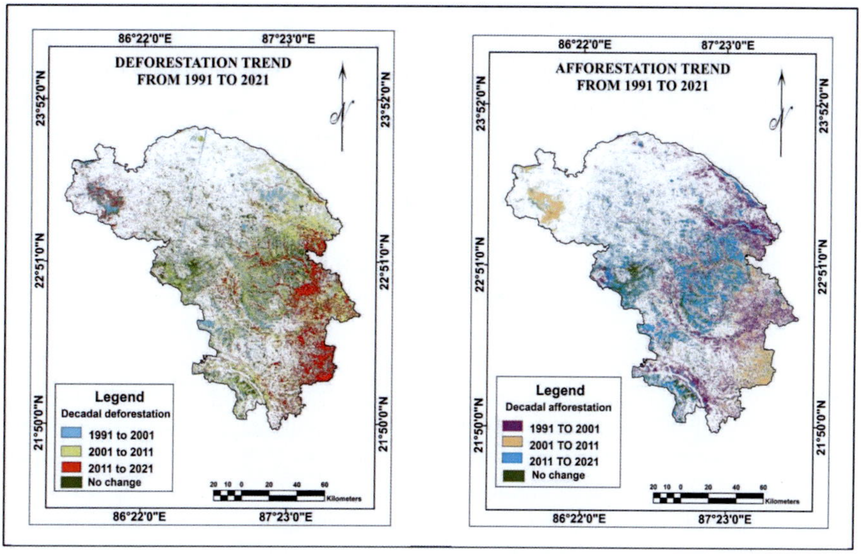

Fig. 6.4 Deforestation and afforestation status on a temporal scale

study (Fig. 6.4), the deforestation process was its full swing during 2001–2011, which was a time of political instability in Jangalmahal because of extremist activity.

This afforestation process maintained its pace during a time of stable state government, which authorized several forestry projects. As much as 10.51% of the total forest area was afforested during 2011–2021. This has helped to counter the fast rate of deforestation (8.97%) and reflects an increase of 1.54% in dense vegetative area over the past decade. Therefore, human interference can not only shrink natural vegetation cover but also expand it under certain socioecological conditions. When communities buy into the coexistence of ecosystems and their way of life for socioeconomic upgradation, they thoroughly integrate into natural environments. Indeed, the communities that live in or around the Jangalmahal area are the most important components of this ecosystem.

5.3 Delineating Socioecological Spaces

Being the most active components of an ecosystem, tribes both directly and indirectly control the health and energy flow of that ecosystem. When a forest is heathier and has more foliage cover, this indicates a surplus of production from the ecosystem, which is bound to offer more life-supporting opportunities to those who live in them. Our study identified five core forest areas in the Jangalmahal area: the Nayagram forest in Jhargram district, the Salboni forest in West Medinipur district, the Ranibundh and Jaypur forests in Bankuradistrict, and the Bagmundi forest in Purulia district. These five forests were delineated according to the overall size of their polygon area of coverage and their NDVI-based forest class. For the demarcation of forest cores, a maximum of 200 m of linear stretches of land without forests were merged with forest shapes taken from GIS images of their environment conditions (Fig. 6.5). The forest with the largest polygon area was Salboni (33.2 km^2), and the forest with the smallest polygon area was Nayagram (3.8 km^2).

By applying the same technique, forest cores were also developed for the year 2021. With the help of a combination raster technique, locational changes to five forest cores in Jangamahal were assessed. According to Table 6.2, in 1991, the largest forest core area was in the Ranibundh area, spanning 238.84 km^2, and the smallest forest core area was in the Jaypur area, spanning 28.36 km^2.

Among the forests that suffered deforestation from 1991 to 2021, Bagmundi was the highest. Less change in deforestation was observed in Salboni and Ranibundh (Fig. 6.6). The highest level of afforestation among forest core areas occurred in Nayagram. At the micro level, each forest core can be assigned to a corresponding

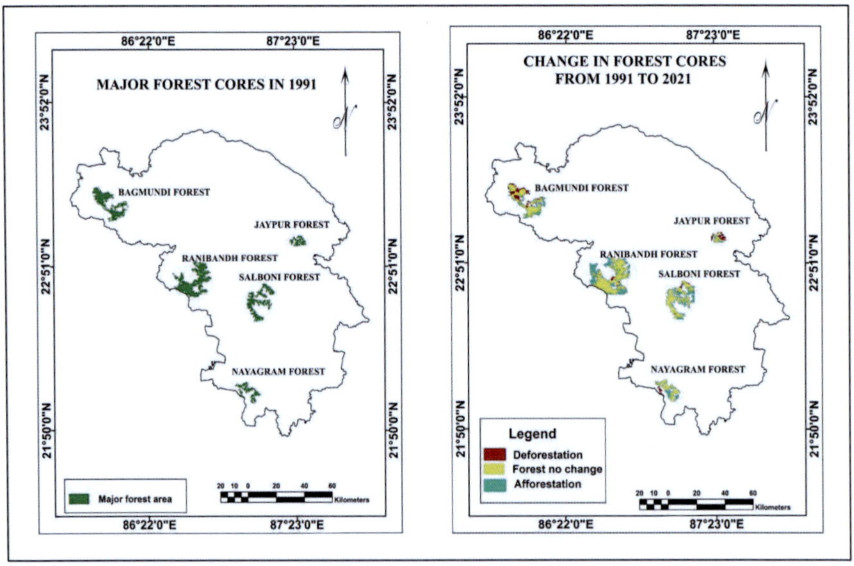

Fig. 6.5 Spatiotemporal change in forest cores in Jangalmahal

Table 6.2 Qualitative changes in forest core areas, 1991–2021

Forest	Forest core area in km^2 (1991)	No change in area, in km^2 (2021)	Deforested area, in km^2 (2021)	Afforested area, in km^2 (2021)
Bagmundi	175.61	120.02	55.59	41.69
Jaypur	28.36	21.92	6.44	7.13
Raninundh	238.84	128.62	10.25	71.36
Salboni	138.87	221.54	17.30	51.48
Nayagram	64.63	54.59	10.04	19.56

Source: Authors' estimation

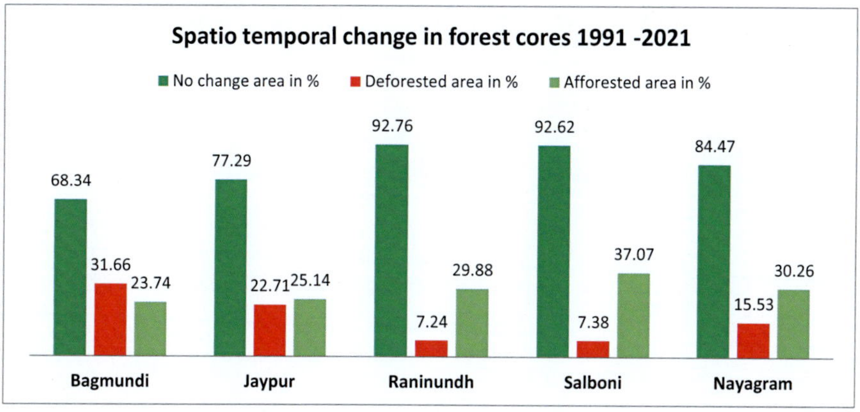

Fig. 6.6 Changes in vegetative landscape in or around forest cores

tribal village. Both deforestation and afforestation in any village are the consequences of socioecological factors. Every tribal forest village uses a great deal of ecosystem services from its forest. Deforestation usually reduces the flow of ecosystem services, whereas afforestation increases all ecosystem services.

In this study, villages were identified in forest areas to determine the status of an ecosystem after changes to its forest cover. Figure 6.7 gives a micro-level demonstration of forest dynamics. The Lodha tribe who live in Ambisole village were primitive, and the Santal people in Tilia village were not very concerned about the preservation of forest land. Additionally, Tilia villagers were forced to minimize their use of ecosystem services because of mismanagement. To understand the synergetic relationship between tribes and their ecosystems, this study was organized according to two groups of stakeholders and their responses to deforestation, afforestation, and trends in stability: (a) villagers living in forest core areas and (b) villages living in and around the buffers of forest cores.

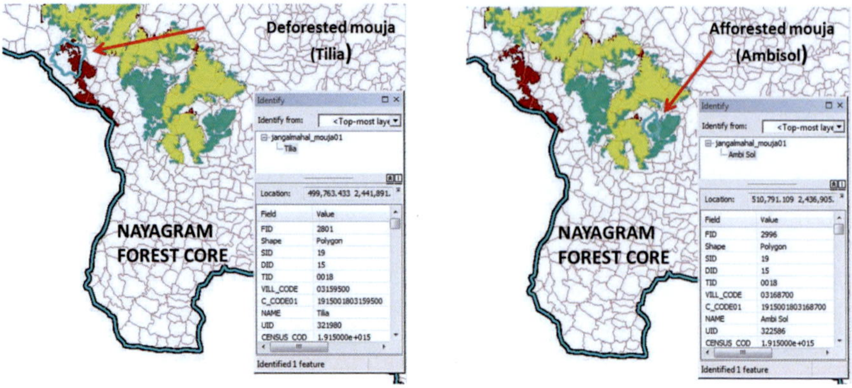

Fig. 6.7 Micro-level demonstration of forest dynamics

6 Synergetic Relationship Between Tribes and Their Ecosystems

In general, tribes in the Jangalmahal area still believe that they are the children of the forest. Hence, they have every right to use the entire resource base of the forest and have within their remit to take every possible initiative for the protection of forests. Tribes acquire food, house-building materials, earnings, and even land for shelters from forests; in return, they protect forests and concerned ecosystems by traditional indigenous knowledge, such as minimizing the use of resources to as little as is needed for survival, maintaining the flow of resources, imposing certain taboos in favor of conservation, and reusing natural resources. The role of Indigenous people in the Jangalmahal ecosystem is quite complex as we found in our observation during field visits. We have decided to study tribal forest association by including the four pillars of human–ecosystem relationships e.g. society, the environment, livelihood, and food.

6.1 Contribution of Ecosystem to Tribes

In the context of human–ecosystem interdependencies, a complex socioecological system must be highlighted. To do that, various socioeconomic considerations must be factored in. In this study, we used various ecosystem service indicators to explore and determine the status of services provided by a local ecosystem to the tribes living in it.

6.1.1 Framework for Ecosystem Service Delineation

The monitoring of ecosystem services for the study area was carried out by following the experience-based Rapid Assessment of Wetland Ecosystem Services (RAWES) model (Ramsar Convention, 2018). Later, many ecologists used the ecosystem service technique to evaluate the degradation of ecosystems other than wetlands (Körner et al., 2005; Mcinnes & Everard, 2017). The RAWES method makes a semiquantitative assessment of the significance of an ecosystem service (Everard, 2020). With the help of the RAWES method, four categories of ecosystem services, namely provisions, regulations, cultural services, and support services, were explored. In the Millennium Ecosystem Assessment (2005), 36 ecosystem service indicators were included.

However, in this study, only 18 ecosystem service indicators were used instead of 36 indicators; it also used other indicators to assess Liang village, Arunachal Pradesh, from Everard et al. (2021). Out of the 18 indicators, five were selected from provisions and five from regulations, and four were selected from cultural services and four from support services. The quantification of each indicator used respondents' perceptions of an ecosystem service. In this study, responses were obtained either on a weighted scale or as a precise value. According to Table 6.3, the indicators were normalized to between 0 and 1 for processing. The importance of each service was assessed according to the scale established by Everard et al. (2021).

With the help of four categories of ecosystem services, Ecosystem Service Index (ESI) values for the study area (Davidson et al., 2019) were assigned to the ecosystem services in the Jangalmahal area. To calculate the total ESI score for all the relevant services, only values within the positive range were used. The ESI values can range from +1 to −1 for compound values. The more positive that a score is, the better off the beneficiaries of the ecosystem service will be. The findings of this study are used to verify the high dependency of local people on ecosystem services, reflecting the mutual interactions between natural resources and local stakeholders.

Table 6.3 Numerical expressions of ecosystem service values

Importance of ecosystem service	Description code	Numerical value
Significantly positive	++	1.0
Positive	+	0.5
Neutral	(..)	0.0
Negative	-	−0.5
Significantly negative	- -	−1.0
Unknown	xx	Excluded

Source: Millennium Ecosystem Assessment (2005)

6.1.2 Selected Ecosystem Service Indicator Scores

The obtained responses to questions on evaluating ecosystem services according to the 18 selected indicators from four categories (provisions, regulations, cultural services, and support services) were complex. The score for the regulation category was 0.79, which was the highest among the categories and was within the significantly positive range (++). The score for support services was 0.44 and was within the neutral range (..). The score for the provision category and that for cultural services were within the significantly positive (+) and significantly neutral (..) ranges, respectively.

Of the five provision services, freshwater and food production were assessed by calculating the months during which they are available per year. According to the results, the respondents were on average able to access freshwater for 7 months and food for 6 months per year. For the assessment of fiber and fuel production, the average household-level earning from collecting forest products was calculated. It included the market value of collected hard wood, fuel wood, leaves, and fruits at INR 5982 per household for the study year. The most effective provision indicator emerged as genetic resources, with an indicator score of 0.76. Major genetic resources found in the area are products of cultivation and forest collection, which directly help tribes to earn their livelihoods. Food adequacy for households was surveyed according to months per year, finding adequacy in 9 months per year. By assessing tribal dependency on natural medicines, the biochemical and natural medicine indicator was calculated, yielding a value of 0.59. The availability of various herbs, such as neem (*Azadirachta indica*), amlaki (*Phyllanthus emblica*), gulancha (*Tinospora cordifolia*), ishwari mul (*Aristolochia indica*), and satamul (*Asparagus racemosus*), accounted for this high score.

In terms of the regulation services in the Jangalmahal area, all the selected indicators had very high scores, with the maximum positive response reached for natural hazard regulation (0.89) and air-quality regulation (0.84). For the assessment of air-quality and natural hazard regulation, respondents were asked whether the incidence of droughts and that of poor air quality are regulated by forest cover. Unlike other indicators, most of the negative responses to water regulation (hydrological) services were assessed according to household's proximity to a water source. Unfavorable hydrogeological conditions are prominent throughout Jangalmahal, which emerged as the cause of the water scarcity in the region.

Out of the four selected cultural services, two —cultural heritage and education/research—were especially relevant, with indicator scores of 0.55 and 0.61, respectively. The score for recreation and tourism was particularly poor, having a value of only 0.40. The score for spiritual and religious services was high (0.50) due to the participation of tribes in several religious practices throughout the year. In fact, various natural products are used as key ingredients and substances in offerings during religious tribal festivals, such as Baha, Sankranti, and Saharay.

Among the supportive services, respondents mostly stressed primary production (0.51) and provision-of-habitat (0.56) services. The nutrient-cycling indicator also yielded a high value (0.47), which was assessed according to the average calorie consumption per day for an individual stakeholder. Consumed calories were assessed according to the standard guideline on consumed food (Gopalan et al., 1971, 2002) in kilocalories, or kcal. In 2009, the government of India (ICMR, 2009) set a standardized norm of 2400 kcal per consumer per day for rural people, in accordance with the definition of the poverty line. However, in the present case, the obtained value was 2017.07 kcal, which would fulfill almost 85% of the daily energy requirements of the respondents.

6.1.3 Results of Ecosystem Service Index (ESI) in Jangalmahal

On the basis of four categories of ecosystem services, an overall ESI score of 0.57 was calculated (Table 6.4), which is positive (+). Everard et al. (2021) found an ESI score of 0.60 in the case of a hilly village in Uttarakhand. It reflects the moderate dependency of the tribe on their exploitation of forest resources. The value found in the Sundarbans of West Bengal (Everard et al., 2019a, b) was even lower, due to the

Table 6.4 Ecosystem Service Index (ESI) scores for indicators in Jangalmahal

Ecosystem service	Selected indicators	Indicator score	Category score	ESI
Provision	Freshwater resources	0.64	0.55	**0.57**
	Food production	0.51		
	Fiber and fuel production	0.23		
	Genetic resources	0.76		
	Biochemicals and natural medicines	0.59		
Regulation	Air-quality regulation	0.84	0.79	
	Local climate regulation	0.72		
	Water regulation (hydrological)	0.73		
	Natural hazard regulation	0.89		
	Erosion regulation	0.77		
Cultural	Cultural heritage	0.55	0.52	
	Spiritual and religious values	0.50		
	Recreation and tourism	0.40		
	Educational and research	0.61		
Supportive	Soil fertility	0.25	0.44	
	Primary production	0.51		
	Nutrient cycling	0.47		
	Provision of habitat	0.56		

Source: Everard et al. (2021)

limited interaction of dwellers with forests under strict forest-protection laws. The present ESI score was intermediate—between scores found in studies on hilly regions and those on flat regions. Nevertheless, that the forest ecosystem in the selected study area yielded an ESI score of 0.57 was significantly relevant to its financial and environmental stability.

6.2 *Contribution of Tribes to Their Ecosystems*

Forests have many ecosystem services that help to provide food and opportunities for tribal livelihoods, form societies, and stabilize the environment. Due to the high availability of renewable resources with good-quality nutrient components in forest ecosystems, tribes in the area heavily concentrated around these ecosystems. Over the years, Indigenous people have maintained the forest ecosystem and have done their best to maintain ecological balance, instead of acting purely in self-interest, according to the four pillars of the SELF method: society, the environment, livelihood, and food (Balasubramaniana & Sangha, 2021).

6.2.1 Integrated Framework for Linking Livability with Ecosystem Services

In the absence of empirical studies using the livability approach to assessing ecosystem services in the Indian tribal context, where dependence on natural systems is relatively high, this study makes unique links between Indigenous peoples and their natural ecosystems. To improve policymaking to better understand these complex connections, we propose a framework where tribal livability forms an integral part of the two-way human–ecosystem relationship (Sangha & Russell-Smith, 2017). We provide evidence that is based on the responses of tribal peoples in the Jangalmahal area. The study analyzes the links between ecosystem services and Indigenous peoples' livability via the SELF method.

6.2.2 Linking Indicators of Ecosystem Services with Livability Approach

Livability simply represents quality of life. In the area at hand, tribal livability depends mostly on the quality of the ecosystem services that tribal people avail themselves of. Accordingly, livability deals with the needs of the habitants of an area and the fulfillment of those needs. Elements such as housing, food, safety, security, and environmental stability are regarded as the basic needs of human life. In this study, only four important components of livability are addressed. Many tribal

Table 6.5 Ecosystem services related to tribal livability

	Livability pillars			
Ecosystem services	Society	Environment	Livelihood	Food
Provision (freshwater, food, fiber and fuel, genetic & medicinal resources)	Medicinal	Freshwater	Fiber and fuel, genetic	Food
Regulation (air quality, local climate, hydrological, natural hazards, erosion)		Air quality, local climate	Natural hazards	Hydrological cycle, soil erosion
Cultural (cultural heritage, tourism, spiritual & religious, educational)	Cultural heritage, spiritual and religious values	Educational	Recreation and tourism	
Supportive (soil fertility, primary production, nutrient cycling, habitat)	Habitat		Soil fertility	Primary production, nutrient cycling

Source: Sangha and Russell-Smith (2017)

communities have knowledge of cultural norms and practices that can successfully use and manage ecosystem services. Understanding the value of people's interactions with their natural resources by using four pillars of livability, as shown in Table 6.5, can help policymakers better understand the true value of ecosystem services for Indigenous wellbeing.

6.2.3 Ecosystem Services and Livelihoods of Tribal Households

This section presents the relationships between four livability pillars and ecosystem service indicators. Regarding the society pillars, medicinal provisioning, cultural heritage, spiritual and religious values, and habitat support services were found to be closely related. To maintain these goods and services, tribes make various contributions to their ecosystems.

Pillars of Society

Some tribes restrict entry to forests in rainy seasons and enter forests only after Bandna Parab has finished being celebrated, in late October or early November, which facilitates the growth and reproduction of medicinal plants. These plants are easily available and regularly used by tribes, which they maintain and often nurture year-round. With their knowledge, passed down through cultural heritage, tribes help to maintain their natural environments. In almost every folk song and story featuring tribes, forest dependency is highlighted. Tribal life is full of uncertainty but never devoid of cultural integrity. Participation in cultural activities related to forests was agreed on by almost all the respondents. Thanks to spiritual and religious values, several species of flora and fauna survive in Jangalmahal. In the case of vegetation,

Sal, Mahua, and Karam trees are considered sacred trees; they are used to protect such plants in grooves around their sacred place of *Jaher Than*. Tribes gather forest products to use as house-building materials. Many tribal households also have bamboo or wood gardens, which they maintain for house-building materials, but at the same time, these gardens help to stabilize local ecosystems.

Pillars of the Environment

In the Jangalmahal area tribal community was found to know how to live in a clean environment. On freshwater provision, 72% of respondents agreed that forests help to stave off acute water crises. At the same time, better water availability supports more life and a greater diversity of life in an ecosystem. To achieve good air quality and local climate regulation, tribes often maintain trees around their houses. They band together in several "Bon committees" to take part in social forestry schemes. Of the respondents, 87% want flexibility in forest resource collection laws and an increase in their share of profits from forest resources. Indigenous peoples' knowledge and skills in using of forest resources have over the years improved with education. With the help of several awareness campaigns and protection committees, deforestation has largely been controlled.

Pillars of Livelihood

Within this study's framework, livelihood and provision services were linked to the collection of various forest products. Wood is the major product, but forests also offer minor resources, such as Sal leaves, fuel wood, Kendu leaves, and Mahua seeds, among many others. Tribes always prefer to maintain mother trees whenever the forest is cleared for wood collection. The cutting techniques of tribes are also suitable for regrowth because they allow 1–1.5 feet of a tree to remain, from which new growth emerges. Natural hazards, especially conditions such as droughts, are very common in this region, due to which they lose resources and workdays. Therefore, they rely heavily on planting indigenous trees instead of foreign trees like eucalyptus or akashmoni. This study also suggests a shift in paradigm: from perceiving a forest as a local resource base to perceiving it as a public benefit. New generations are advocating for better public attractions and more tourism development, which may open up a new gateway for livelihood opportunities. According to the supportive service components, soil fertility was found to be the most important. Tribal communities have mostly refrained from excessive chemical use on soil and have instead relied on biofertilizer for crop production. The use of compost, biofertilizer, and natural pesticides is an important contribution of tribal people toward maintaining the stability of their ecosystems.

Pillars of Food

To secure a sustainable supply of food, tribal people make the most of every food source (Jelliffe, 1966). A large amount of food is collected directly from nature. With the help of natural irrigation and rainwater, they produce their own vegetables. Of the tribal households surveyed, 43% have been found to have a kitchen garden. Food is also produced by maintaining livestock in large numbers. The scarcity of water and soil erosion are regarded as the two major challenges to tribal prosperity. Tribes take utmost care in the preservation of water and soil. Of the tribal farmers surveyed, 92% cultivated crops on "Tokra", or fragmented piece of land, which helps to minimize the flow of water and control soil erosion. Primary production and nutrient cycling are two other major aspects among the pillars of food. Due to the high amount of nutrient generation, respondents favored access to nutritious food. Various fruits, roots, leaves, and animal products are regularly collected, the collection of which is managed by several tribal laws and state forest acts.

6.3 *Interrelationships Between Tribes and Their Ecosystems*

This study analyses and integrates both ecosystem services and livability aspects with empirical evidence to establish interconnections between tribes and their ecosystems. In the absence of theoretical and empirical studies on ecosystem services and the livability approach in the Indian context, where tribal people's dependence on natural systems is relatively high, it offers a multidimensional perspective of various links between people and their natural systems that can directly inform policy decision-making. All the interconnections can be represented by using a holistic system model. Such a holistic understanding can help to improve the wellbeing of local tribal communities and better manage rapidly depleting natural resources.

6.3.1 Selection of Ecosystem Service Indicators

Principal component analysis (PCA) was conducted on household responses to determine which relevant indicators have eigenvalues above 1 (Kaiser's rule) (Jolliffe, 2002). The presence of many correlations above 0.3 in the correlation matrix of the selected indicators suggests the appropriateness of the indicators. No correlation value was found above 0.9, which may raise the problem of singularity.

According to the correlation matrix, the whole dimension cannot be represented by any single indicator. This fact is well supported by the scree plot, which shows that six of the 18 indicators have eigenvalues above 1. The first two components represented 70.896% of the total variance. Bartlett's test of sphericity was also significant, at a value of 0.00. Anything less than 0.05 is considered as significant, which suggests that the correlation matrix is not an identity matrix.

Table 6.6 shows the component matrix of the selected indicators, from which six components were extracted. The higher the absolute value of the loading, the more the factor contributes to the variable. In our study, 18 indicators were divided into four factors, in which cultural factors were the most important, showing high loading for cultural heritage (0.942) in component 1 and for spiritual and religious values

Table 6.6 Component matrix[a]

Ecosystem services	Sl. no.	Indicators	Component					
			1	2	3	4	5	6
Provisioning	1	Freshwater	−0.046	−0.320	0.717	0.103	−0.042	0.132
	2	Food production	0.241	0.462	−0.494	0.214	0.032	0.139
	3	Fiber and fuel production	−0.083	0.594	−0.140	−0.095	0.021	0.199
	4	Genetic resources	0.338	0.198	0.498	0.241	−0.133	0.094
	5	Biochemicals, natural medicines	0.054	0.613	−0.129	−0.148	−0.377	0.107
Regulating	6	Air-quality regulation	0.483	0.272	0.026	−0.090	−0.082	−0.039
	7	Local climate regulation	0.245	−0.023	−0.304	0.209	−0.297	0.058
	8	Water regulation (hydrological)	−0.339	−0.272	−0.212	0.482	−0.298	0.256
	9	Natural hazard regulation	0.007	0.010	0.027	−0.256	0.037	−0.004
	10	Erosion regulation	0.082	−0.254	0.283	−0.182	−0.047	−0.055
Cultural	11	Cultural heritage	0.942	−0.185	0.027	−0.101	0.122	−0.031
	12	Spiritual and religious values	−0.113	0.793	0.165	0.244	0.064	−0.118
	13	Recreation and tourism	−0.055	0.150	0.369	0.611	−0.088	0.037
	14	Education and research	0.789	−0.258	0.069	0.148	−0.159	0.112
Support	15	Soil fertility	0.184	0.059	0.154	0.780	0.434	−0.238
	16	Primary production	0.052	0.181	−0.038	−0.019	−0.085	0.967
	17	Nutrient cycling	−0.071	−0.063	−0.100	−0.044	0.923	−0.043
	18	Provision of habitat	0.126	0.063	0.685	0.097	0.105	−0.165

Source: SPSS output
Extraction method: principal component analysis
[a]Six components extracted

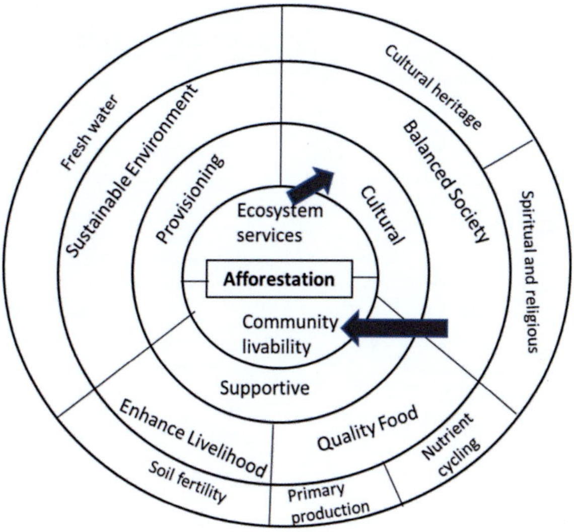

Fig. 6.8 Holistic presentation of the interdependence between ecosystem services and tribal livability

(0.793) in component 2. The freshwater indicator under provision services was found to be an important indicator in component 3. However, the maximum number for high loading was seen in the following supportive services: soil fertility (0.780) in component 4, nutrient cycling (0.923) in component 5, and primary production (0.967) in component 6. Therefore, respondents were most satisfied with the cultural and supportive services of their ecosystems.

6.3.2 Interlinking Extracted Indicators with Tribal Livability

In this study, the respondents listed three relevant ecosystem services in Jangalmahal. In order of importance, they are cultural, supportive, and provision services (Fig. 6.8). Another interconnection was found between cultural heritage and society. Tribal people want to maintain the stability of their ecosystems because they believe in the importance of those ecosystems to their cultures. Their culture is best maintained within their societies. Ecosystems also play roles in their spiritual knowledge. Indeed, they locate their existence within their Indigenous sociocultural setups.

Another important interconnection was found between provision services and the environment. From an environmental perspective, tribes are most concerned about the water crisis. They know the value of water, which is essential to grow crops and support livestock. Most of the tribal settlements in this region have developed around a source of water. Tribes use every possible source of water available, and they rate water availability as their foremost concern among all provision services. A

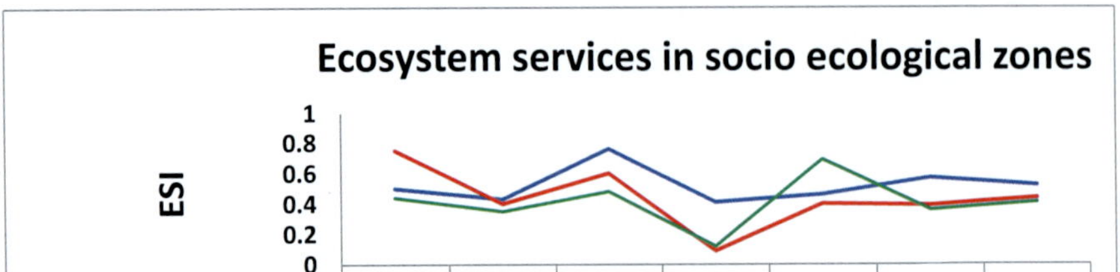

	Cultural heritage	Spiritual and religious value	Fresh water	Soil formation	Primary production	Nutrient cycling	ESI
No forest cover change	0.5	0.43	0.76	0.41	0.46	0.57	0.52
Deforested area	0.75	0.4	0.6	0.09	0.4	0.39	0.44
Afforested area	0.44	0.35	0.48	0.12	0.69	0.36	0.41

Fig. 6.9 ESI scores for socioecological zones as per, Field Survey, 2020

clear link among the supportive services was found between food and livelihood opportunities. The availability of food and livelihood opportunities depend on primary production, nutrient cycling, and soil fertility. Increasing species diversity and forest production in the area are connected to the level of primary production.

6.3.3 Ecosystem Services in Socioecological Zones

Several PCA-extracted ecosystem service indicators have been studied by assessing forest dynamics. The entirety of our sample survey data was classified into the following groups: no change in forest cover, degraded areas, and afforested areas ($n = 16$, 16, and 28, respectively). The obtained values were scaled from 0 to 1, and the results are shown in Fig. 6.9.

Analyzing the important ecosystem service indicators in different socioecological zones is essential for understanding the dynamics between forest ecosystems and tribal livability. The highest ESI values were found in the areas of Jangalmahal showing no forest cover change, with value of 0.52, which was even higher than the overall sample ESI value of 0.49. An ESI value of 0.41 was the smallest and was found in the afforested zone of Jangalmahal, whereas the ESI was intermediate in deforested areas. This study suggests that due to deforestation, ESI values can drastically drop, especially in soil production, primary production, and nutrient cycling. In afforestation areas, primary production services remain high for all socioecological zones. Thus, reduced forest health generally indicated poor primary production, which led to a diminishing relationship between a tribe and their ecosystem.

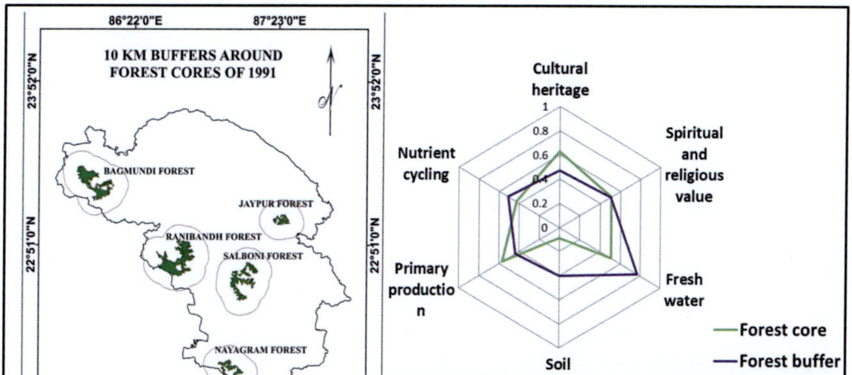

	Cultural heritage	Spiritual & Religious	Fresh water	Soil formation	Primary production	Nutrient cycling	ESI
Forest core	0.63	0.51	0.51	0.09	0.57	0.43	0.46
Forest buffer	0.47	0.5	0.77	0.4	0.44	0.51	0.51

Fig. 6.10 Variation in ecosystem services between forest cores and forest buffers

6.3.4 Changing Ecosystem Services in Forest Core and Buffer Areas

The study of ecological boundaries has played an important role in human–ecosystem relationships. To explore it, we arranged ecosystem indicator values according to their spatial forest zones. The two zones were separated by forest core boundaries. A buffer area of 10 km was constructed along each of the five forest cores (Fig. 6.10). During the survey, a sample was collected each from forest core area and each buffer area to identify the roles that ecotones play in generating ecosystem services.

ESI values were comparatively high in the forest buffer areas (10 km) of Jangalmahal. Here tribes enjoy various opportunities from nature to fulfill their requirements live better social and economic lives. Primary production services from ecosystems were much higher in forest cores (0.57), whereas nutrient cycling was comparatively lower (0.43), due to the presence of more forest land than agricultural land. Tribes living in forest cores enjoy more services for their traditions and customs. However, no significant change was observed in religious services between the two selected spaces.

7 Conclusion

Tribes that live in forests collect resources from them and take measures to maintain their renewability. Over the past 30 years, the forest area of Jangalmahal has decreased from 17.29% (1991) to 16.45% (2021), yet over the past decade, the rate of deforestation was lower than that of afforestation. Whenever an area has faced economic or social instability, deforestation has increased, but restoration of forests begins with the active participation of tribes. According to this study, the hilling of natural forest is the most vital role played by Indigenous peoples in the Jangalmahal ecosystem.

Traditionally, tribes in Jangalmahal have lived in and around the five forest cores of Jangalmahal. These cores are not always safe, as has been found in the Bagmundi forest of Purulia. The micro-level assessment in this study suggests that the self-awareness of tribes living in forest cores has significantly controlled the dynamics of those forest cores. The ecosystem services in this area were measured via the RAWES method, yielding a satisfactory value of 0.57. Out of the selected components, cultural heritage emerged as the most influential on generating ecosystem services for tribes. Among the provision services, freshwater availability was significant, with a good score according to PCA extraction, suggesting the high value of water in the semidry areas of Jangalmahal.

The study showed that the livability of the Jangalmahal ecosystem rested on the active contributions of Indigenous tribes. They interact with their ecosystems to achieve the four pillars of livability: society, the environment, livelihood, and food. The most important link between ecosystem services and tribal livability was for cultural heritage; an important interconnection was found between provision services and the environment, for which Indigenous people take a holistic approach. Food and livelihood generation for tribes were found to largely depend on supportive services.

Socioecological factors were found to play two roles in ecosystem service generation. They increased with stable forest cover and decreased in deforested areas. The ESI values for forest buffers were higher than those for forest core areas due to transition effects and increased opportunities, especially in supportive services. Insufficient essential resources and earning opportunities have stagnated the lives of tribal people living at the age of technological expansion in Jangalmahal because such tribal life in Jangalmahal beats to the rhythm of nature.

References

Badland, H., Whitzman, C., Lowe, M., Davern, M., Aye, L., Butterworth, I., Hes, D., & Giles-Corti, B. (2014). Urban livability: Emerging lessons from Australia for exploring the potential for indicators to measure the social determinants of health. *Social Science and Medicine, 111*, 64–73.

Balasubramaniana, M., & Sangha, K. K. (2021). Integrating capabilities and ecosystem services approaches to evaluate indigenous connections with nature in a global biodiversity hotspot of Western Ghats, India. *Global Ecology and Conservation, 27*, 1–16.

Bhattarjee, S., & Siddique, G. (2016). Significance of forests in the tribal habitation of Jangalmahal area in West Bengal. *International Journal of Research on Social and Natural Services, 1*, 1–13.

Bijl, R. (2011). Never waste a good crisis: Towards social sustainable development. *Social Indicators Research, 102*(1), 157–168.

Burger, J., Gochfeld, M., & Greenberg, M. (2008). Natural resource protection of buffer lands: Integrating resource evaluation and economics. *Environmental Monitoring and Assessment, 142*, 1–9. [PubMed: 17899418].

Chazal, J. D. (2010). A systems approach to livability and sustainability: Defining terms and mapping relationships to link desires with ecological opportunities and constraints. *Systems Research and Behavioral Science, 27*(5), 585–597.

Clements, F. E. (1905). *Research methods in ecology*. University of Nebraska Publishing Co.

Cruz-Garcia, G. S., Sachet, E., Blundo-Canto, G., Vanegas, M., & Quintero, M. (2017). To what extent have the links between ecosystem services and human well-being been researched in Africa, Asia, and Latin America? *Ecosystem Services, 25*, 201–212.

Davidson, N. C., Dinesen, L., Fennessey, S., Finlayson, C. M., Grillas, P., Grobicki, A., et al. (2019). Wetland ecological status and changes: A meta-analysis. *Marine and Freshwater Research, 71*(1), 127– 138. https://doi.org/10.1071/MF18329.

Everard, M., Gupta, N., Scott, C. A., Tiwari, P. C., Joshi, B., Kataria, G., et al. (2019a). Assessing livelihood-ecosystem interdependencies and natural resource governance in Indian villages in the Middle Himalayas. *Regional Environmental Change, 19*(1), 165–177. https://doi.org/10.1007/s10113-018-1391-x.

Everard, M., Kangabam, R., Tiwari, M. K., McInnes, R. J., Kumar, R., Talukdar, G. H., et al. (2019b). Ecosystem service assessment of selected wetlands of Kolkata and the Indian Gangetic Delta: Multibenefcial systems under diferentiated management stress. Wetlands Ecology and Management. https://doi.org/10.1007/s11273-019-09668-1.

Everard, M. (2020). *Rebuilding the earth: Regenerating our planet's life support systems for a sustainable future*. Palgrave Macmillan.

Everard, M., Kataria, G., Kumar, S., & Gupta, N. (2021). Assessing livelihood-ecosystem interdependencies and natural resource governance in a tribally controlled region of India's north-eastern middle Himalayas. *Environment, Development and Sustainability, 23*, 7772–7790.

Gopalan, C., Sastri, B. V. R., Balasubramanium, S. C. (1971). *Nutritive value of Indian foods, National Institute of Nutrition, Indian Council of Medical Research*, Hyderabad.

Gopalan, S. (2002). Nutrition security. In P. Chaturvedi (Ed.), *Food security in South Asia*. Indian Association for the Advancement of Sciences.

Guha, R. (1983). Forestry in British and Post-British India: A Historical Analysis, Economic and Political Weekly, Nov. 5–12, Vol. 18, No. 45/46, pp. 1940–1947.

Herrman, T., & Lewis, R. (2015). *Research initiative 2015–2017: Framing livability – What is livability?* University of Oregon.

Hunter, W. W. (1897). Statistical account of the district of Midnapur, A Statiscal Account of Bengal, Volume III, Part I. Government of West Benal.

ICMR. (2009). *Nutrient requirements and recommended dietary allowances for Indians*. National Institute of Nutrition, Indian Council of Medical Research. Annx. 2.1 and 2.2.

Jelliffe, D. B. (1966). *The assessment of the nutritional status of the community* (WHO Monograph No. 53). WHO.

Jolliffe, I. T. (2002). Principal Component Analysis, Springer. https://link.springer.com/book/10.1007/b98835

Kaal, H. (2011). A conceptual history of livability. *City, 15*(5), 532–547.

Körner, K., Ohsawa, M., Spehn, E., Berge, E., Bugmann, H., Groombridge, B., et al. (2005). Chapter 24: Mountain Systems. In R. Hassan, R. Scholes, & N. Ash (Eds.), Ecosystems and human wellbeing. Current state and trends: Findings of the condition and trends working group. Millennium ecosystem assessment (Vol. 1, pp. 681–716). Washington: Island Press.

Li, F., Liu, X., Hu, D., Wang, R., Yang, W., Li, D., & Zhao, D. (2009). Measurement indicators and an evaluation approach for assessing urban sustainable development: A case study for China's Jining City. *Landscape and Urban Planning, 90*(3–4), 134–142.

Luck, G. W. (2007). The relationships between net primary productivity, human population density and species conservation. *Journal of Biogeography, 34*, 201–212.

McCann, E. J. (2004). Quality of life and popular media discourse. *Urban Studies, 41*(10), 1909–1929.

Millennium Ecosystem Assessment. (2003). *Ecosystems and human well-being: A framework for assessment*. Island Press.

Millennium Ecosystem Assessment. (2005). *Ecosystems and human well-being* (p. 563). The UN Island Press.

Ninan, K. N., & Kontoleon, A. (2016). Valuing forest ecosystem services and disservices – Case study of a protected area in India. *Ecosystem Services, 20*, 1–14.

Notess, L., Veit, P. G., Monterroso, I., Sulle, A. E., Larson, A. M., Gindroz, A.-S., Quaedvlieg, J., & Williams, A. (2018). *The scramble for land rights: Reducing inequity between communities and companies*. World Resources Institute (WRI).

McInnes, R. J., & Everard, M. (2017). Rapid assessment of wetland ecosystem services (RAWES): An example from Colombo, Sri Lanka. *Ecosystem Services, 25*, 89–105

Polishchuk, Y., & Rauschmayer, F. (2012). Beyond "benefits"? Looking at ecosystem services through the capability approach. *Ecological Economics, 81*, 103–111.

Ramsar Convention. (2018). Resolution XIII.17: Rapidly assessing wetland ecosystem services. In *13th meeting of the conference of the contracting parties to the Ramsar Convention on Wetlands*. https://www.ramsa r.org/about/cop13–resolutions

Rouse, J. W., Haas, R. H, Schell, J. A. & Deering, D. W. (1974). Monitoring vegetation systems in the great plains with ERTS, Greenbelt, NASA. Goddard Space Flight Center 3d ERTS-1 Symp, Vol. 1, Sect. A(35): 3010–3017.

Sanford, E. (2013). *What is the difference between livability and sustainability?* [Online]. Available from http://www.camsys.com/kb_experts_livability.htm

Sangha, K., & Russell-Smith, J. (2017). Towards an indigenous ecosystem services valuation framework: A north Australian example. *Conservation and Society, 15*, 255–269.

Sangha, K. K., Preece, L., Villarreal-Rosas, J., Kegamba, J. J., Paudyal, K., Warmenhoven, T., & RamaKrishnan, P. S. (2018). An ecosystem services framework to evaluate Indigenous and local peoples' connections with nature. *Ecosystem Services, 31*(Part A), 111–125.

Sangha, K. K., Maynard, S., Pearson, J., Dobriyal, P., Badola, R., & Hussain, S. A. (2019a). Recognising the role of local and Indigenous communities in managing natural resources for the greater public benefit: case studies from Asia and Oceania region. *Ecosystem Services, 39*, 100991.

Sangha, K. K., Russell-Smith, J., & Costanza, R. (2019b). Mainstreaming indigenous and local communities' connections with nature for policy decision-making. *Global Ecology and Conservation, 19*, e00668.

Sen, A. (1993). Capability and wellbeing. In: Nussbaum, M., Sen, A. (Eds.), The Quality of Life. Clarendon Press, Oxford.

Sen, A. (1999). Development as Freedom. New York.

Soderqvist, T., Eggert, H., Olsson, B., & Soutukorva, A. (2005). Economic valuation for sustainable development in the Swedish coastal zone. *Ambio, 34*, 169–175. [PubMed: 15865316].

Stumpff, L. M. (2006). Reweaving the earth: An indigenous perspective on restoration planning and the National Environmental Policy Act. *Environmental Practice, 8*, 93–103.

Verma, M., Negandhi, D., Khanna, C., Edgaonkar, A., David, A., Kadekodi, G., Costanza, R., Gopal, R., Bonal, B. S., Yadav, S. P., & Kumar, S. (2017). Making the hidden visible: Economic valuation of tiger reserves in India. *Ecosystem Services, 26*, 236–244.

Part II
Application of Geospatial Technologies

Chapter 7
SENTINEL-Derived Aerosol Optical Depth and Land Surface Temperature of a Newly Developed Urban Metropolitan Area in Eastern India

Surajit Chakraborty, Rohit Basu Dhar, and Pradip K. Sikdar

Abstract Aerosols in the atmosphere act as a significant meteorological parameter whose presence help us to develop models for long-term prediction of environmental degradation and understand the effects of climate change. Technologies such as remote sensing help us to retrieve aerosol data from space more efficiently and reliably. Although ground level measurement techniques are readily available for determination of aerosol loads over a localized area, aerosol optical depth (AOD) data obtained from satellites provide us reliable information on particulate concentration over a large area. This study portrays the AOD distribution over an urban metropolitan town (Rajarhat) in West Bengal, India, using ESA Sentinel 3 OLCI sensor data. We have used AOD data for four different wavelengths (400_{nm}, 778.75_{nm}, 865_{nm}, and 940_{nm}) and tried to establish the relationship between AOD and land surface temperature (LST) in 4 months (January, April, October, and November) of three consecutive years (2017, 2018, and 2019). Results show that out of the four wavelengths, $AOD_{778.75\ nm}$ shows best correlation between aerosol load and LST data in the 4 months of the 3 years. A positive correlation between $AOD_{778.75\ nm}$ and LST data obtained for the months of January 2017 ($R^2 = 0.984$), April 2017 ($R^2 = 0.938$), October 2017 ($R^2 = 0.856$), and November 2017 ($R^2 = 0.695$) reflects the rise in aerosol load more in winter (January) than in summer (April) or post-monsoon (October–November) which have a direct effect on the LST. This study also helps to understand the spatiotemporal distribution of AOD over various land use types. Therefore, aerosol data retrieved from Sentinel 3 satellite are useful to understand the temporal variation of LST in a regional scale. This temporal variation will help the planners to take effective decisions to control the rise in LST with increasing urbanization.

S. Chakraborty · R. B. Dhar · P. K. Sikdar (✉)
Department of Environment Management, Indian Institute of Social Welfare and Business Management, Kolkata, India

S. Hazra et al. (eds.), *Geographies of the Indian Subcontinent*,
https://doi.org/10.1007/978-3-031-74813-4_7

1 Introduction

The total radiation budget of the Earth and atmospheric interactions are strongly related with the total content of aerosol present in the atmosphere through direct and indirect effects. Being short-lived with low half-life, they have a large spatial and temporal variation which is difficult to monitor. To quantify the radiative effects and climatic impacts on Earth's atmosphere, the accurate estimation of physical and optical properties of aerosols is extremely important (Ramanathan et al., 2001). The challenges for retrieval of aerosol data are ascertaining their concentration in atmosphere, its chemical composition, size distribution, and characteristics. Remote sensing techniques have served as one of the major methods for characterizing the large spatial and temporal heterogeneity of aerosol distribution on a regional to global scale (Chin et al., 2009). Techniques involved in aerosol data retrieval from ground-based and satellite data observations have been used worldwide to improve quantification and characterization of their physical and optical properties on a global scale (Dubovik et al., 2002; Tanré et al., 2001; Hsu et al., 1999; Remer et al., 2005; Ma et al., 2015; Van Donkelaar et al., 2010) and regional scale (Misra et al., 2008;Bibi et al., 2015; Xiao & Weng, 2007).

The ground-based aerosol retrieval method, especially AERONET, has its network covered across the world in almost 450 locations (Hsu et al., 1999) providing information on aerosol conditions since 1993. Satellite-based aerosol retrieval products have been quite useful in obtaining aerosol data globally. Among most of the available aerosol retrieval satellites, the Moderate Resolution Imaging Spectroradiometer (MODIS) collects aerosol data over terrestrial and aquatic regions at multispectral wavelengths (Kaufman et al., 2002; Levy et al., 2015; Remer et al., 2005). Certain uncertainty still exist as various physical parameters such as surface reflectance, aerosol models, and cloud screening needs to be assumed (Remer et al., 2005). Therefore, the retrieved aerosol data is either quantified or validated by ground-based measurements, aiding integrated efforts of accurate aerosol characterization.

The major challenge for long-term retrieval of aerosol optical depth (AOD) data from satellite observation is generation of correct AOD data from diverse and sophisticated instruments having limited lifeline requiring mature algorithms which can derive aerosol data based upon platforms. Popp et al. (2016) have used the Visible and Infra-red Imaging Spectroradiometer Suite (VIIRS) of MODIS sensor algorithms for upgrades and improvements, whereas the Advanced Along-Track Scanning Radiometer (AATSR) has been applied for the observations of sea and land surface temperature (SLST). This suggests that both the algorithms need to be more robust for AOD retrieval. In spite of MODIS data availability for aerosol retrieval, the onboard Environmental Satellite (ENVISAT) instrument called Medium Spectral Resolution Imaging Spectroradiometer (MERIS) provides sufficient and credible data for various applications. Two major algorithms, namely, Bremen AErosol Retrieval (BAER) (von Hoyningen-Huene et al., 2011) and European Space Agency (ESA) standard aerosol retrieval method (Lenoble et al.,

2007) provide AOD data. But considering many factors like dark-vegetated surfaces and for non-absorbing aerosols the BAER algorithm has limited accuracy and mixed success (de Leeuw et al., 2015), and the ESA AOD retrieval method tends to overestimate AOD (de Leeuw et al., 2015).

Ocean and Land Color Instrument (OLCI) onboard ESA's Sentinel 3A takes the heritage of MERIS as it contains all MERIS channels. The MERIS algorithms can be transferred onto the OLCI instrument. Theoretically, it is possible to transfer the mature MERIS retrieval algorithms to the OLCI instrument. Therefore, the first objective of the present research is to explore which wavelength among four different wavelengths (400 nm, 778.75 nm, 865 nm, 940 nm) of ESA's Sentinel 3A OLCI sensor is best suited for AOD retrieval. The second objective is to understand the relationship between AOD and land surface temperature (LST) using LANDSAT OLI.

2 Methods

2.1 *Study Area*

Rajarhat block is a semi-metropolitan neighborhood of Kolkata located between 22° 40′54.79"N–22°30′50.25″N Latitude and 88°23′42.96″E–88°34′18.42″E Longitude in North 24-Parganas district of West Bengal which includes the entire Saltlake block, Rajarhat New Town, and East Kolkata Wetlands falling under the Bidhannagar Municipal Corporation with an area of 165 km^2 (Fig. 7.1). It also includes the Netaji Subhash Chandra Bose International Airport as one of the key landmarks. There are two important canals or *khals* which drain the area. They are Nowai *khal* and Kestopur *khal* which meet River Bidyadhari outside the eastern boundary of Rajarhat block (Fig. 7.1). Rajarhat has the characteristics of a deltaic land. According to the Census of India (2011), the total population of Rajarhat block is 189,893, and the population density has increased from 701 people/km^2 in 1991 to 1151 people/km^2 in 2011. The climate of Rajarhat is tropical similar to the rest of Gangetic West Bengal and characterized by south-west monsoon. The monsoon sets in the first week of June and continues till mid-September. Winters are usually dry ranging from mid-November to mid-February while summers (March to June) are humid. The average annual rainfall for the period 1990–2020 is around 1756 mm and falls under Zone-3 (semi-humid where annual rainfall ranges from 1500 to 1800 mm). Maximum and minimum temperatures recorded are 42 °C in May and 11 °C in January (IMD, 2020). Quite recently, in the past 10 years, the area is under the focus of large-scale real estate development. During the 1990s, Rajarhat was dominantly rural with sprawling agricultural and open land. The area experienced rapid urbanization post-2000 with several habitat encroachments with possible changes in land surface temperature and aerosol load.

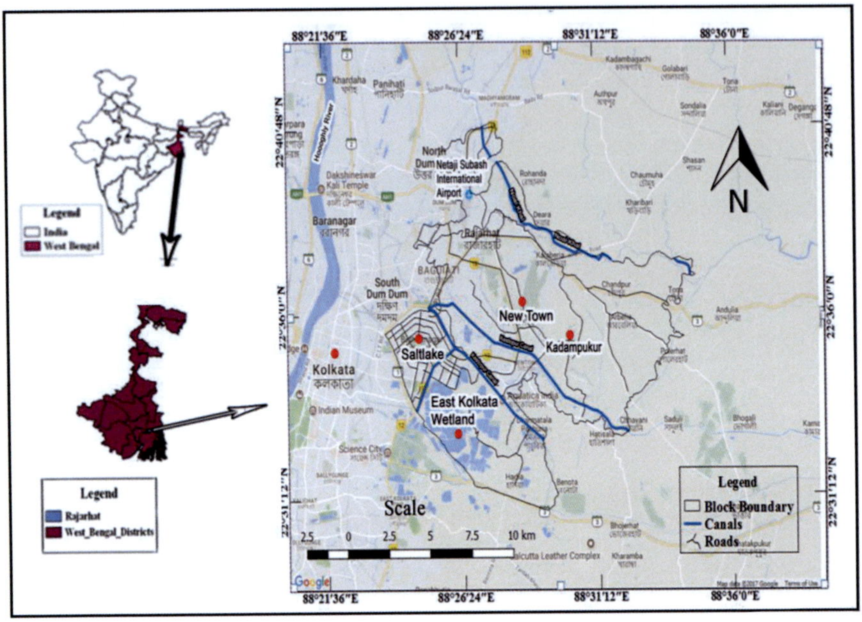

Fig. 7.1 Location map of Rajarhat block

2.2 Satellite Data

Satellite data used are Landsat 8OLI (January, April, October and November, 2017–19) and ESA Sentinel 3A sensors (January, April, October, and November, 2017–19) which are downloaded from the USGS Earth Explorer website as a GeoTIFF data product. These are then geo-referenced to UTM zone 45 North projection using WGS-84 datum. The thermal bands of Landsat 8 OLI are used for the estimation of LST, and ESA Sentinel 3A OLCI is used for the retrieval of AOD data. The data acquisition is enlisted in Table 7.1. The downloaded satellite images are pre-processed by an image processing system which includes both atmospheric and geometric corrections.

2.3 Aerosol Data Retrieval from Sentinel 3 OLCI

2.3.1 Sentinel 3 Instrument Payload

The OLCI Instrument Payload (Sentinel 3 Handbook, 2017) onboard ESA's Sentinel 3 satellite is particularly designed just like ENVISAT MERIS's opto-mechanical and imagery machinery built-in. Few important features of the satellite are as follows:-

Table 7.1 Satellite data acquisition showing time, path/row, available bands, and spatial resolution

Satellite	Time	Path/row	Available bands	Spatial resolution
Landsat 8	January 2017–19, April 2017–19, October 2017–19, November 2017–19, 2018	138/044	Visible: Bands 1, 2, 3, 4	30 m × 30 m
Sensor—Operational Land Imager/Thermal Infrared Sensor			NIR: Band 5	
			SWIR: Band 6, 7, 9	
			TIR: Bands 10 and 11	
ESA Sentinel 3A	January 2017–19, April 2017–19, October 2017–19, November 2017–19	Orbit Direction: **Descending**	Visible: Bands 1[a], 2, 3, 4, 5, 6, 7, 8, 9, 10	300 m × 300 m
Sensor—OLCI		Relative Orbit No: **9802**	NIR: Bands 11, 12, 13, 14, 15, 16[a], 17[a], 18	
			SWIR: Bands 19, 20[a], 21	

[a]Aerosol retrieval bands

(a) Swath width: 1270 km.
(b) Spectrometer imaging: Five cameras equipped with push-broom scanners preventing contamination of sun-glint by camera tilts in western axes.
(c) Spatial sampling: 300 m @ SSP.
(d) Spectrum: 21 bands (0.4–1.02) μm.
(e) Radiometric accuracy: 2% abs, 0.1% rel.
(f) Launch mass: 153 kg.
(g) Size: 1.3 m^3.
(h) Design lifetime: 7.5 years.

2.3.2 Band Designations

Among the available 21 MSI bands in Sentinel 3 LFR Level 2 products, the bands which give the quantification of the AOD data are given in Table 7.2.

2.3.3 Processing Levels of Sentinel 3 OLCI

Usually Sentinel 3 OLCI data products are subjected to three layers of processing which are as follows:

Table 7.2 Band designations of Sentinel 3 satellites

Band	λ center (nm)	Width (nm)	Function
Oa01[a]	400	15	Aerosol correction, improved water constituent retrieval
Oa13	761.25	2.5	O_2 absorption band/aerosol correction
Oa16[a]	778.75	15	Atmospheric correction/aerosol correction
Oa17[a]	865	20	Atmospheric correction/aerosol correction, pixels, cloud co-registration
Oa20[a]	940	20	Water vapor absorption, atmospheric correction/aerosol correction
Oa21	1020	40	Atmospheric correction/aerosol correction

[a]Indicates AOD retrieval bands

(a) Level 0 includes reassembled instrument at full space-time resolution which is a time-sorted Instrument Source Packet (ISP). Communication artifacts such as time frames of synchronization and data duplication along with invalid packets are removed during dispatch of data. The OLCI data attains full resolution (300 m) at this particular stage.
(b) Level 1 includes Top of Atmosphere (TOA) corrected radiometric measurements along with radiometrically corrected with preservation of spectral characteristics. The data is checked for quality control and is ortho-geolocated (latitude and longitude, coordinates and altitude) with marked satellite positioning and pointing. The landmarks for preliminary pixel classification (land/water/cloud masks) are also generated and products are generated and dispatched in Full Resolution (FR—300 m) and Reduced Resolution (RR—1 km) for the whole globe with same coverage.
(c) Level 2 products include products pertaining to geophysical parameters derived from the processing of measurement data provided in the Level 1 product. Level 2 products generally include both land and marine applications and also have a capability to retain water vapor and integrated water vapor column data through both data streams, respectively.

The availability of Sentinel 3 data is made available through NetCDF product files and usually is classified further into Near Real Time (NRT) and Non-Time Critical (NTC). The product delivery time to users in case of NRT is less than 3 h after acquisition of data by sensors while NTC products have a scheduled delivery period of 1 month or later after acquisition of data from sensor directly or either from available data archives. The different radiometric scaling techniques, which are used by the Sentinel 3 OLCI prior data acquisition at both NRT and NTC, are as follows:

(a) Radiometric scaling—Derivation of TOA radiance from previously recorded numerical counts.
(b) Stray light correction.
(c) Geo-referencing.
(d) Pixel classification.

2.3.4 Preprocessing and Rayleigh Correction

The AOD data is retrieved from the OLCI sensor onboard Sentinel 3 which captures both at NRT and NTC. The two modes of data capture provide users to access both present and historical data. In our study, both NRT and NTC types of data are used based on the availability. From the Level 0, calibrated images available at Enhanced Full Resolution (EFR) are selected. A full resolution data product is downloaded from scihub.copernicus.eu (ESA web-based platform to download imagery) and is opened in Sentinel Applications Platform Toolbox (SNAP). It is to be mentioned that the AOD data retrieval is restricted to the bands as referred in Table 7.2. The preprocessing steps in SNAP are as follows:

(a) The downloaded data product is opened in SNAP and at first a subset mask is prepared by the shapefile of the study area.
(b) The data products are then individually converted to TOA reflectance from TOA radiance using the algorithm as expressed in the Eq. (7.1)

$$R_{TOA}(\lambda) = \frac{\pi\, L_{TOA}(\lambda)}{E_0(\lambda)\cos\theta} \tag{7.1}$$

where E_0 and θ are the solar spectral irradiance and the angle corresponding to the sun zenith, respectively. The sensors, which are supported by the radiance to reflectance processor, are provided within the algorithm along with the sun zenith angle value in case of L1b products.

The irradiance values of the solar spectral region are provided per pixel in case of OLCI L1b products. For SLST Radiometer, the values are taken from L1 product metadata, and for MERIS the metadata is collected from available auxiliary files.

(c) After successful completion, the atmospherically corrected data product is selected for Rayleigh correction. The necessary processing for Rayleigh correction is carried out in several steps. The Rayleigh reflectance is first computed for every 4 × 4 pixel window and then the Rayleigh transmittance and Rayleigh albedo are computed for every 4 × 4 pixel window. The available TOA apparent reflectance is corrected for gaseous absorption (ρ_{ng}) for obtaining corrected Rayleigh contributions for every pixel and finally deriving the TOA reflectance (r_{top}). It is ensured that there is no pixel overlap during Rayleigh correction.
(d) Thus, the final bands are obtained containing necessary information about AOD over the study area.

2.4 LST Retrieval from Landsat 8 OLI

2.4.1 Conversion of Digital Number (DN) Values to Radiance at the Sensor's Aperture

The metadata file within the satellite imagery bundle is used to convert the digital number (DN) values into reflectance. Before the data is made available to users, the Landsat images have individual radiance values tagged within them with proper scaling factor.

The spectral radiance at sensor's aperture for Landsat satellites (L_λ) after NASA (2016) is expressed by the equation

$$L_\lambda = M_L{}^* Q_{\mathrm{cal}} + A_L - O_i \quad (7.2)$$

where

L_λ = Spectral radiance at sensor's aperture in (W m $^{-2}$ sr^{-1} μm^{-1}).
M_L = A multiplicative rescaling factor which is particularly band specific obtained from Landsat metadata (RADIANCE_MULT_BAND_x, where the band number is denoted by *x*).
A_L = an additive rescaling factor which is particularly band-specific obtained from Landsat metadata.
(RADIANCE_ADD_BAND_x, where the band number is denoted by *x*).
Q_{cal} = Quantized and standard product pixel values which are pre-calibrated (DN).
O_i = Offset issued by the USGS for Thermal Infrared Sensor (TIRS) band calibration (applicable to Landsat 8 TIRS sensor).

An O_i value of −0.29 is used for Landsat 8 TIRS Band 10 and this value is then subtracted from the radiances as required for the calibration of Landsat 8 TIRS Band 10 radiance. Landsat 8 TIRS has two thermal bands (Bands 10 and 11). The uncertainty of Band 11 values is avoided by restricting the data availability only from Band 10 in LST retrieval (Barsi et al., 2014). Landsat 7 ETM+ and Landsat 5 TM sensors use the conversion factor of DN values to radiance where both "gain and bias" and "spectral radiance" methods are expressed by the Eqs. 7.3 and 7.4.

$$L_\lambda = \mathrm{gain} \times Q_{\mathrm{cal}} + \mathrm{bias} \quad (7.3)$$

$$\begin{aligned} L_\lambda = {} & ((LMAX_\lambda - LMIN_\lambda)/(QCALMAX\text{–}QCALMIN)) \\ & *(QCAL\text{–}QCALMIN) + LMIN_\lambda \end{aligned} \quad (7.4)$$

Where terms have usual meanings.

2.4.2 Conversion to Top of Atmosphere (TOA) Reflectance from Radiance

Top of Atmosphere reflectance values are obtained from radiance values. Top of Atmosphere reflectance forms the summation of atmospheric reflectance and combined surface reflectance to reduce "in between-scene" variability performed by normalization through solar irradiance. This TOA reflectance (ρ_p), which is the unitless ratio of reflected versus total power energy as suggested by NASA is expressed by the equation

$$\rho_p = \left(\pi^* \ L_\lambda{}^* d^2\right) / \left(ESUN_\lambda{}^* \cos \theta_s\right) \tag{7.5}$$

where,

L_λ = Spectral radiance at the sensor's aperture (W m^{-2} sr^{-1} μm^{-1}).
d = Earth-Sun distance in astronomical units (AU).
ESUN$_\lambda$ = The average values of solar exo-atmospheric irradiances in (W m^{-2} μm^{-1}).
θ_s = The zenith angle (solar) in degrees, which is equal to $\theta_s = 90° - \theta_e$, where θ_e is the sun's elevation.

ESUN can be expressed by the equation in case of Landsat 8 as,

$$\text{ESUN} = \left(\pi^* d^2\right) * \text{RADIANCE\_MAXIMUM}/\text{REFLECTANCE\_MAXIMUM} \tag{7.6}$$

where,

RADIANCE_MAXIMUM and REFLECTANCE_MAXIMUM according to the metadata.

2.4.3 Dark Object Subtraction 1 (DOS1) Correction

The Dark Object Subtraction 1 (DOS1) correction forms an additional atmospheric correction based upon the basic assumption about shadowing some pixels whose radiances are received at the sensor due to atmospheric interaction. The assumption behind this might be that few targets are completely black on Earth's surface; hence, 1% reflectance value obtained is after all better than none (Chavez, 1996; Chin et al., 2009). The path radiance after Jiménez-Muñoz et al. (2014) and Sobrino et al. (2004) is expressed by the equation

$$L_p = L_{\min} - L_{\text{DO1\%}} \tag{7.7}$$

$L_{\min}$ = radiance corresponding to a digital count value for which the sum of all the pixels with digital counts lower or equal to this value of the 0.01% of all the pixels considered (Sobrino et al., 2004).

In particular, for Landsat images the radiance is expressed by the equation

$$L_{\min} = M_L * \mathrm{DN}_{\min} + A_L \tag{7.8}$$

The radiance of Dark Object is expressed by the equation after Sobrino et al. (2004)

$$L_{\mathrm{DO1\%}} = 0.01^*[(ESUN_{\lambda}{}^* \cos\theta_s{}^* T_z) + E_{\mathrm{down}}] * T_v/\left(\pi^* d^2\right) \tag{7.9}$$

Currently, there are various DOS corrections available which include (DOS1, DOS2, DOS3, DOS4) based on different assumptions about T_v, T_z, and E_{down}. The simplest technique adopted is the DOS1 where the following assumptions are made (Pielke et al., 1992):

$$T_v = 1, T_z = 1, E_{\mathrm{down}} = 0$$

The Sentinel-2A images are converted to radiance by Eq. 7.1 prior to DOS1 calculation.

The path radiance using the DOS1 correction is expressed by the equation

$$L_p = M_L * \mathrm{DN}_{\min} + A_L\text{–}0.01^* ESUN_{\lambda}{}^* \cos\theta_s/\left(\pi^* d^2\right) \tag{7.10}$$

Therefore, the resulting Land Surface Reflectance is expressed by the equation

$$\rho = \left[\pi^*\left(L_{\lambda}\text{–}L_p\right) * d^2\right]/(ESUN_{\lambda}{}^* \cos\theta_s) \tag{7.11}$$

Incorporation of all these algorithms into the preprocessing of SCP allows the output conversion of the selected bands in Land Surface Reflectance required for LU/LC classification and LST estimation.

In the present study, TIR Band 6 is chosen for Landsat 5 TM with the lowest center wavelength, i.e., 10.60 μm while TIR Band 10 is chosen for Landsat 8 OLI/TIRS with the lowest center wavelength, i.e., 10.80 μm due to uncertainty of using Band 11 (Barsi et al., 2014).

2.4.4 Conversion of TOA Reflectance to Surface Reflectance

The TOA reflectance is converted to surface reflectance (ρ) using the formula of Pielke et al. (1992) which is expressed by the following equation:

$$\rho = \left[\pi^*\left(L_{\lambda}\text{–}L_p\right) * d^2\right]/[T_v{}^*((ESUN_{\lambda}{}^* \cos\theta_s{}^* T_z) + E_{\mathrm{down}})] \tag{7.12}$$

where,

L_p = path radiance (W m^{-2} sr^{-1} μm^{-1}).
T_v = atmospheric transmittance in the viewing direction.
T_z = atmospheric transmittance in the illumination direction.
E_{down} = down-welling diffused radiance (W m^{-2} sr^{-1} μm^{-1}).

2.4.5 Estimation of NDVI (Normalized Difference Vegetation Index) to Calculate the Land Surface Emissivity (LSE) Using the Land Surface Temperature (LST) Estimation Plugin

Thermal Infrared Sensor involves dealing with emitted thermal energy and therefore the knowledge of Land Surface Emissivity (LSE) is essential for estimating LST to reduce error on estimation of surface temperature from satellite-derived data. Out of the various methods available to estimate LSE (Zhang et al., 2006; Sobrino et al., 2004; Ma et al., 2015; Jiménez-Muñoz et al., 2014), the Normalized Difference Vegetation Index (NDVI) is widely used for estimation of the LSE which is expressed by the following equation (Zhang et al., 2006):

$$\text{NDVI} = \left(\frac{\text{NIR} - \text{R}}{\text{NIR} + \text{R}}\right) \tag{7.13}$$

where,

NIR = Near infrared band (Band 4 of TM and 5 of OLI) and R = Red band (Band 3 of TM and 4 of OLI).

The value of NDVI ranges from −1 to +1. Values toward +1 indicate highest possibility of vegetation density and lower values toward −1 indicate lesser vegetation cover.

According to Zhang et al. (2006), LSE of an area can be estimated if the NDVI values of that particular area is known. In a NDVI raster, if the NDVI value of a pixel is below −0.185, then the LSE value of 0.995 is assigned to a newly formed LSE raster. When the NDVI value is ≥ − 0.185 and < 0.157, a pixel of the NDVI raster gets embedded with a LSE value of 0.985, when the NDVI value is >0.157–0.727, a logarithmic relationship between NDVI and LSE is used to assign the values to the raster and lastly when the NDVI is >0.727, a value of 0.990 is assigned to a pixel in the raster (Table 7.3).

Table 7.3 The DN values of LSE been assigned accordingly against the corresponding NDVI values according to Zhang et al. (2006)

NDVI	LSE
< −0.185	0.995
≥ −0.185–0.157	0.985
>0.157–0.727	1.009 + 0.047 × ln(NDVI)
>0.727	0.990

2.4.6 Computation of LST from LSE and at-Satellite Brightness Temperature (T_B)

After the estimation of LSE, LST is calculated using the modified Planck's equation (Artis & Carnahan, 1982) which is as follows:

$$\mathrm{LST} = \frac{T_B}{\left[1 + \left(\lambda * \frac{T_B}{c_2}\right) * \ln \varepsilon\right]} \tag{7.14}$$

Where,

LST = land surface temperature in Kelvin.
λ = wavelength of the emitted radiance, i.e., 10.6 for Landsat 5 TM (1990) and 10.8 for Landsat 8 TIRS (2016).
$c_2 = h * c/s = 1.4388 \times 10^{-2}$ m K = 14,388 μm K,
where, h = Planck's constant (6.626×10^{-34} J/s), c = speed of light (3×10^8 m/s), s = Stefan's-Boltzmann's constant (1.38×10^{-23} J/K).
ε = Land surface emissivity.

After obtaining the LST values in Kelvin, they are converted to LST values in°C using band–calculator tool as provided in QGIS.

3 Results and Discussions

3.1 AOD Retrieval Scenarios

The extracted AOD data at specific wavelengths 400, 778.75, 865, and 940 nm provide important and relevant information about the net aerosol content for the months of January, April, October, and November in three consecutive years (2017–2019) over Rajarhat area. A comparative analysis between AOD and LST is shown in Table 7.4 which states the different R^2 values that are obtained during these three-year periods and also at three different months. Results show that out of the four wavelengths, R^2 values of $\mathrm{AOD}_{778.75\ \mathrm{nm}}$ show the highest goodness of fit and high positive correlation between aerosol load and LST data in the 4 months of the 3 years. A possible reason for this change is because of the influence of various meteorological parameters like humidity, total water vapor content, rainfall, incoming solar energy flux that have a direct control over the incoming long wave radiation. This leads to the inference that land processes, which control AOD such as urbanization, land use changes, and other physical dynamics, are retrieved better in $\mathrm{AOD}_{778.75}$ compared to the other center wavelengths. This means that if we intend to use Sentinel 3 OLCI data for AOD retrieval exclusively for land biophysical processes and to know whether LST has some pronounced effect on AOD variation, it is preferable to use $\mathrm{AOD}_{778.75}$.

Table 7.4 R^2 values for AOD of all four wavelengths versus LST over 3 year series and also in three different seasons

		Wavelengths			
		400 nm	**778.75 nm**	**865 nm**	**940 nm**
Months	**January-17**	0.960	0.984	0.977	0.963
	January-18	0.979	0.990	0.986	0.978
	January-19	0.726	0.710	0.654	0.665
	April-17	0.944	0.938	0.921	0.880
	April-18	0.919	0.984	0.976	0.975
	April-19	0.921	0.9801	0.901	0.912
	October-17	0.972	0.856	0.835	0.849
	October-18	0.953	0.923	0.923	0.930
	October-19	0.682	0.9806	0.956	0.822
	November-17	0.973	0.695	0.749	0.769
	November-18	0.586	0.972	0.980	0.978
	November-19	0.837	0.979	0.862	0.849

Figures 7.2, 7.3, 7.4 and 7.5 show the aerosol load at 778.75 nm for all the seasons. A perusal of Fig. 7.2 shows concentrations of aerosol load over the entire Rajarhat area in 3 years: January 2017, January 2018, and January 2019. The $AOD_{778.75\ nm}$ shows a maximum value of 0.402 and a minimum value of 0.160 during January 2017 (Table 7.5). The northern fringes of Airport, Chinar Park, and other adjoining areas show higher aerosol load compared to the northwestern region. Waterbodies show less aerosol load compared to terrestrial areas. In the next 2 years, there is an overall decline of aerosol load (0.288 maximum and 0.114 minimum for 2018 and 0.266 maximum and 0.105 for 2019), but the distribution is more concentrated in areas where urban development is high. Thus, it can be said that the range of aerosol load decreases with time, but the concentration tends to show increased levels in urban areas. The entire East Kolkata Wetlands (EKW) and in some aquaculture farms in the northern part show lower AOD.

Figure 7.3 shows the distribution of aerosols over Rajarhat in April 2017, 2018, and 2019. The $AOD_{778.75\ nm}$ shows a maximum value of 0.351 and a minimum value of 0.150 during April 2017 (Table 7.5). For April 2018, the range of AOD decreases (maximum being 0.331 and minimum being 0.145). The concentration of AOD load is more over Salt Lake area and East Kolkata Wetlands (EKW) in 2018. For the month of April 2019, there is a variable fall of AOD concentration in the Salt Lake area, whereas an increase is noticeable in the Hatisala and Rajarhat-Chandpur region.

The possible reasons for this variable AOD values each year are mainly due to variation in concentration of water vapor present in the upper layers of the atmosphere (Román et al., 2014; Clifford, 1991). This variation is also due to the pollution load over the particular area and other factors mainly in winter season (January 2017–19 and November 2017–19). In January, the maximum AOD concentration remains over the Salt Lake, Rajarhat New Town, and Chinar Park area due

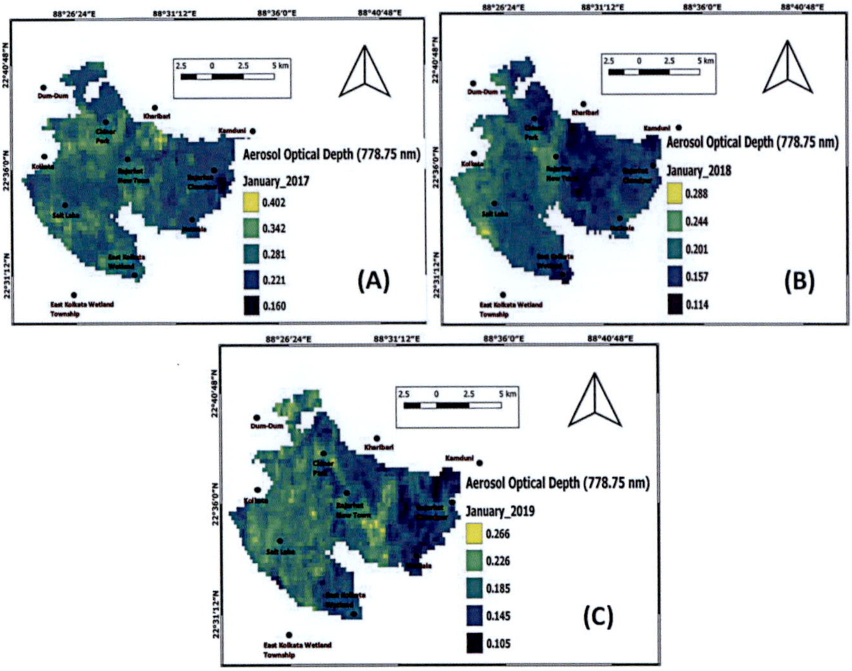

Fig. 7.2 $AOD_{778.75}$ of Rajarhat area as captured by S3A-OLCI: (**a**) January 2017, (**b**) January 2018, and (**c**) January 2019

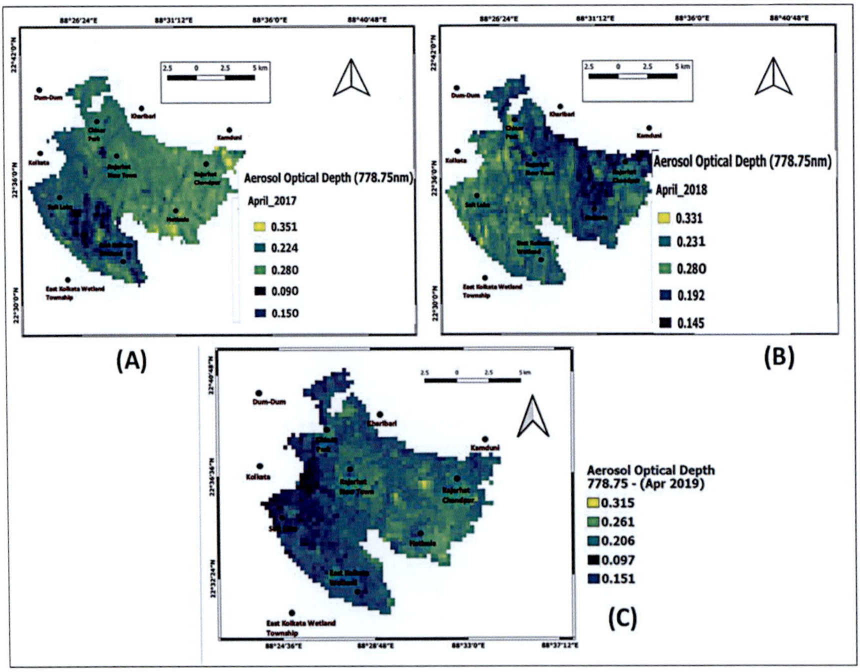

Fig. 7.3 $AOD_{778.75}$ of Rajarhat area as captured by S3A-OLCI: (**a**) April 2017, (**b**) April 2018, and (**c**) April 2019

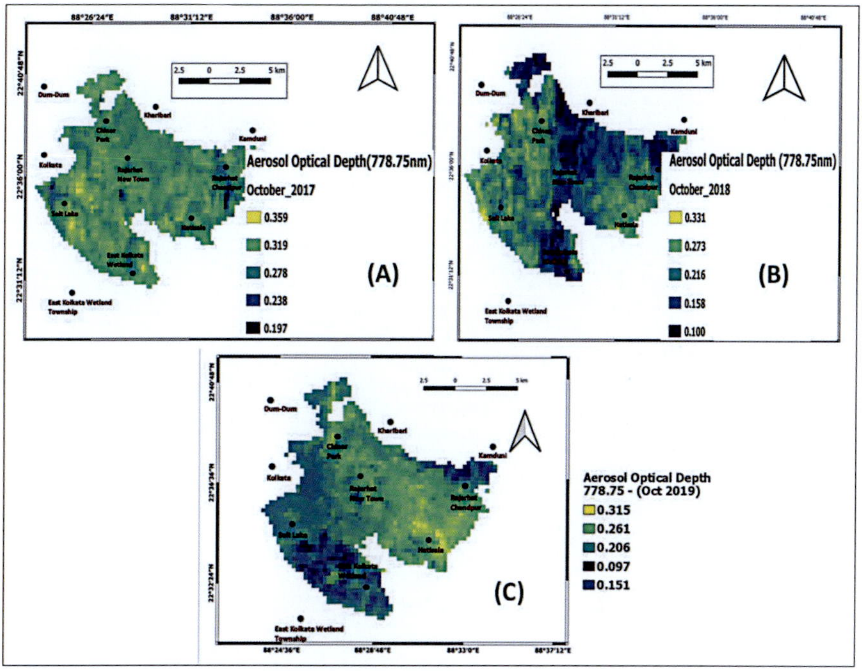

Fig. 7.4 $AOD_{778.75}$ of Rajarhat area as captured by S3A-OLCI: (**a**) October 2017, (**b**) October 2018, and (**c**) October 2019

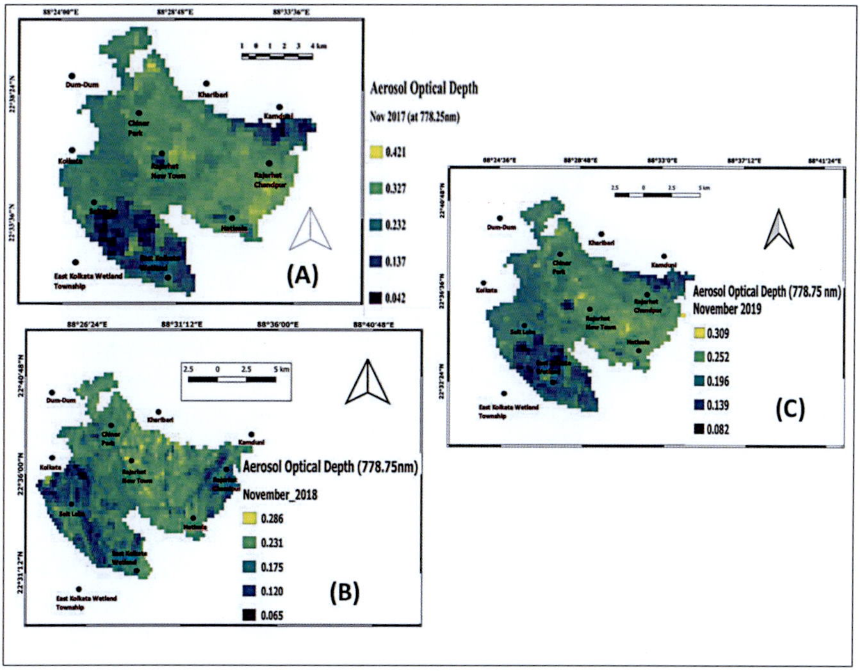

Fig. 7.5 $AOD_{778.75}$ of Rajarhat area as captured by S3A-OLCI: (**a**) November 2017, (**b**) November 2018, and (**c**) November 2019

Table 7.5 Descriptive statistics for $AOD_{778.75}$ of all the three-year periods of different seasons

		Descriptive statistics for $AOD_{778.75}$			
		Min	**Max**	**Mean**	**SD**
Months	**January-17**	0.160	0.402	0.274	0.033
	January-18	0.114	0.288	0.179	0.028
	January-19	0.105	0.266	0.175	0.031
	April-17	0.150	0.351	0.239	0.044
	April-18	0.145	0.331	0.228	0.034
	April-19	0.151	0.315	0.232	0.023
	October-17	0.197	0.359	0.309	0.017
	October-18	0.100	0.331	0.203	0.055
	October-19	0.151	0.315	0.262	0.042
	November-17	0.042	0.421	0.254	0.052
	November-18	0.065	0.286	0.198	0.036
	November-19	0.082	0.309	0.186	0.039

to the vehicular pollution (Fig. 7.2). In November 2017–19, the AOD concentrates more over to the adjoining areas of Hatisala, Rajarhat-Chandpur regions also. This aerosol dispersion occurs mostly because of the urbanization and increasing air pollution load in winter.

Figures 7.2, 7.3, 7.4 and 7.5 reveal that AOD load is more in winter months than in summer or post-monsoon months since the pollution load in winter is maximum. The range of AOD decreases from 2017 to 2019, but the spatial extent of AOD concentration increases. Thus, it may be concluded that AOD increases significantly over the period 2017–2019.

3.2 *Correlation Between AOD and LST*

The LST of Rajarhat during the period 2017–2019 of four different seasons as retrieved from Landsat 8 OLI sensor in the form of descriptive statistics is given in Table 7.6. For the month of January, the mean temperature of Rajarhat is found to be 24.72 °C in 2017, 23.77 °C in 2018, and 21.29 °C in 2019. This suggests that the LST decreases by 0.95 °C during the year 2017–18 and again decreases by 2.48 °C during 2018–19. The decrease in LST can be linked to increased average rainfall in 2018 (+18.72% increase from 2016–17, 146.5 mm excess) and 2019 (+15.12% increase, 107 mm excess from 2017–18) and also to other factors like El-Nino Southern Oscillation events in the Southern Pacific. The direct impact of urbanization-coupled LST increase is noticeable in the month of April 2017–2019 where the LST first decreases from 34.51 °C to 28.95 °C in 2017–18 but then increases from 28.95 °C to 35.91 °C in 2018–19 suggesting that weather anomalies like El-Nino has major effects in controlling LST. Air pollution serve as a contributing factor to increasing AOD and thereby leading to progressive rise in LST.

Table 7.6 Descriptive statistics for LST of the 3 years for different seasons

		Descriptive statistics for LST			
		Min	Max	Mean	SD
Months	**January-17**	21.61	31.11	24.72	1.451
	January-18	20.51	28.21	23.77	1.269
	January-19	19.77	27.78	21.29	1.438
	April-17	30.81	40.03	34.51	1.446
	April-18	24.34	34.49	28.95	1.796
	April-19	30.68	44.15	35.91	1.987
	October-17	28.24	36.33	31.07	1.304
	October-18	24.6	39.02	31.72	1.833
	October-19	23.91	30.53	26.69	1.043
	November-17	23.25	32.3	25.8	1.491
	November-18	22.72	23.93	23.6	0.216
	November-19	23.76	37.35	28.01	2.14

To understand the relationship and to formulate the hypothesis that LST may have an effect on the aerosol load over Rajarhat area or vice versa, a multivariate correlation was performed using Data Analyzer tool's in-built Microsoft Excel considering all the goodness-of-fit values of all the wavelengths in 3 years. It is interesting to note that in all the years, LST increased with increasing AOD suggesting that aerosols over land surface have a direct effect over temperature. The increase in AOD leads to smaller atmospheric window which does not allow NIR waves to exit thereby causing rise in temperature. In the NIR region, 778.75 and 865 nm wavelength shows positive goodness-of-fit values, but maximum is for $AOD_{778.75}$ (Table 7.4). This serves as more statistically significant than other bands. A perusal of Figs. 7.2, 7.3, 7.4 and 7.5 states that the AOD value is significantly less in case of waterbodies which causes the LST to remain low. For areas where there are urban settlements, construction projects, and housing projects, AOD is high which in turn leads to the rise in LST.

The correlation graphs between AOD and LST (Figs. 7.6, 7.7, 7.8 and 7.9) give a better picture on the extent of increase in LST with rise in AOD over Rajarhat area. At each circumstance (season and year), it is seen that the LST shows a rising trend with increase in AOD values. The linear trend line shows variable slope due to variation in AOD values with LST. For January 2017 (Fig. 7.6c), it is observed that both LST and AOD values become constant after 25 °C but again increases post 27 °C. Such variations can be noticed again in the month of April 2019 (Fig. 7.7c) where both AOD and LST values remain constant post 40 °C.

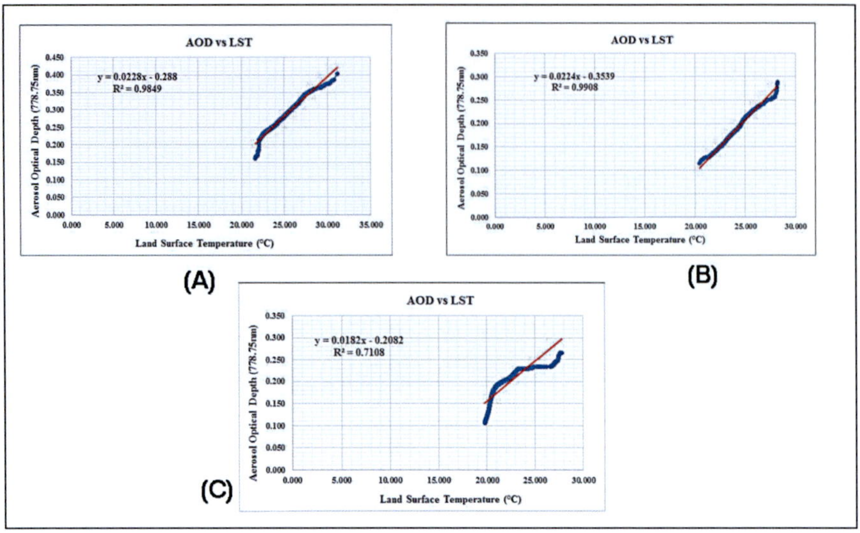

Fig. 7.6 $AOD_{778.75}$ versus LST graphs of (**a**) January 2017, (**b**) January 2018, and (**c**) January 2019 showing R^2 values and trend lines

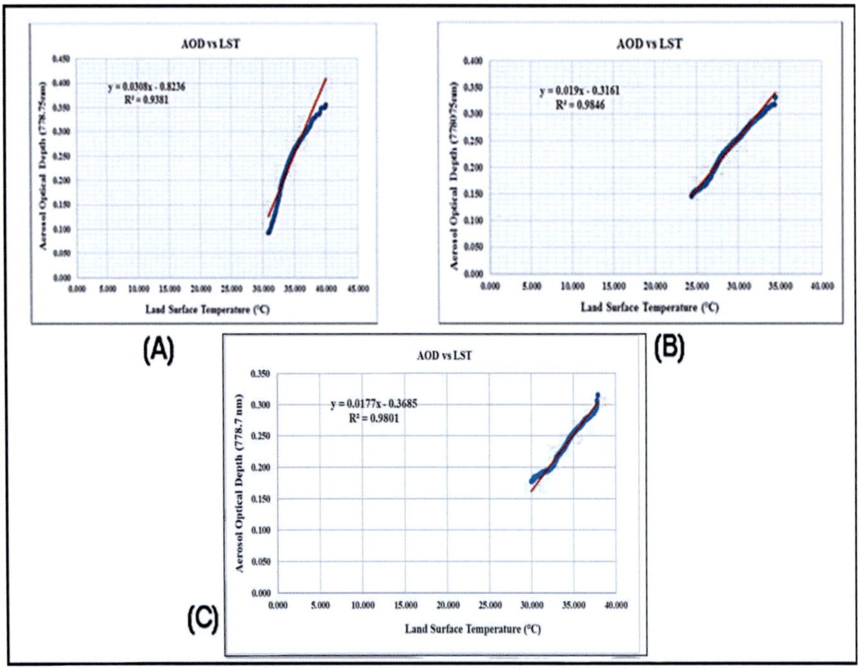

Fig. 7.7 $AOD_{778.75}$ versus LST graphs of (**a**) April 2017, (**b**) April 2018, and (**c**) April 2019 showing R^2 values and trend lines

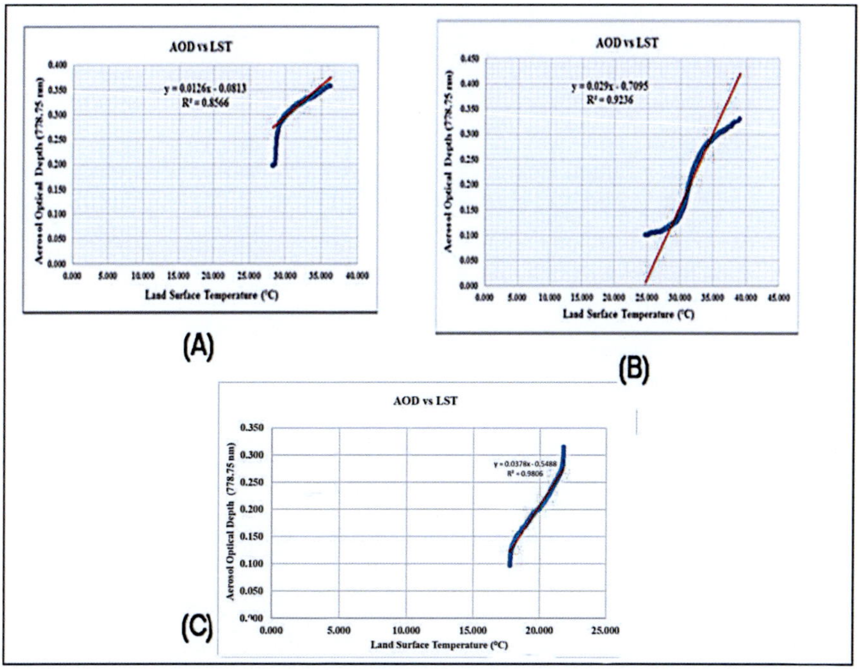

Fig. 7.8 $AOD_{778.75}$ versus LST graphs of (**a**) October 2017, (**b**) October 2018, and (**c**) October 2019 showing R^2 values and trend lines

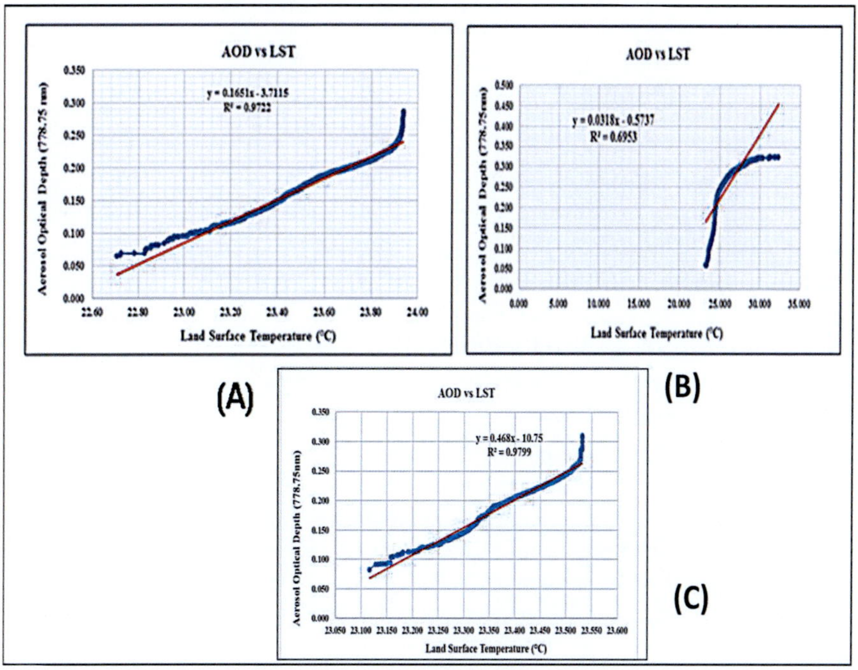

Fig. 7.9 $AOD_{778.75}$ versus LST graphs of (**a**) November 2017 and (**b**) November 2018 showing R^2 values and trend lines

Table 7.7 R^2 values for AOD versus PM_{10} of all the three-year periods of different seasons

		R^2 values of AOD vs. PM_{10}
Months	**January-17**	0.931
	January-18	0.798
	January-19	0.903
	April-17	0.896
	April-18	0.708
	April-19	0.739
	October-17	0.778
	October-18	0.725
	October-19	0.785
	November-17	0.823
	November-18	0.863
	November-19	0.691

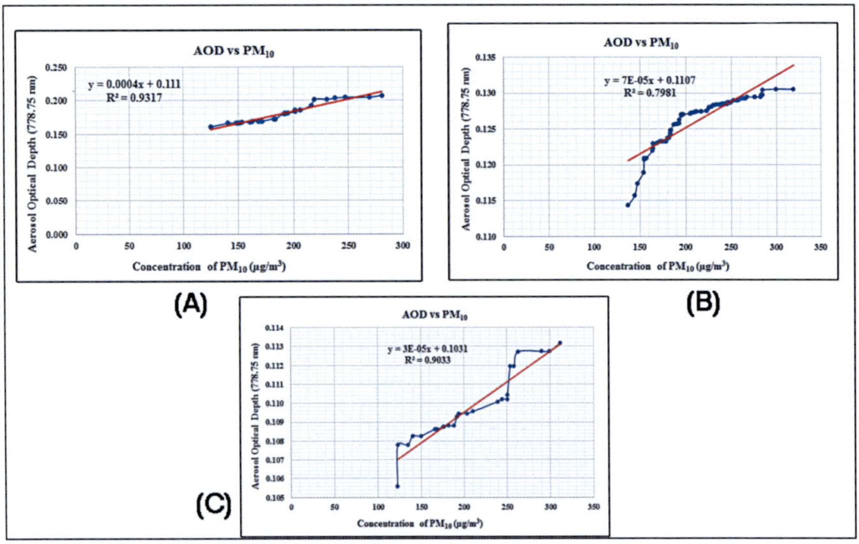

Fig. 7.10 $AOD_{778.75}$ versus PM_{10} concentration graphs of (**a**) January 2017, (**b**) January 2018, and (**c**) January 2019 showing R^2 values and trend lines

3.3 *Validation of AOD with PM_{10}*

Concentration of PM_{10} in the lower atmosphere due to atmospheric inversion results in increase of AOD values. In order to understand this, we validate $AOD_{778.75}$ with PM_{10} data provided by the West Bengal Pollution Control Board (WBPCB) for the months of January, April, October, and November for 2017–19 (Table 7.7) from their air quality monitoring stations. The data are then correlated with AOD values falling over these areas (values are chosen arbitrarily within a defined radius of the monitoring stations) and the results are given in Figs. 7.10, 7.11, 7.12 and 7.13.

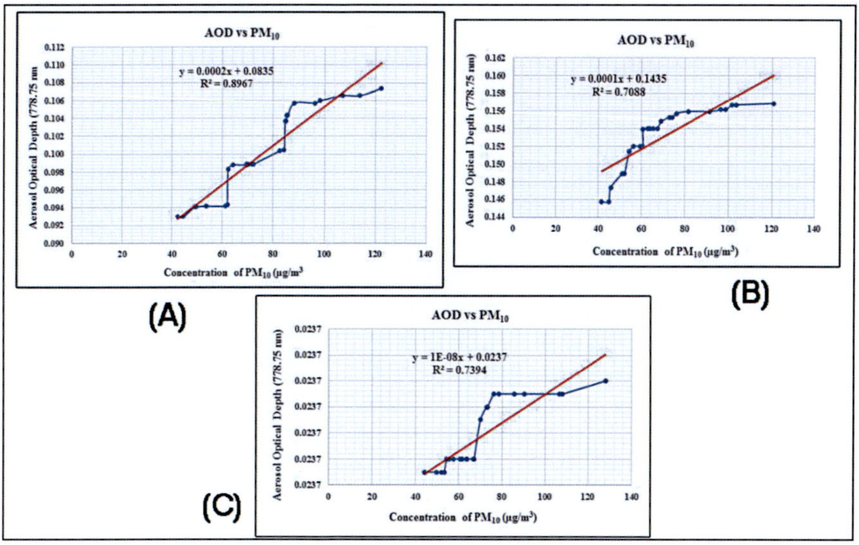

Fig. 7.11 $AOD_{778.75}$ versus PM_{10} concentration graphs of (**a**) April 2017, (**b**) April 2018, and (**c**) April 2019 showing R^2 values and trend lines

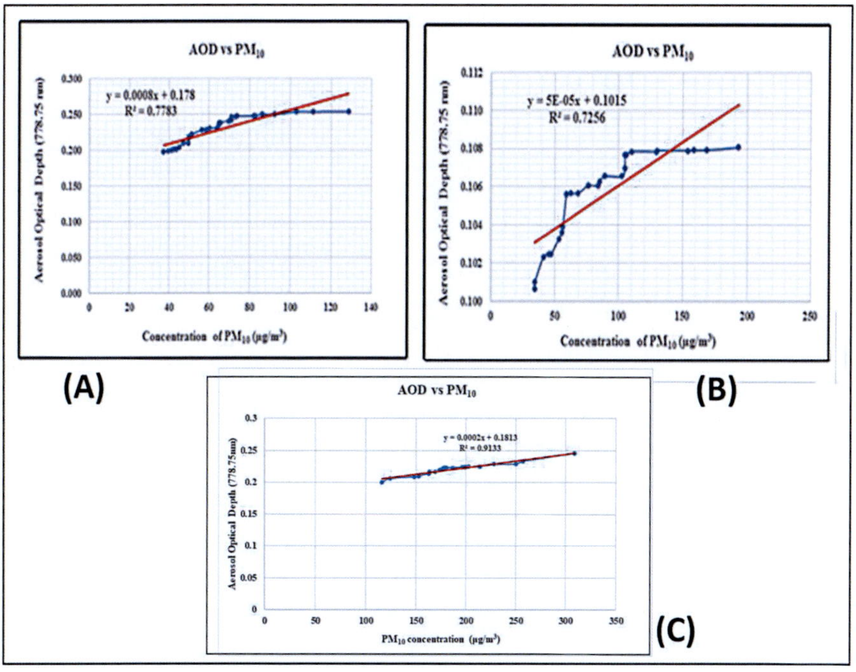

Fig. 7.12 $AOD_{778.75}$ versus PM_{10} concentration graphs of (**a**) October 2017, (**b**) October 2018, and (**c**) October 2019 showing R^2 values and trend lines

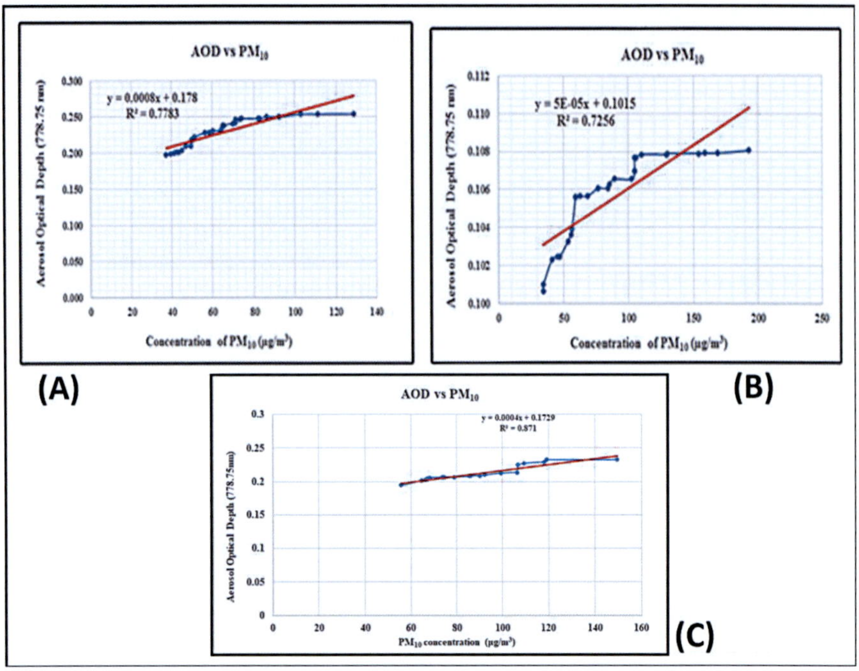

Fig. 7.13 $AOD_{778.75}$ versus PM_{10} concentration graphs of (**a**) November 2017, (**b**) November 2018, and (**c**) November 2019 showing R^2 values and trend lines

From the correlation graphs (Figs. 7.10, 7.11, 7.12 and 7.13), it is quite clear that AOD values increase with increase in PM_{10} values. Thus, this confirms that increase in PM_{10} concentration results in increasing trend of aerosols.

3.4 *Relation Between AOD and LULC of Rajarhat*

To understand the spatial extent of aerosol distribution over Rajarhat more efficiently, a spatiotemporal relationship is obtained between land use/land cover (LULC) classes of Rajarhat and the AOD data. Using the LULC map (Fig. 7.14) of Rajarhat for the year 2016, the eight land use classes (Aquaculture, City/Town, Open Fields, Periurban, Scrubland, Tree Cover, Lakes/Ponds, and Croplands) are extracted and then the individual cluster of AOD values are obtained for each land use types. AOD data chosen are for April 2017 and January 2019 to show the temporal variation. Table 7.8 shows the descriptive statistics of AOD values for April 2017 for each LULC type. The change in AOD values are observed in both the year periods April 2017 and January 2019 (Table 7.9).

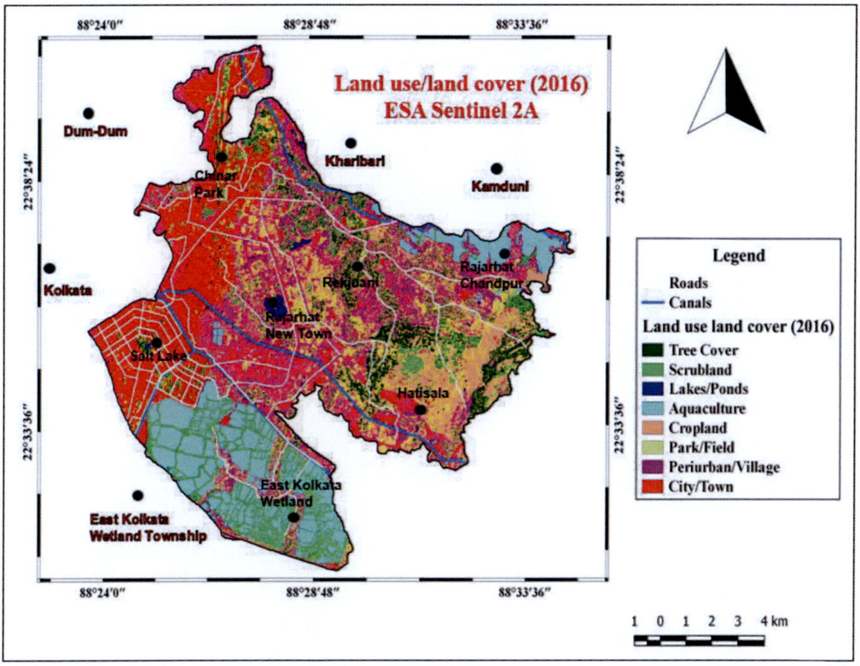

Fig. 7.14 Land use land cover map of 2016 of Rajarhat area as derived from Sentinel 2A MSI satellite imagery. (Source: Dhar et al., 2019)

Table 7.8 $AOD_{778.75}$ (April 2017) values of different LULC types

AOD (LULC wise)—April 2017				
LULC type	Max	Min	Mean	SD
Aquaculture	0.351	0.093	0.216	0.059
City town	0.348	0.094	0.229	0.036
Cropland	0.330	0.094	0.269	0.028
Lakes	0.311	0.107	0.232	0.043
Open field	0.354	0.099	0.251	0.040
Periurban	0.354	0.094	0.243	0.046
Scrubland	0.345	0.093	0.239	0.046
Tree cover	0.354	0.100	0.247	0.043

So far as urban areas like city/town or periurban landscapes are concerned, the minimum AOD value is around 0.094 (Table 7.8) for April 2017, whereas for January 2019 it is around 0.108 (Table 7.9). Again the same trend can be observed in case of aquaculture ponds, lakes, or tree cover where the minimum value of AOD increases from 2017 to 2019. One of the possible reasons for this variable AOD change is because during the summer period there are contributions to the AOD due to water vapor concentration, other atmospheric factors, and increase in air pollution load (Clifford, 1991). The maximum values for nearly all the classes show a

Table 7.9 $AOD_{778.75}$ (January 2019) values of different LULC types

AOD (LULC wise)—January 2019				
LULC type	Max	Min	Mean	SD
Aquaculture	0.252	0.106	0.174	0.024
City town	0.265	0.108	0.194	0.030
Cropland	0.264	0.113	0.162	0.019
Lakes	0.246	0.124	0.184	0.024
Open field	0.264	0.109	0.173	0.031
Periurban	0.252	0.109	0.176	0.032
Scrubland	0.261	0.121	0.187	0.027
Tree cover	0.266	0.108	0.173	0.030

decreasing trend in the January 2019, whereas the minimum values increase particularly for LULC classes like city/town and periurban areas. The same can also be observed in Figs. 7.2 and 7.3 where in January 2019 the AOD gets concentrated more over the urban landscapes. This increase in AOD values for cities/towns definitely suggests that aerosols concentrate and increase in the winter months due to vehicular pollution, burning of stubbles, and other anthropogenic reasons. Therefore, there is an urgent need to introduce better and proper management practices, to reduce the aerosol load over various land use types.

4 Conclusion

This chapter highlights the aspects of AOD retrieval from Sentinel 3 OLCI sensor which has a huge potential to deliver accurate aerosol results at a much better much spatial resolution. The NRT products are easily available and are efficient to process compared to other satellite sensors. Since the aerosol content is exclusively dependent upon both physical and anthropogenic factors, we use multiple wavelengths to verify which center wavelength has better results in obtaining AOD correctly. $AOD_{778.75\ nm}$ is found to be the best suited wavelength for obtaining AOD. Since Rajarhat is an area which is a center of continuous land use change, it is obvious that the aerosol load is contributed largely due to the developmental activities. AOD values show high positive correlation with LST. Rise in LST results in increase of AOD over Rajarhat. To control the rise in LST, measures like planting of trees, construction of aquaculture ponds, fountains, reduction of vehicular load by introduction of mass transport system, and introduction of electric vehicles should be implemented immediately. This study will definitely help to carry out better research on AOD with Sentinel 3 satellite data.

Acknowledgments The authors acknowledge the Director, IISWBM, for providing the infrastructural facilities for this research work.

References

Artis, D. A., & Carnahan, W. H. (1982). Survey of emissivity variability in thermography of urban areas. *Remote Sensing of Environment, 12*(4), 313–329.

Barsi, J. A., Schott, J. R., Hook, S. J., Raqueno, N. G., Markham, B. L., & Radocinski, R. G. (2014). Landsat-8 thermal infrared sensor (TIRS) vicarious radiometric calibration. *Remote Sensing, 6*(11), 11607–11626.

Bibi, H., Alam, K., Chishtie, F., Bibi, S., Shahid, I., & Blaschke, T. (2015). Intercomparison of MODIS, MISR, OMI, and CALIPSO aerosol optical depth retrievals for four locations on the Indo-Gangetic plains and validation against AERONET data. *Atmospheric Environment, 111*, 113–126.

Census of India. (2011). *Directorate of census operations West Bengal* (Series-20, Part XII-B) (p. 296). District Census Handbook.

Chavez, P. S. (1996). Image-based atmospheric corrections-revisited and improved. *Photogrammetric Engineering and Remote Sensing, 62*(9), 1025–1035.

Chin, M., Diehl, T., Dubovik, O., Eck, T., Holben, B., Sinyuk, A., & Streets, D. G. (2009). Light absorption by pollution, dust, and biomass burning aerosols: A global model study and evaluation with AERONET measurements. *Annales de Geophysique, 27*, 3439–3464. http://www.ann-geophys.net/27/3439/2009/

Clifford, M. B. (1991). Effects of water vapor and anisotropic scattering on aerosol optical depth retrieval. *Naval Postgraduate School Monterey CA*. http://hdl.handle.net/10945/28152

de Leeuw, G., Holzer-Popp, T., Bevan, S., Davies, W. H., Descloitres, J., Grainger, R. G., Griesfeller, J., Heckel, A., Stefan, K., Klüser, L., Kolmonen, P., Litvinov, P., Martynenko, D., North, P., Ovigneur, B., Pascal, N., Poulsen, C., Ramon, D., Schulz, M., Siddans, R., Sogacheva, L., Tanré, D., Thomas, G. E., Virtanen, T. H., von Hoyningen Huene, W., Vountas, M., & Pinnock, S. (2015). Evaluation of seven European aerosol optical depth retrieval algorithms for climate analysis. *Remote Sensing of Environment, 162*, 295–315.

Dhar, R. B., Chakraborty, S., Chattopadhyay, R., & Sikdar, P. K. (2019). Impact of land use/land cover change on land surface temperature using satellite data: A case study of Rajarhat Block, North 24-Parganas District, West Bengal. *Journal of the Indian Society of Remote Sensing, 47*(2), 331–348.

Dubovik, O., Holben, B., Eck, T. F., Smirnov, A., Kaufman, Y. J., King, M. D., Tanré, D., & Slutsker, I. (2002). Variability of absorption and optical properties of key aerosol types observed in worldwide locations. *Journal of the Atmospheric Sciences, 59*(3), 590–608.

Hsu, N. C., Herman, J., Torres, O., Holben, B., Tanre, D., Eck, T., Smirnov, A., Chatenet, B., & Lavenu, F. (1999). Comparisons of the TOMS aerosol index with Sun-photometer aerosol optical thickness: Results and applications. *Journal of Geophysical Research: Atmospheres, 104*(D6), 6269–6279.

IMD (India Meterological Department), Pune. (2020). *Observed rainfall variability and changes over West Bengal state*. ESSO/IMD/HS/Rainfall Variability/29(2020)/53(7).

Jiménez-Muñoz, J. C., Sobrino, J. A., Skoković, D., Mattar, C., & Cristóbal, J. (2014). Land surface temperature retrieval methods from Landsat-8 thermal infrared sensor data. *IEEE Geoscience and Remote Sensing Letters, 11*(10), 1840–1843.

Kaufman, Y. J., Tanré, D., & Boucher, O. (2002). A satellite view of aerosols in the climate system. *Nature, 419*(6903), 215.

Lenoble, J., Herman, M., Deuzé, J., Lafrance, B., Santer, R., & Tanré, D. (2007). A successive order of scattering code for solving the vector equation of transfer in the earth's atmosphere with aerosols. *Journal of Quantitative Spectroscopy and Radiative Transfer, 107*(3), 479–507.

Levy, R. C., Munchak, L. A., Mattoo, S., Patadia, F., Remer, L. A., & Holz, R. E. (2015). Towards a long-term global aerosol optical depth record: Applying a consistent aerosol retrieval algorithm to MODIS and VIIRS-observed reflectance. *Atmospheric Measurement Techniques, 8*(10), 4083–4110.

Ma, Z., Hu, X., Sayer, A. M., Levy, R., Zhang, Q., Xue, Y., Tong, S., Bi, J., Huang, L., & Liu, Y. (2015). Satellite-based spatiotemporal trends in PM2. 5 concentrations: China, 2004–2013. *Environmental Health Perspectives, 124*(2), 184–192.

Misra, A., Jayaraman, A., & Ganguly, D. (2008). Validation of MODIS derived aerosol optical depth over Western India. *Journal of Geophysical Research: Atmospheres, 113*(D4).

Pielke, R. A., Cotton, W., Walko, R. E. A., Tremback, C. J., Lyons, W. A., Grasso, L., Nicholls, M. E., Moran, D. A., Wesley, T., Lee, J., & Copeland, J. H. (1992). A comprehensive meteorological modeling system-RAMS. *Meteorology and Atmospheric Physics, 49*(1–4), 69–91.

Popp, T., De Leeuw, G., Bingen, C., Brühl, C., Capelle, V., Chedin, A., & Xue, Y. (2016). Development, production and evaluation of aerosol climate data records from European satellite observations (Aerosol_cci). *Remote Sensing, 8*(5), 421.

Ramanathan, V., Crutzen, P., Kiehl, J., & Rosenfeld, D. (2001). Aerosols, climate, and the hydrological cycle. *Science, 294*(5549), 2119–2124.

Remer, L. A., Kaufman, Y., Tanré, D., Mattoo, S., Chu, D., Martins, J. V., Li, R. R., Ichoku, C., Levy, R. C., Kleidman, R. G., Eck, T. F., Vermote, E., & Holben, B. N. (2005). The MODIS aerosol algorithm, products, and validation. *Journal of the Atmospheric Sciences, 2*(4), 947–973.

Román, R., Bilbao, J., & de Miguel, H. A. (2014). Uncertainty and variability in satellite-based water vapor column, aerosol optical depth and Angström exponent, and its effect on radiative transfer simulations in the Iberian Peninsula. *Atmospheric Environment, 89*, 556–569.

Sobrino, J. A., Jimenez-Munoz, J. C., & Paolini, L. (2004). Land surface temperature retrieval from LANDSAT TM 5. *Remote Sensing of Environment, 90*(4), 434–440.

Tanré, D., Kaufman, Y. J., Holben, B. N., Chatenet, B., Karnieli, A., Lavenu, F., Blarel, L., Dubovik, O., Remer, A., & Smirnov, A. (2001). Climatology of dust aerosol size distribution and optical properties derived from remotely sensed data in the solar spectrum. *Journal of Geophysical Research: Atmospheres, 106*(D16), 18205–18217.

Van Donkelaar, A., Martin, R. V., Brauer, M., Kahn, R., Levy, R., Verduzco, C., & Villeneuve, J. P. (2010). Global estimates of ambient fine particulate matter concentrations from satellite-based aerosol optical depth: Development and application. *Environmental Health Perspectives, 118*(6), 847.

von Hoyningen-Huene, W., Joon, Y., Vountas, M., Istomina, G., Rohen, G., Dinter, T., Kokhanovsky, A. A., & Burrows, J. P. (2011). Retrieval of spectral aerosol optical thickness over land using ocean color sensors MERIS and SeaWiFS. *Atmospheric Measurement Techniques Discussions, 4*, 151–171.

Xiao, H., & Weng, Q. (2007). The impact of land use and land cover changes on land surface temperature in a karst area of China. *Journal of Environmental Management, 85*(1), 245–257.

Zhang, J., Wang, Y., & Li, Y. (2006). A C++ program for retrieving land surface temperature from the data of Landsat TM/ETM+ band6. *Computers & Geosciences, 32*(10), 1796–1805.

Chapter 8
Flood Monitoring Using Synthetic Aperture Radar Data and Google Earth Engine: A Case Study of the 2020 Monsoon Flood in Assam, India

Deb Kumar Maity, Sukla Hazra, Dipanjan Das Majumdar, Mampi Pal, and Tapan Pramanick

Abstract Floods wreak havoc in the state of Assam in India every year; indeed, the year 2020 has been no different. Torrential monsoonal rain from May to September 2020 produced a severe flood that became catastrophic for Assam. Changes in the distributional patterns of monsoonal rainfall result in irregularities in the state of Assam's tropical flood occurrence, duration, and intensity. This chapter assesses the flood situation in Assam using the Sentinel-1A and Sentinel-1B SAR images in combination with other ancillary datasets such as precipitation, ground-based river gauge data, and gridded population data. Seventy-five pre-flood and 106 post-flood images were processed within the Google Earth Engine platform for developing the flood model. Difference and pixel-by-pixel ratio change method are used to highlight the flood occurrence patterns and trends across various land use and landforms within the state. Seasonality of flood and its duration were the main concerns. Three long and heavy spells of rain triggered three distinct flood phases. The river Brahmaputra and its tributaries flowed above danger level for almost all the flood

D. K. Maity (✉)
Department of Geography, East Calcutta Girls' College, West Bengal State University, Kolkata, West Bengal, India

S. Hazra
East Calcutta Girls' College, West Bengal State University, Kolkata, West Bengal, India

D. Das Majumdar
Department of Geography, Netaji Satabarshiki Mahavidyalaya, Ashoknagar, West Bengal, India

M. Pal
Department of Geography, Kazi Nazrul University, Asansol, West Bengal, India

T. Pramanick
Department of Geography, The University of Burdwan, Bardhaman, West Bengal, India

S. Hazra et al. (eds.), *Geographies of the Indian Subcontinent*,
https://doi.org/10.1007/978-3-031-74813-4_8

phases. In the second phase, all experienced the most noticeably awful flooding scenarios. The result suggests that Synthetic Aperture Radar (SAR) data is more effective for mapping floods from space than optical data because SAR data can capture images in any weather situation. The developed flood model indicates that around 12.27 lakh hectares (ha) of land were flooded, over 3650 villages were under water, and more than 20 lakh people were directly affected. This chapter will be helpful in accurately delineating the spatial extent of flood footprints and major hotspot areas and putting in place the right flood prevention actions for the affected regions of any part of the globe.

1 Introduction

Extreme floods are one of the greatest natural catastrophes, and Brahmaputra Valley is hit hard by them annually. The regional topography is defined by a variety of grasslands, low wetlands, low-altitude forests, and croplands that are relatively vulnerable to flooding due to the proximity of a river (Kushwaha et al., 2000). The monsoon usually delivers two or more intense flood waves in the Brahmaputra Valley over the entire 6-month rainy season, from June to October (Bhatt et al., 2013). The river gauge indicates that floods are becoming more frequent because of intense rain events (Bhatt & Rao, 2016). Information on flood hazards at both spatial and temporal scales is essential for quantifying risk and developing flood management strategies. Satellite-based flood assessment must be undertaken before, during, and immediately after a flood to accurately predict the extent and intensity of the damage done. The extensive collection of remote sensing data, especially about floods, has expanded the number of observations, allowing for near-real-time inundation mapping and evaluation. Real-time assessment of floods is essential for policymakers during a flood situation.

Satellite-based flood extent and severity assessments are critical to understanding the different flood phases. Flood inundation can be mapped and monitored in near real time with the help of the vast array of remote sensing data provided by the growing number of satellites. Several researchers have examined remote sensing datasets and methods for flood inundation mapping and monitoring. Several academics have assessed the potential of diverse remote sensing datasets and methodologies for accurate flood inundation mapping and monitoring (Pulvirenti et al., 2011; Gao et al., 2018). The Normalized Difference Water Index (NDWI), which utilizes reflected near-infrared radiation and visible green light, was introduced by McFeeters in 1996 to evaluate both the quantity and quality of water resources. Xu (2006) substituted middle infrared for near-infrared in NDWI to effectively improve open water from other feature classes.

For a very long time, optical remote sensing data has been extensively utilized to monitor floods'extent. Cloud cover prevents optical data from being used for continuous flood monitoring. Synthetic Aperture Radar (SAR) data is the best way to find flooded areas and get accurate information during a flood because it can work in any weather and may be able to see through thick vegetation (Patel et al., 2009). In a radar-related study, the fallen signal goes through an identified reflection, which leads to low backscatter when there is more than 50% water in the soil (Srivastava et al., 2006). Based on the histogram of the radar signal over open water-rich land, several experts have suggested methods to automatically separate open water (Fan & Lei, 2012). Sudd wetland in South Sudan was investigated by Wilusz et al. (2017) using SAR imagery to detect flood periods. Sivasankar et al. (2019) explored the capabilities of VV and VH polarizations and thresholding methods to detect the extent of rivers. It was then utilized to pinpoint embankment breach spots. Rahman and Thakur (2017) employed HH polarization and a thresholding technique to map floodwaters in the Kendrapara area of the Indian state of Orissa. In most research, flood maps are produced using proper settings and adjustments of radar data in conjunction with various change detection techniques, like log ratio (Bazi et al., 2005), ratio index (Rignot & van Zyl, 1993), and polarimetric difference image (Marinelli et al., 1997). These techniques are made possible by the availability of pre- and post-flood images.

Support vector machines (Tehrany et al., 2015; Han et al., 2007), fuzzy logic (Vanama & Rao, 2019; Costache, 2019), artificial neural networks (Hashemi-Beni & Gebrehiwot, 2021; Ichim & Popescu, 2019), and texture-based (Zhang & Lu, 2005) classification algorithms are some of the machine learning methods already investigated in flood mapping. Users can avoid the hassle of downloading and processing large amounts of Sentinel-1 data locally by doing it via a cloud-based system like Google Earth Engine (GEE). Data processing is done in parallel on Google's computer infrastructure, drastically enhancing processing efficiency and opening up new potential for end users. The GEE cloud platform has been significantly used in various recent remote sensing applications. The GEE platform has various advantages, including the availability of multitemporal EO datasets, a parallel processing design, and the capacity to handle large datasets efficiently.

Through mapping, monitoring, and an assessment of flood catastrophe footprints in several Assam districts using the year 2020 as an example, this study demonstrates both the inundation and persistence of floods. In addition, the initiative seeks to utilize the Google Earth Cloud Platform in order to monitor and evaluate bigger flood-affected regions.

2 Study Area

We have examined the Brahmaputra in Assam only. The state of Assam is known for its diverse landscape, which includes a range of elevations, slopes, rivers, physiography, drainage, terrain, and flood conditions. The state of Assam has a range of elevations, with the highest point being the peak of Bor Madhuri in the Karbi Anglong Hills, which has an elevation of 1220 meters above sea level. The lowest point is in the Brahmaputra River delta, which has an elevation of around 56 m

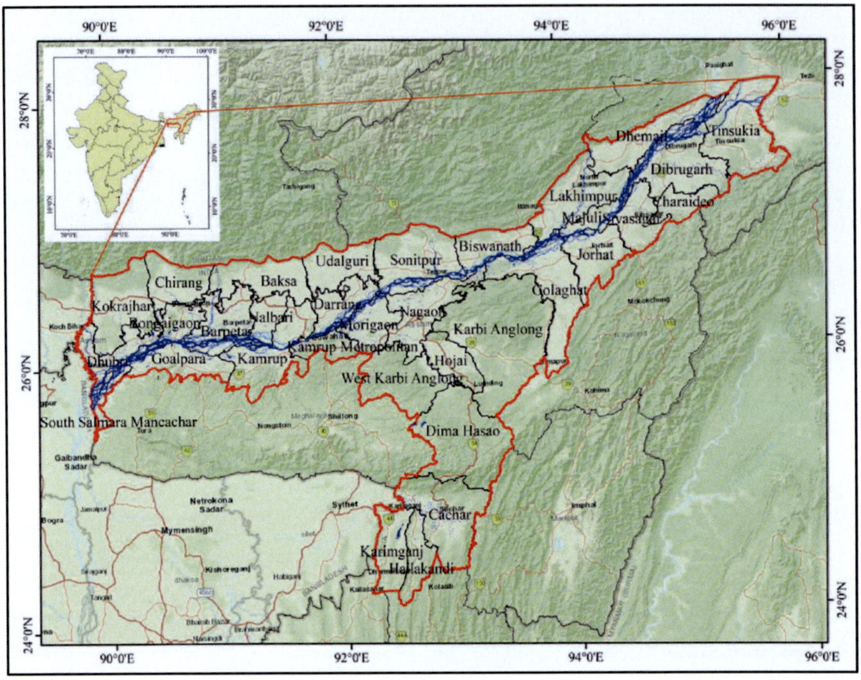

Fig. 8.1 Location map showing the satellite view depicting the course of the Brahmaputra River as it flows through the several districts of Assam

above sea level. The state also has several hill ranges, including the Barail Range, Garo Hills, and Karbi Anglong Hills. Assam has varying slopes, the highest in the hill ranges and the lowest in the Brahmaputra River delta. The hill ranges have steep slopes with inclinations of over 40° (Figs. 8.1, 8.2 and 8.3).

On the other hand, the Brahmaputra River delta has gentle slopes with inclinations of less than 2°. The state of Assam has a diverse physiography, with several distinct regions. The topography is dominated by riverine lakes, wetlands, grasslands, and elevated woodland-covered plains (Figs. 8.4 and 8.5) that offer animals refuge and shelter during floods. The main river that traverses the state of Assam is the Brahmaputra. It is one of the biggest rivers in Asia and is well known for the cliff-side gorges and rapids in Arunachal Pradesh's higher regions. The river gets braided when it enters Assam, forming a flood plain that is between 50 and 60 miles broad and 600 miles long. Another significant river, the Barak, rises in the Barail Range, passes through the Cachar area, and then enters Bangladesh as the Surma River. The state also has several other rivers, including the Dibang, Lohit, Kameng, Subansiri, and many others. Every year as it rains, the rivers rise, which causes the water to overflow their banks and flood the nearby area. Floodwaters may damage bridges, railroad lines, and highways in addition to damaging homes and killing cattle, which disrupts communication. Natural disaster also causes mortality in several places of the state. Flooding in 2020 was brought on by heavy rains in

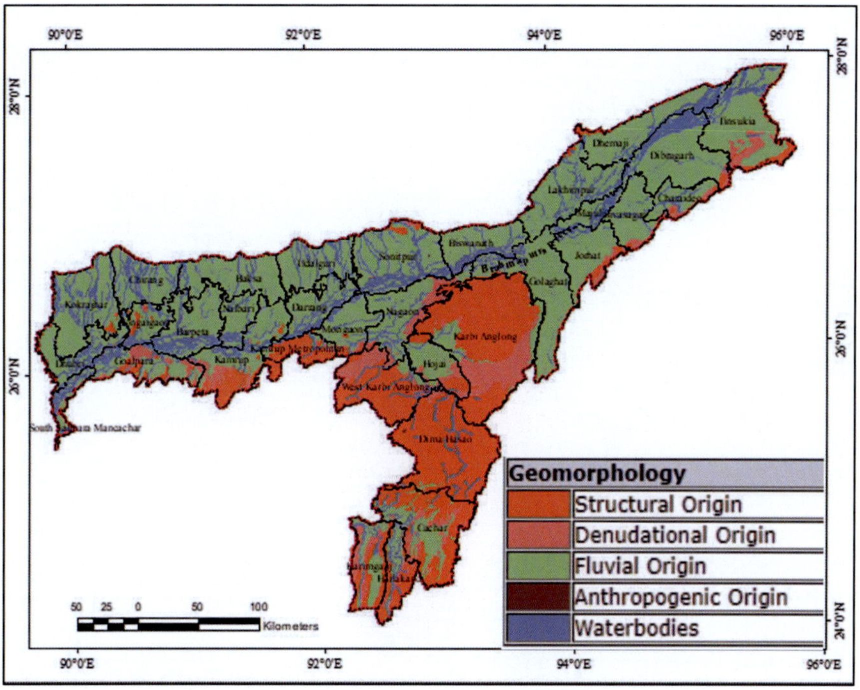

Fig. 8.2 A map showcasing the origin of various geomorphological landforms in the state of Assam. (Source: Map prepared from Bhuvan Thematic Layer)

Assam and the Brahmaputra's rising water level. Around 30 thousand people across five districts were impacted, and crops were ruined. Additionally, a COVID-19 epidemic that was still active in the Indian state of Assam coincided with the floods; 87,000 hectares of farmland were devastated, and 2200 settlements were affected by the flooding. The water level rose above the danger line in several cities. More than 50% of the animals in Kaziranga National Park have been pushed out by floods.

3 Materials and Methods

The following stages would make up the flowchart for the full procedure (Fig. 8.4). The initial step would be to use the Google Earth Engine platform to access and analyze SAR data. To analyze the data for flood mapping utilizing ratio and difference thresholding methods, the suitable dataset, such as Sentinel-1 data, would need to be located. The evaluation of flood proxy indicators, such as rainfall and river gauge data, would come next. Predicting future flood occurrences often entails looking at previous data to find patterns and trends in precipitation and river gauge data. Once the flood proxy indications have been evaluated, the flood's

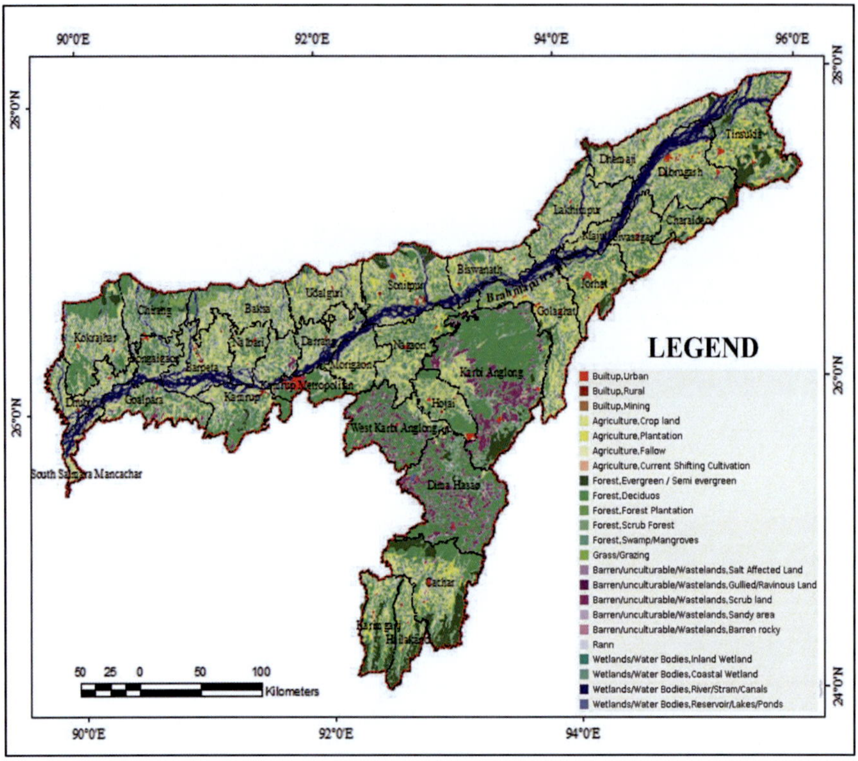

Fig. 8.3 Land use/land cover map of Assam providing valuable insights into the state's geomorphological, ecological, and socioeconomic landscape. (Source*:* Map prepared from Bhuvan Thematic Layer)

damage has to be evaluated as the following stage. Typically, it would entail gathering statistics on the disrupted transportation systems, agricultural loss, and impacted communities. To map the extent of the flood, several Sentinel-1 SAR imageries from the monsoon season were employed. The Indian Meteorological Data (IMD) of the Indian Water Portal was utilized to investigate the precipitation. Additional river water level information utilized during the monsoon season was given by the Central Water Commission (CWC). The effects of the various flood stages on the populace, agricultural land, and transportation network are assessed using data from the Bhuvan site, population statistics, land use/land cover data, and open street maps (OSM).

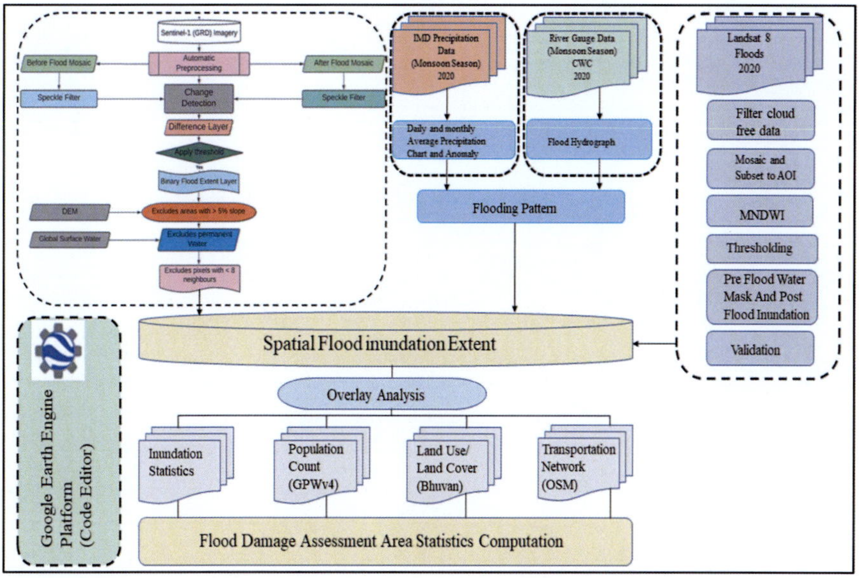

Fig. 8.4 A graphical representation of the systematic procedure used to conduct this research is provided in the "methodology flowchart."

4 Mapping of Flood Hazard

4.1 Sentinel-1 SAR Data Preprocessing

The GEE Code Editor gives you a way to run custom scripts and access data stored on the GEE platform through a web-based interface. It has powerful tools for preprocessing, analyzing, and visualizing large geospatial datasets, such as Sentinel-1 SAR data. For mapping flood risk, the GEE platform can be used to process and analyze Sentinel-1 SAR records. Within GEE, images are saved that have been fixed in terms of their position and their radiometrics. Once the data is accessed, it needs to be preprocessed to make the geometrically corrected backscattering coefficient (r_) images. This would involve three steps: radiometric correction, geometric correction, and co-registration. Using an improved Lee speckle filter with a 3 × 3 pixel window size, speckle noise that lowers image quality is removed from the images. In the last step, the backscatter intensity is turned into a backscatter coefficient (r_) that is given in decibels (dB). Users can quickly and easily process and analyze a lot of Sentinel-1 SAR data and get the flood inundation layer by running a custom script through the GEE Code Editor. It will let us find the flooded areas, figure out how big the flood is and how high the water is, and make high-resolution maps of flood hazards that can be used for flood management and prevention.

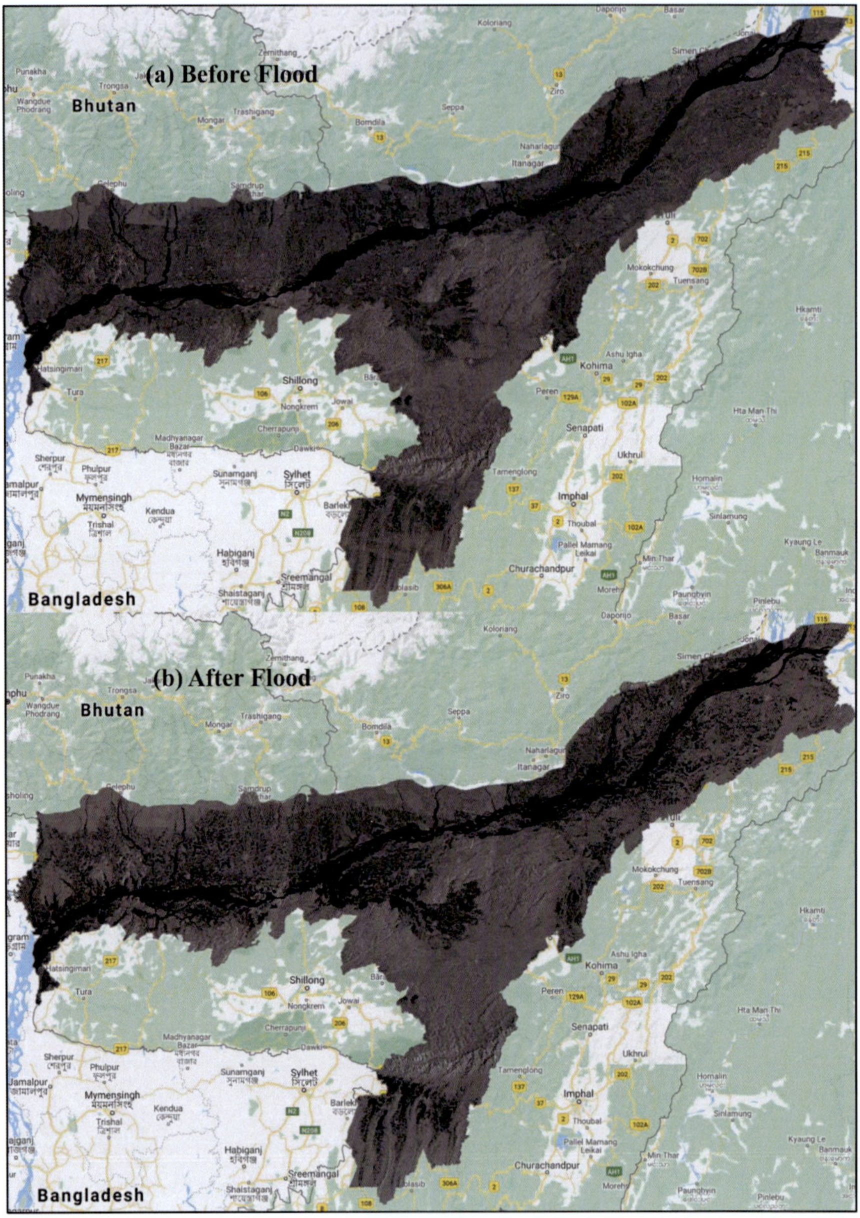

Fig. 8.5 Pre-flood (**a**) and post-flood (**b**) maps of Assam highlighting the extent of the flooding in the year 2020

4.2 Analysis of SAR Backscatter Data

Water appears black in the radar data. For the pre-flood area, the backscatter intensity values ranged from _7 to _14 dB polarization and from _14 to _19 dB for the VH polarization. Backscatter intensity values for the same places during the flood period ranged for VH polarization roughly between _24 and _32 dB and for VV polarization between _17 and _24 dB. As a result of flooding, the backscattering intensity tends to decrease, leading to a decline toward lower negative values.

4.3 Mapping of Flooded Area

Utilizing multi-dated SAR images, difference and ratio change detection techniques are frequently used in mapping and monitoring flooded regions. These techniques have proved useful in estimating the severity of floods (Rignot & van Zyl, 1993). Change detection methods are more effective in removing stationary bodies of water as compared to single-image methods. Singh, 1989 divided the intensity values of the mosaic picture from the flood period (Fig. 8.5b) by the mosaic images from before the flood period to calculate the pixel-by-pixel changes (Fig. 8.5a). By comparing the two dates, it is possible to determine the degree of change (inundated regions), which is shown by an increment in pixel brightness. The pixels with greater brightness are then separated using the threshold operator (.gt(threshold) operator). The flood extent is then determined by adjusting the best threshold through trial and error. The range of threshold values used in the current experiment was 1.15 to 1.30. "Flooded" pixels were those whose values were within the specified threshold, while the remaining pixels received the value 0, eventually dubbed "non-flooded pixels."

4.4 Refining of Flood Inundation Layer

At a resolution of 30 m, the Joint Research Centre (JRC) Global Surface Water Explorer "Seasonality" dataset offers details on the onset and duration of surface water presence. For more than 10 months, it was employed to keep the submerged areas hidden under water. The identification of flooded regions might also be impacted by hill shadows brought on by the landscape. The operator "ee.Algorithms.Terrain" was used to remove the shadows from a 30-meter DEM obtained from the HydroSHEDS and the Shuttle Radar Topography Mission (SRTM). Last but not least, it could occasionally be required to remove isolated pixels that do not actually depict flooded regions. To determine how the flooded pixels are connected, utilize the connected Pixel Count' operation within GEE. Each pixel's linked pixel count is determined by this operator, and any pixels with a low connected pixel count may be eliminated. The final flood hazard footprint layer is then applied once again to gather information about floods and estimate damage. This layer may be used to compare the changes in land cover, structures, and infrastructure before and after a flood event in order to determine the size and depth of the flood as well as the damage it caused.

4.5 Validation of Flood Hazard Layer

The validation procedure makes use of both direct techniques (from the field) and indirect methods (from secondary sources). Landsat-8 and Sentinel-2 optical imageries, IMD rainfall data, flow accumulation, water level data, and a hydrologically corrected digital elevation model were used in the study (DEM). To confirm the flood region identified by Sentinel-1, it was compared to the optical data from Sentinel-2 that were available. The flood region from Sentinel-2, which was discovered using the Normalized Difference Water Index (NDWI) approach, was then placed on top of the flood area from Sentinel-1 radar image analysis (McFeeters, 1996). Additionally, by contrasting it with the areas that were supposedly inundated by a number of different publicly accessible sources, flood inundation was double validated. The GEE platform now has the ability to determine daily precipitation by adding together the daily totals. In addition, a few photographs were acquired from web sources. For comparative purposes, the real flood area was digitally represented on the GEE platform using photos and survey data.

5 Results and Discussion

5.1 Spatiotemporal Pattern of Precipitation

From May 1 to October 15, this graph shows how much rain the Indian Meteorological Department measured every day. A close look at the rainfall numbers shows that about 24% (152 mm) of the total rainfall happened in the 19th and 20th weeks (the last 2 weeks of the month of May) of 2020. Analysis of the second wave of rainfall data shows that Assam got the most rain between the 24th and 28th weeks of the year (the last week of June and the entire month of July 2020). It was responsible for 62% of the annual rainfall in the state (340 mm). Analysis of daily rainfall data showed that more than half of July's rain fell during the 7 days of heavy monsoon rain that never stopped (June 24 to July 30, 2020) (Fig. 8.6). The geographical location shows that it rains more in the north, east, central, and southern districts of Assam that are close to the Brahmaputra River during the first two weeks of July. In the last week of September, people notice the third rainy spell.

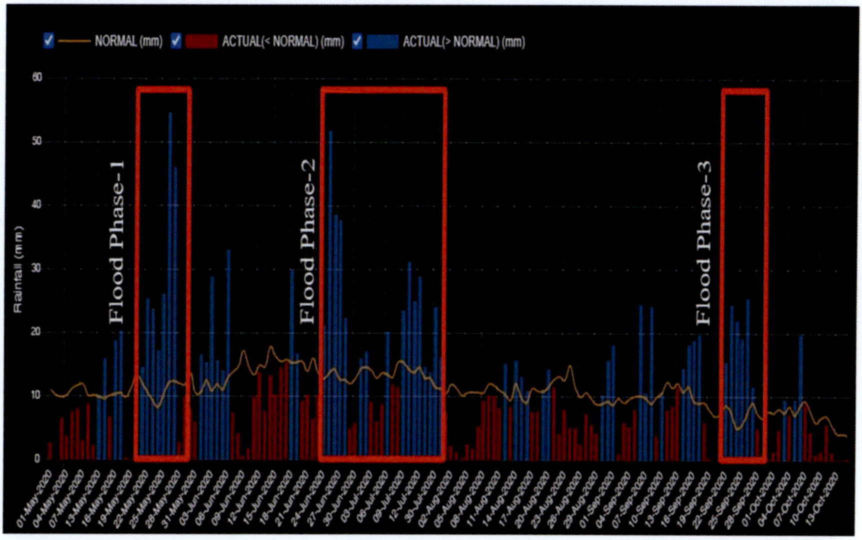

Fig. 8.6 Graph showing how the monsoon rains are distributed throughout the several phases of flooding in Assam. (Source: Indian Meteorological Department)

5.2 Spatiotemporal Changes in River Gauge Level

The Central Water Commission (CWC), which has developed several categories of water levels at key points all around the nation, is in charge of forecasting floods on India's major rivers. A river is stated to be flooding if the water level is at or over the danger line at that place; if the water level is at least one meter above the danger line, the flooding is considered a big flood for the respective year. As shown in Table 8.1, until June 20, 2020, every area along the Brahmaputra's tributaries and main canal will be below the danger line. On June 23 and 24, 2020, the water level started to rise and a few areas crossed the DL, albeit the rise was sporadic. Starting on June 25, 2020, the water level started to steadily rise, going above the WL mark and then the DL mark (Fig. 8.7). On June 26, 2020, the river's water level was over DL at nine locations, at fifteen stations by June 28 and at seventeen stations by June 30. For three days (June 27–29, 2020), the main river ran continuously over DL at seven locations, reaching their maximum flood levels. It is also noteworthy to mention that during this event, the Brahmaputra River's water level peaked. The data shows that it took four days for the water in Dhubri to reach its peak flood level on June 30, 2020, after the Brahmaputra River at Dibrugarh reached its greatest flood level on June 26, 2020.

Table 8.1 Result on the assessment of the flood hazard at the village and district levels in Assam State and rivers flowing above danger levels

Date	Details	Remarks
25-May	51.06 hectares (ha) of cropland in Lakhimpur, one village in Dhemaji, 1957 hectares (ha) in Nagaon, and 28 hectares (ha) in Darrang were flooded	Jia Bharali (Sonitpur) & Puthimari (Kamrup) above danger level
26-May	Cropland areas of 112 ha in Dhemaji, 3 ha in Lakhimpur, 4.01 ha in Baksa, 16.66 ha in Bongaigaon, and 1590 ha in Nagaon were all flooded	Jia Bharali (Sonitpur) & Brahmaputra at Neamatighat (Jorhat) above danger level
27-May	There are 31 villages in Hojai. There are 4 villages in Golaghat, 10 ha in Jorhat, 428 ha in Majuli, 292.31 ha in Darrang, 173 ha in Bongaigaon, and 57 ha in Nagaon. 3030.50 ha in Golaghat, 63.60 ha in Lakhimpur	Jia Bharali (Sonitpur), Puthimari (Kamrup) & Brahmaputra at Neamatighat (Jorhat) above danger level
28-May	816 ha at Majuli 558 ha at Darrang, 1823.69 ha Barpeta 183.50 ha Bongaigaon, 887 ha Kamrup	Brahmaputra at Neamatighat (Jorhat) & Kopili (Kampur) above danger level
29-May	5 ha in Lakhimpur, 869.50 ha in Darrang, 1420 ha in Nagaon, 262.02 ha in Golaghat, 132 ha in Majuli, 1823.69 ha in Barpeta, 32.5 ha in Bongaigaon, 120.6 ha in Kamrup, 20 ha in Morigaon	Kopili (Nagaon) above danger level
05-June	521 ha in Nagaon, 10.26 ha in Golaghat, 140 ha in Golaghat croplands affected	No evidence of river water exceeding danger levels has been found
06-June	125 ha in Bis wanath, 167 ha in Nagaon, 160 ha in Morigaon croplands affected	No evidence of river water exceeding danger levels has been found
07-June	61 ha at Lakhimpur, 85 ha at Bis wanath, 262 ha at Nagaon, 160 ha at Morigaon, 800 ha at Goalpara, 677 ha at South Salmara croplands affected	No evidence of river water exceeding danger levels has been found
08-June	45 ha in Bis wanath, 675 ha in Nagaon, 160 ha in Morigaon, 677 ha in South Salmara	Jia Bharali (Sonitpur) & Brahmaputra at Neamatighat (Jorhat) above danger level
18-Jun	131 ha croplands affected in Morigaon	Jia Bharali (Sonitpur) & Brahmaputra at Neamatighat (Jorhat) above danger level
19-June	131 ha crop affected in Morigaon	Jia Bharali, Dikhow & Des ang (Sibs agar), Dans hiri (S) at Golaghat, Beki at Barpeta & Brahmaputra at Neamatighat (Jorhat) above danger level
20-June	131 ha in Morigaon and 70 ha in Majuli	Jia Bharali, Des ang (Sibs agar), Dans hiri (S) at Golaghat, Brahmaputra at Neamatighat (Jorhat) above danger level
21-June	11.6 ha in Lakhimpur, 131 ha in Morigaon, 3 ha in Jorhat & 12 ha in Sib s agar	Jia Bharali, Dikhow & Des ang (Sibs agar), Dans hiri (S) at Golaghat, Brahmaputra at Neamatighat (Jorhat) above danger level

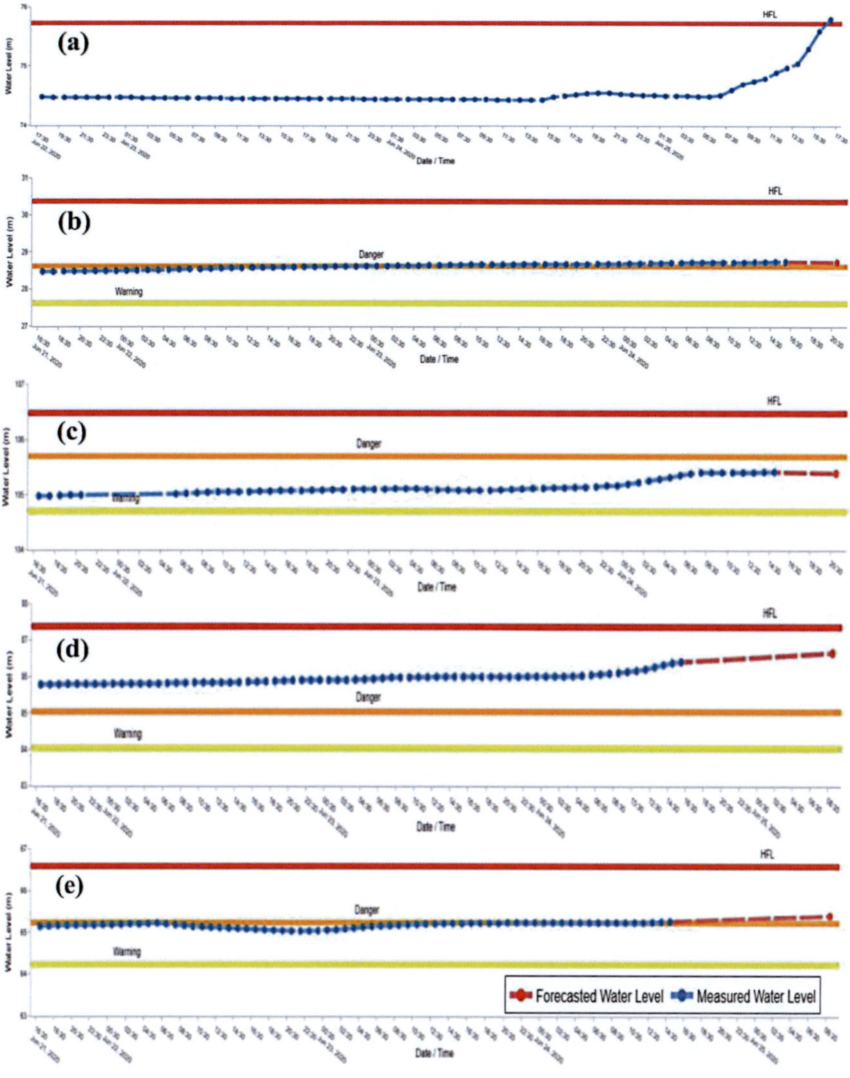

Fig. 8.7 Hydrograph showing observations of water levels from a variety gauging stations in Assam. (**a**) Suklai, Kamrup District (Brahmaputra Basin). (**b**) Dhubri, Dhubri District (Brahmaputra Basin). (**c**) Dibrugarh, Dibrugarh District (Brahmaputra Basin). (**d**) Neamatighat, Jorhat District (Brahmaputra Basin). (**e**) Tezpur, Sonitpur District (Brahmaputra Basin). (Source: Central Water Commission (CWC)

5.3 *Spatiotemporal Assessment of Flood Damage*

The maximum area that will be inundated during this flood event may be determined by merging the flood footprint layers that were created by examining satellite data on June 27, 29, 30, and July 2, 2020. Approximately 4.65 lakh hectares, or roughly 6% of the state's total gain area, was inundated, according to the integrated study of flood layers from several dates. Observations of the initial flood wave were made between May 22 and June 2, 2020 (Fig. 8.8a). During this time, 3 lakh hectares of land were inundated. Between June 21 and August 10, 2020 (Fig. 8.8b), the second phase was detected. During this period, floods affected roughly 12.27 lakh hectares of land. The most severe flooding conditions that have occurred during the phases have, by far, been experienced in the second phase. In 2020, between September 25 and October 4 (Fig. 8.8c), the third phase was seen. 3.45 lakh hectares of land were flooded at this time. Districts heavily hit by floods in Assam include Golaghat, Dibrugarh, Lakhimpur, Nagaon, Barpeta, and Morigaon. Rivers running above the danger threshold are specifically covered by the flood footprint assessment (Fig. 8.9). Additionally, it evaluates flood damage (Fig. 8.10), which includes submerging crops and cattle as well as causing harm to roads, rail lines, and residences. Figure 8.11 shows the district-wise impacts of flooding in the year 2020. Table 8.1 also shows the amount of supplemental relief that was given out after floods.

6 Accuracy Assessment and Validation

On June 2, 2020, identical images from Sentinel 2 and Landsat 8 were acquired; 200 random samples of water pixels have been generated. Cross-referencing the samples with water bodies obtained by Sentinel 1 was done. Since 173 out of 200 samples returned correct results, accuracy was calculated to be 86.5%. The Sentinel-1 SAR data revealed that, in 27 areas identified as water bodies in the Landsat 8 images, the boundaries did not align with the actual water body locations. This discrepancy indicates that the flood footprint area may be overestimated by approximately 13.5%. Sentinel 1 and Landsat 8 images of flood scenarios were compared on June 2, 2020 (Fig. 8.12).

7 Novelty of Flood Assessment Using SAR Data in GEE

In comparison to conventional methodologies, the integration of SAR data and GEE offers a novel strategy for flood assessment. SAR data has the advantage of being able to see through clouds, delivering a steady signal in all-weather circumstances, and providing excellent accuracy in flood monitoring. The incorporation of SAR

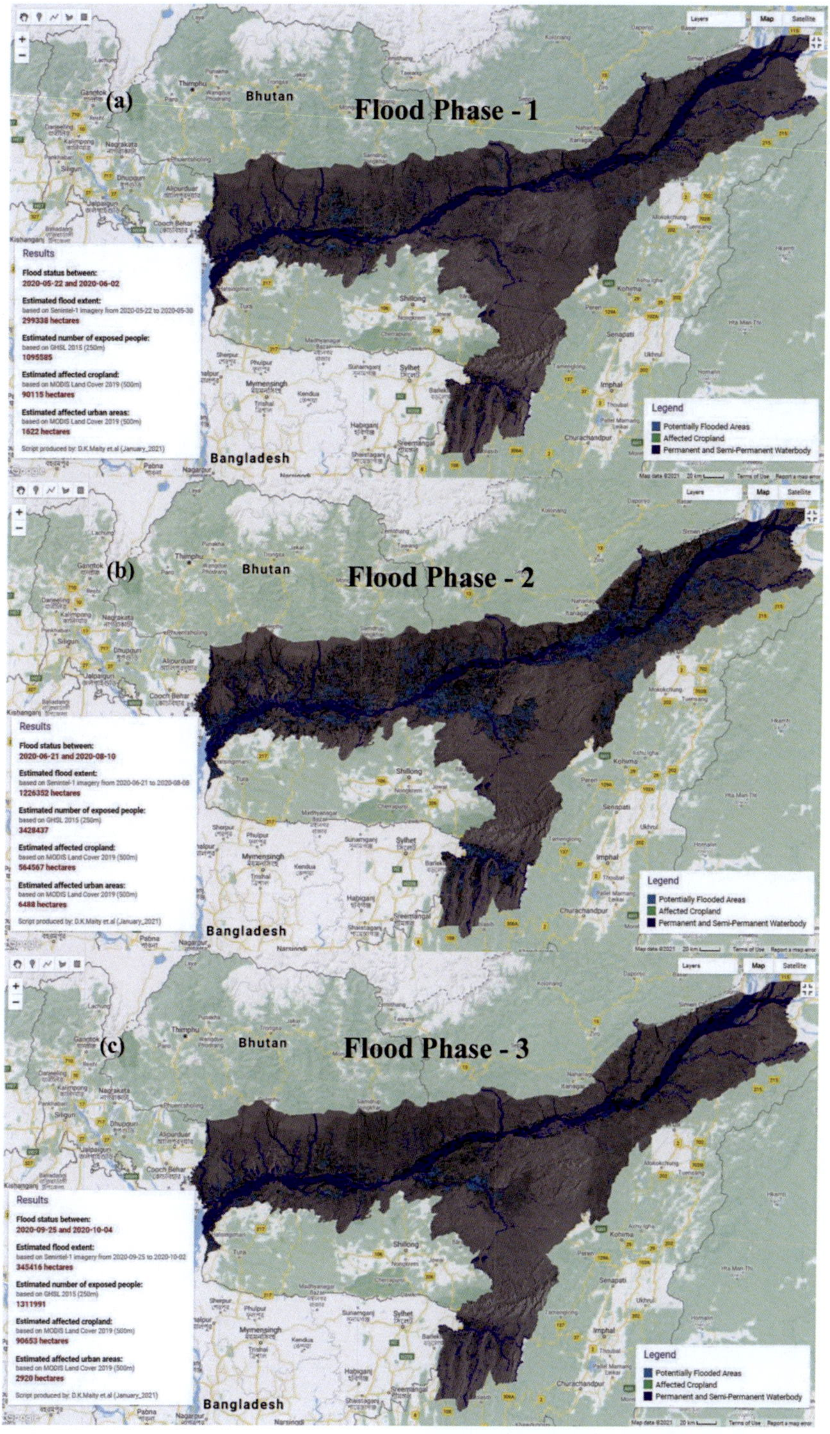

Fig. 8.8 Analysis of different flood phases of Assam. (**a**) Flood Phase 1, (**b**) Flood Phase 2, and (**c**) Flood Phase 3

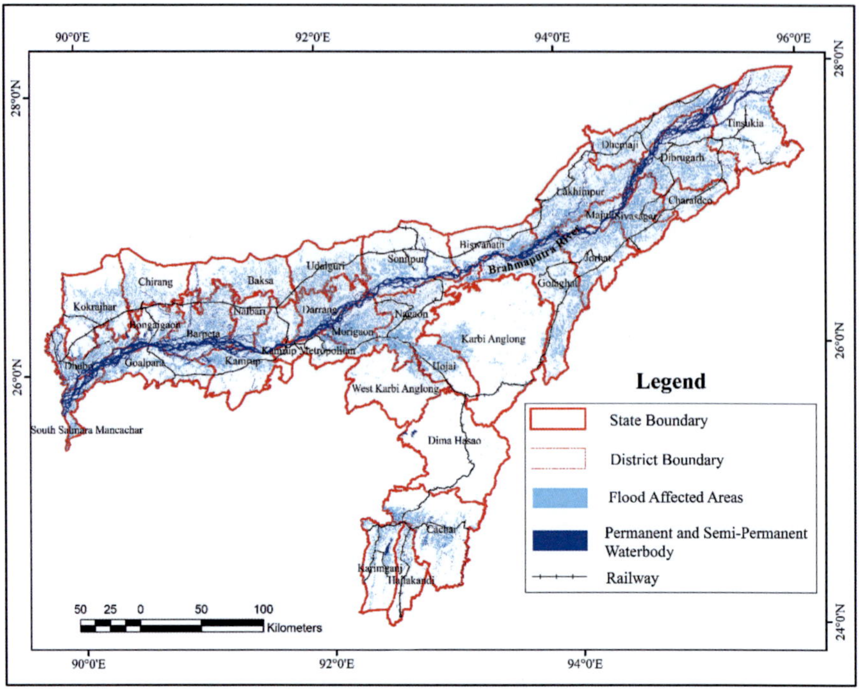

Fig. 8.9 Flood-affected areas of Assam in the year 2020

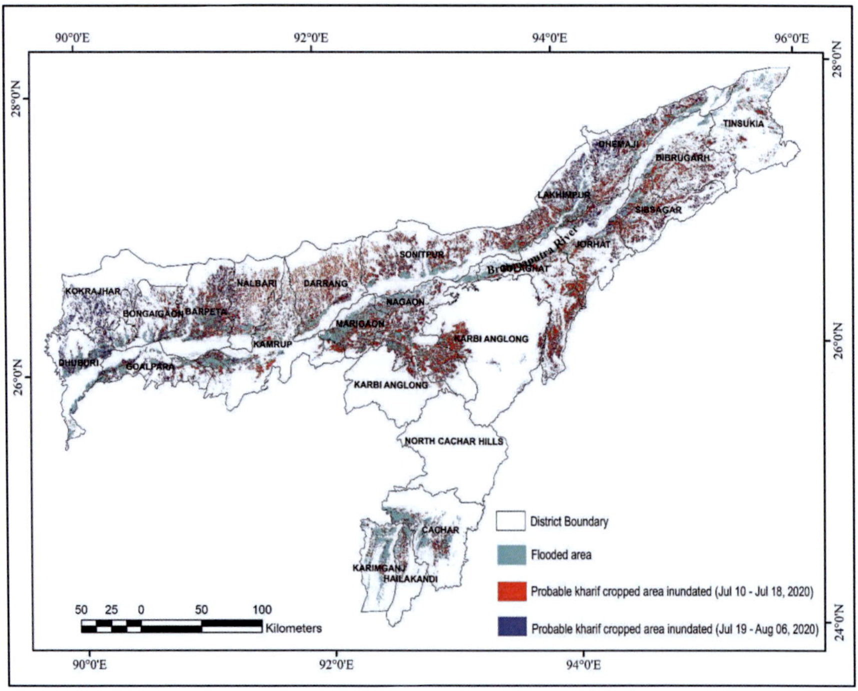

Fig. 8.10 Cropped area affected due to flooding in the state of Assam in the year 2020

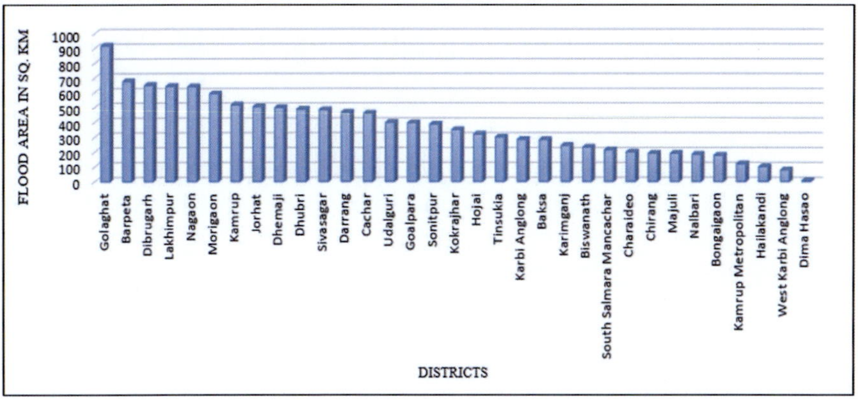

Fig. 8.11 A breakdown, by district, of the areas of Assam that were impacted by flooding in the year 2020

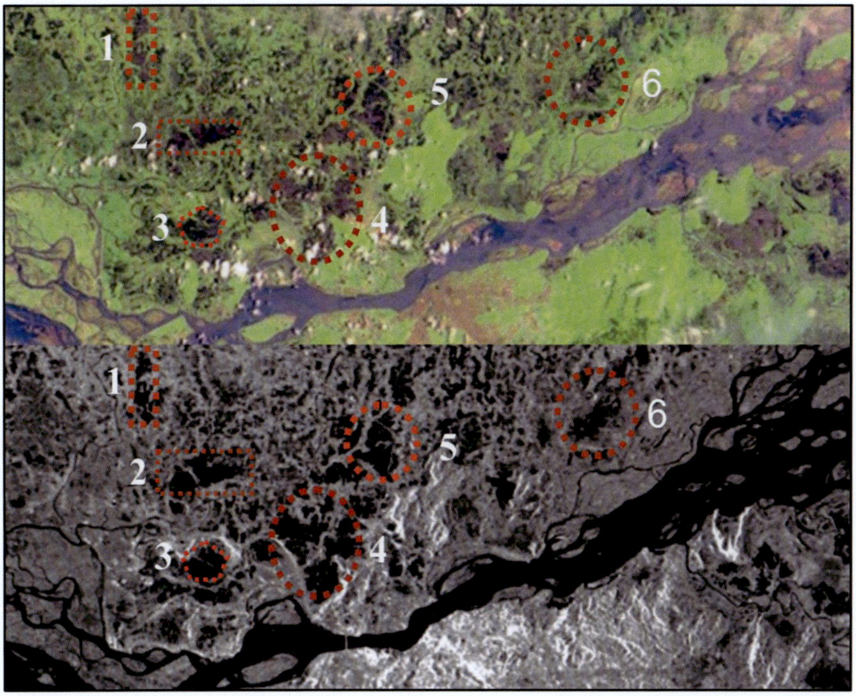

Fig. 8.12 A comparison of SAR and Optical Data to better understand flood events in different districts of Assam

data in GEE offers a quick and effective method for assessing flood risk. GEE is a cloud-based platform with quick processing power. SAR data may be swiftly processed to produce flood maps and gauge flood depths. This makes it possible to accurately analyze flooding on a broad scale in nearly real time, giving timely information for evacuation and rescue attempts. In areas like Assam that are prone to flooding, GEE‘s low cost and simple access to SAR data (Sentinel 1A and 1B) make it a desirable alternative for disaster management. SAR data and GEE integration is a huge step forward in flood assessment and has the potential to reduce flood damage and save casualties.

8 Conclusions

It is crucial to map the flood‘s footprint and evaluate the impacted region before making any choices or taking any actions to lessen the damage. The Brahmaputra River’s active river dynamics have resulted from human intervention to change land use and river morphology, which has led to the emergence of new environmental issues. The difficulties of developing sustainable flood control systems are highlighted through studies of previous precipitation events, water level data, evaluations of flood inundation, and ground verification. By continually monitoring floods using a combination of satellite-based precipitation record and flood inundation area approaches, it would be able to build AI-based region-growing models for flood inundation forecasts. About 2/3 of the entire yearly flooded area will be covered by the flooding in May and June 2020. Even though the season has just recently started, significant high-intensity floods are predicted for the months of July, August, and September. During the COVID-19 crisis, it is essential to anticipate floods in advance and to prepare the necessary actions. For restoring damage, formulating future plans, and transferring people to safer regions, flood inundation mapping and scientific evaluation of rainfall products are both helpful. Integrated studies are used to estimate how much of a region would be inundated based on real-time satellite data on precipitation to see where high-intensity rainfall is falling. Through the integration of satellite data, on-the-ground observations, and mathematical models, flood inundation mapping may be used to inform the public about the potential severity of flooding in any particular location. Local and state governments can utilize the data from these integrated studies to create strategies for flood management. This mapping may also be used to create early warning systems that alert communities to the possibility of floods and provide them with the tools and guidance they need to be safe in such circumstances. SAR-based flood mapping using Google Earth Engine is an important tool for disaster management and preparedness, and its continued development and implementation in Assam and other flood-prone regions around the world can help save lives and reduce the damage caused by floods.

Acknowledgments The authors thank NASA and the Indian Metrological Department (IMD) for daily rainfall products. The authors would also like to express gratitude to the European Space Agency and the Sentinel 1 Science Hub for their assistance sharing Sentinel 1 C Band S.A.R. Data. Assam State Disaster Management Authority's daily flood reports from 16 May to 15 November are also appreciated. Flood forecast bulletins were shared by C.W.C. H.O. Circle Guwahati, which the authors would also like to thank.

References

Bazi, Y., Bruzzone, L., & Melgani, F. (2005). An unsupervised approach based on the generalized Gaussian model to automatic change detection in multitemporal SAR images. *IEEE Transactions on Geoscience and Remote Sensing, 43*(4), 874–887.

Bhatt, C. M., Srinivasa Rao, G., Begum, A., Manjusree, P., Sharma, S. V. S. P., Prasanna, L., & Bhanumurthy, V. (2013). Satellite images for extraction of flood disaster footprints and assessing the disaster impact: Brahmaputra floods of June–July 2012, Assam, India. *Current Science, 104*(12), 1692–1700.

Bhatt, C. M., & Rao, G. S. (2016). Ganga floods of 2010 in Uttar Pradesh, North India: A perspective analysis using satellite remote sensing data. *Geomatics, Natural Hazards and Risk, 7*(2), 747–763.

Costache, R. (2019). Flood susceptibility assessment by using bivariate statistics and machine learning models – A useful tool for flood risk management. *Water Resource Management, 33*(12), 3239–3256. https://doi.org/10.1007/s11269-019-02301-z

Fan, J.-L., & Lei, B. (2012). A modified valley-emphasis method for automatic thresholding. *Pattern Recognition Letters, 33*(6), 703–708.

Gao, W., Shen, Q., Zhou, Y., & Li, X. (2018). Analysis of flood inundation in ungauged basins based on multi-source remote sensing data. *Environmental Monitoring and Assessment, 190*(3), 129. https://doi.org/10.1007/s10661-018-6499-4

Han, D., Chan, L., & Zhu, N. (2007). Flood forecasting using support vector machines. *Journal of Hydroinformatics, 9*(4), 267–276. https://doi.org/10.2166/hydro.2007.027

Hashemi-Beni, L., & Gebrehiwot, A. A. (2021). Flood extent mapping: An integrated method using deep learning and region growing using UAV optical data. *IEEE Journal of Selected Topics in Applied Earth Observations and Remote Sensing, 14*, 2127–2135.

Ichim, L., & Popescu, D. (2019). Flooded areas evaluation from aerial images based on convolutional neural network. In *Proceedings of the International Geoscience and Remote Sensing Symposium (IGARSS), Yokohama, Japan, 28 July–2 August 2019* (pp. 9756–9759).

Kushwaha, S. P. S., Roy, P. S., Azeem, A., Boruah, P., & Lahan, P. (2000). Land area change and rhino habitat suitability analysis in Kaziranga National Park, Assam. *Tiger Paper, 27*(2), 9–17.

Marinelli, L., Michel, R., Beaudoin, A., & Astier, J. (1997). Flood mapping using ERS tandem coherence image: A case study in South France. In *Proceedings of the 3rd ERS symposium, Florence, Italy* (pp. 531–536) Retrieved with permission from www.researchgate.net

McFeeters, S. K. (1996). The use of the normalized difference water index (NDWI) in the delineation of open water features. *International Journal of Remote Sensing, 17*(7), 1425–1432. https://doi.org/10.1080/01431169608948714

Patel, P., Srivastava, H. S., & Navalgund, R. R. (2009). Use of synthetic aperture radar polarimetry to characterize wetland targets of Keoladeo National Park, Bharatpur, India. *Current Science, 97*(4), 529–537.

Pulvirenti, L., Chini, M., Pierdicca, N., Guerriero, L., & Ferrazzoli, P. (2011). Flood monitoring using multi-temporal COSMOSkyMed data: Image segmentation and signature interpretation. *Remote Sensing of Environment, 115*(4), 990–1002.

Rahman, M. R., & Thakur, P. K. (2017). Detecting, mapping and analyzing of flood water propagation using synthetic aperture radar (SAR) satellite data and GIS: A case study from the Kendrapara district of Orissa state of India. *Egypt Journal of Remote Sensing and Space Science, 21*, S37–S41. https://doi.org/10.1016/j.ejrs.2017.10.002

Rignot, E. J. M., & van Zyl, J. J. (1993). Change detection techniques for ERS-1 SAR data. *IEEE Transactions on Geoscience and Remote Sensing, 31*, 896–906. https://doi.org/10.1109/36.239913

Singh, A. (1989). Review article: Digital change detection techniques using remotely-sensed data. *International Journal of Remote Sensing, 10*(6), 989–1003.

Sivasankar, T., Das, R., Borah, S. B., & Raju, P. L. N. (2019). Insight to the potentials of Sentinel-1 SAR data for embankment breach assessment. In P. Rao, K. Rao, & S. Kubo (Eds.), *Proceedings of international conference on remote sensing for disaster management* (Springer Series in Geomechanics and Geoengineering) (pp. 33–41). Springer. https://doi.org/10.1007/978-3-319-77276-9_4

Srivastava, H. S., Patel, P., & Navalgund, R. R. (2006). How far SAR has fulfilled its expectation for soil moisture retrieval. *Proceeding: SPIE, 6410*(641001), 1–12.

Tehrany, M. S., Pradhan, B., & Jebur, M. N. (2015). Flood susceptibility analysis and its verification using a novel ensemble support vector machine and frequency ratio method. *Stochastic Environmental Research and Risk Assessment, 29*, 1149–1165. https://doi.org/10.1007/s00477-015-1021-9

Vanama, V. S. K., & Rao, Y. S. (2019). Change detection-based flood mapping of 2015 flood event of Chennai city using sentinel-1 SAR images. In *In Proceedings of the IGARSS 2019-2019 IEEE international geoscience and remote sensing symposium, Yokohama, Japan, 28 July–2 August 2019*. IEEE.

Wilusz, D. C., Zaitchik, B. F., Anderson, M. C., Hain, C. R., Yilmaz, M. T., & Mladenova, I. E. (2017). Monthly flooded area classification using low resolution SAR imagery in the Sudd wetland from 2007 to 2011. *Remote Sensing of Environment, 194*, 205–218.

Xu, H. (2006). Modification of normalised difference water index (NDWI) to enhance open water features in remote sensed imagery. *International Journal of Remote Sensing, 27*(14), 3025–3033. https://doi.org/10.1080/01431160600589179

Zhang, S., & Lu, H. (2005). Learning texture classifier for flooded region detection in SAR images. In *International conference on computer graphics, imaging and visualization* (pp. 93–98). IEEE. https://doi.org/10.1109/CGIV.2005.51

Chapter 9
Geospatial Modeling for Quick Response to Urban Fire Hazards: A Study in Kolkata Municipal Corporation (KMC) Area

Sandip Ghosh, Ashis Kumar Sen, and Lakshminarayan Satpati

Abstract Hazards are usually the outcome of a combination of extreme natural or man-made events and unplanned development activities affecting increasing population. However, these are manageable through systematic observations and analysis of the hazards to provide measures related to prevention, mitigation, preparedness, emergency response, and recovery. The principal aim of the present research work involves the use of GIS techniques to mitigate the problem of urban fire hazards. This involves identifying the exact location of the victim's place, its connectivity, and accessibility. Clear knowledge about the city layout with the locations and availability of water source, pumping station, nature of the thoroughfares, and types of the fire-management tools, is necessary for this task. The modern techniques of Geomatics are helpful to prepare high quality map, with greater resolution and accuracy, such as transport map, location map of water body, location of fire stations, and urban land use. With the collaboration of these maps, I have constructed a standard geospatial model for the quick response mechanism for fire management of the city. The model is expected to help the administration for managing fire hazards not only in the KMC area but also for similar environmental conditions.

S. Ghosh (✉)
Department of Geography, Government Autonomous College, Rourkela, Odisha, India

A. K. Sen
Department of Geography, Aliah University, Kolkata, India

L. Satpati
UGC-Human Resource Development Centre, University of Calcutta, Kolkata, India

S. Hazra et al. (eds.), *Geographies of the Indian Subcontinent*,
https://doi.org/10.1007/978-3-031-74813-4_9

1 Introduction

In Kolkata, the different land use patterns form a complex mosaic in the landscape and subsequently create varying degrees of fire hazards in the region, particularly in the high-density areas. Fire-prone zones of the KMC area have increased dramatically in many areas, and the past 5 years have witnessed a substantial increase in the number of fires and area burnt, causing loss of crores of rupees due to damage of properties, and even the loss of lives. Kolkata is the pioneering city in India in terms of industrialization (Mukherjee et al., 2018), the most important city of the country in the sense of national economy, polity, and culture (Sen, 2015), characterized by large number of migrants due to communal harmony and political stability (Basu & Sil, 2001), contains major groups of water bodies (Rudra, 2014). Due to growth of population, the land use pattern has been changed with the shrinkage of water bodies and with the increase of residential, commercial, as well as industrial and institutional areas (Chatterjee et al., 2007). For effective fire management, it is crucial to have clear understanding of the physical structure of the city in order to quickly reach to the fire victim locations and minimize the loss of lives and properties (Ghosh, 2017). This is significant, as Kolkata, the present study area, is historically old, congested, and had emerged as an unplanned city (Sikdar & Sahu, 2009) or "Premature Metropolis" due to haphazard settlement from the colonial time. The Kolkata metropolitan city is divided into two clear segments: (a) the older section of the city, spreading over the central and northern parts and (b) the built-up areas, constituting southern and western periphery (Dhar, 2013). These two areas, therefore, have two distinct types of characters. The former section is characterized by poor urban infrastructure, in close proximity to the CBD, while the latter, followed by the "outer crescent of the city," exhibits planned urban development in the form of a purely residential built-up area (Dhar, 2013). In addition, it seems difficult for the quickest arrival at the fire victim locations in Kolkata, because the road space is very congested and vehicular density (500 vehicles/km) is also very high leading to high traffic congestion in the major roads (Chakraborty et al., 2009). The technology can be effectively used to combat and forecast urban fire, as this has both spatial and temporal dimensions. Geomatics, the new technological support of these days, may be immensely beneficial for management of the fire hazard to a great extent (Ghosh, 2018).

The system is capable of helping the managers of urban fire to take the best decision against real-time fire incidents. Such an approach can be considered in view of the existing fire hazard in Kolkata. In general, there is an individual fire extinguishing system in recent times in almost every apartment, hospital, and industry, and other installations against small fires are at the nascent stage (Ghosh & Sen, 2021). But unfortunately, there are major infrastructural lacunae and communication gap between fire events and their management, especially for the large-scale fires. In many cases, it is noticed, that the department of fire and emergency services is not properly informed. Sometimes, narrow roads do not permit to enter the big fire engines, and very often fire fighters have to fetch water from distant places (Ghosh & Sen, 2021).

2 Review of Selected Literatures

Literatures on geography of fire hazards are somehow very limited. However, a number of studies, mainly by the scholars of other disciplines, have been done on the hazard and disasters in the different parts of the world, i.e., (i) the Federal Emergency Management Agency (FEMA), the USA (1999) has prepared a comprehensive profile of the urban fire problem in the United States of America, (ii) De Vligher and Basigos (1995) have prepared fire hazard modeling using remote sensing and GIS with a case study in Greece; (iii) Lentile et al. (2006) have used remote sensing technique in the assessment of fire hazard as an effective technology with the assertion that it has great potential for scientists and managers seeking to map, predict, and assess the ecological effects of fire; (iv) Melesse et al. (2007) have discussed the remote sensing sensors that can be applied for environmental resources mapping and modeling; (v) Picotte and Robertson (2010) have attempted to test the accuracy of remote sensing technology in forest fire damages; (vi) Ghosh Dastidar (2012) has discussed in detail the AMRI hospital fire tragedy of December 9, 2011, while Paul and Ghosh (2014) have analyzed the fire incident of AMRI hospital, which was a very shocking incident for the people of Kolkata.

The major causes of fires in the KMC Area are as follows: fire feeding materials used in building (including slums) and short circuit or illegal hooking; illegal storage of gas cylinder, diesel, motor oil, and wooden furniture; unauthorized chemical industries (paper and plastic materials); breaking of the guidelines issued by Fire Safety and Emergency Department; Spark from the kitchen open stove used for cooking in improper places; and electrical components including loose wiring and computers.

3 The Study Area

The Kolkata Municipal Corporation (KMC) is located at 22° 82′ N of latitudes and 88° 20′ E of longitudes (Fig. 9.1), and it is one of the oldest urban centers developed by the British. The city was not developed in a planned way, so it has a complex character that has resulted in a number of human and environmental problems, including fire hazard. Kolkata (formerly known as Calcutta) is the capital of West Bengal, and it is one of India's oldest urban areas. Historically, the city was the trading and commercial capital of India. It is located on the eastern side of the river Hooghly.

The city is divided into 144 administrative wards that are grouped into 15 boroughs. As per the2011 census data, the city has an area of 185 km^2 and a population of 4.48 million, with a population density of 24,252/km^2.

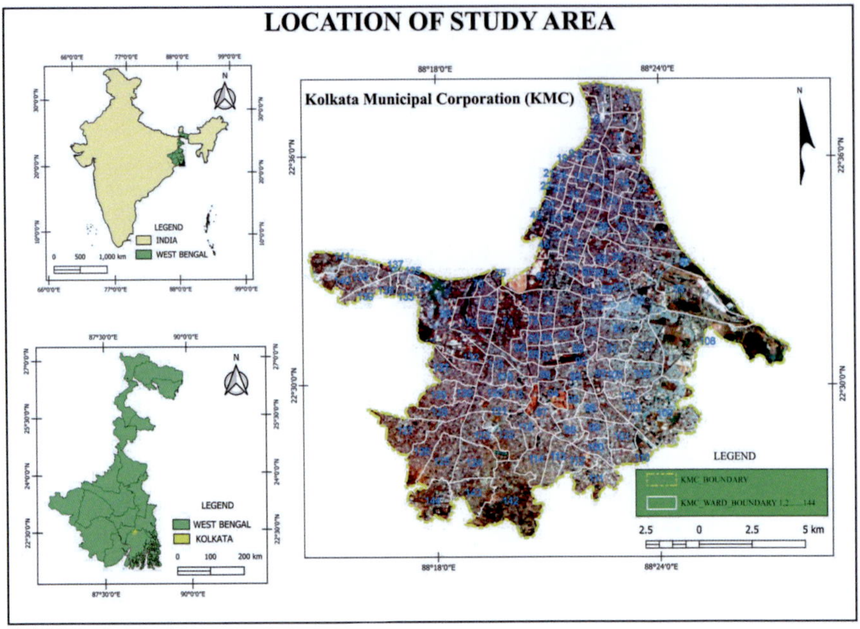

Fig. 9.1 Location of the study area

4 Materials and Methods of Database: Collection and Integration of Remote Sensing and GIS Data

The principal aim of this research work is to effectively manage fire hazards by ensuring accessibility as well as well connectivity, determining the exact location of the victim's place from the nearest fire station, finding out the shortest route to reach the victim place, and obtaining a clear knowledge about the city layout with the water source. The modern techniques of Geomatics may help us to prepare the high-quality map (with greater resolution and accuracy) such as transport map, location map of water body, location of fire station map, and urban land use map, as evident from high-resolution satellite images. With the collaboration of these maps, I have constructed a standard geospatial model for the quick redressal management of city as well as proposed design for the study area: Large Scale LULC map for Fire System Management (FSM) (Anderson et al., 1976), Google Earth pro version Transport map of the KMC Ward maps of KMC (Total ward no. 144), location of water body and pumping station location of fire station (using GPS), and location of slum area, hospitals, markets, and shopping complex using Satellite data (Table 9.1).

ARC GIS (version 10) developed by ESRI Inc. and ERDAS Imagine developed by Leica Geosystems Ltd. Are also used in this study for data processing and preparation of the LULC at the micro-level, i.e., ward level. Selected field verification has made as and when required. The Fire and Emergency Services Department,

Table 9.1 Materials and methods of database

Sl. no.	Data type	Spatial resolution (meter)	Sources	Year	Data format (After image processing)	Application
1.	LANDSAT (4–5) TM	30 × 30	USGS	2016, 2017	Raster (.Img)	Urban LULC map
	LANDSAT (7) ETM+	15 × 15 (pan)				
		30 × 30				
2.	IRS P6 LISS III (Resourcesat-1/ Resourcesat-2)	23.5 × 23.5	NRSC	2014–2015	Raster (.Img)	Inland water body identify
3.	Geo Eye	0.5 m	Google Earth	2005–2018	Raster/vector (.Img/. Shp)	Fire event location identify, networks analysis

Government of west Bengal and various establishments (e.g., Building Department) of the KMC, the Kolkata Environment Improvement Project (KEIP), and the Kolkata Metropolitan Development Authority (KMDA), have also been consulted regularly for the betterment of the study.

5 A Brief Case Study: Bagree Market

The case study has been made on this fire event and can be summed up as follows:

- Bagree market, one of the biggest and busiest retail and wholesale market of Kolkata in close proximity to the Writer's Building and the Reserve Bank of India as well as just few meters from Nandaram market, where a fire damaged nearly 100 houses. The fire started at 2:30 am, and people who stayed in the market rushed out to save their lives. Ultimately, the fire was brought under control after 3 days, i.e., 18 September, 2018 (Fire Type: A, B, C, D, and E).
- The observation revealed that the shops were affected by inflammable materials as mentioned above, and the fire extended rapidly from the ground floor up to fifth floor with continuous damages at every floor.
- Thirty fire engines were engaged into service to retrieve the situation into normalcy at this age-old building with narrow space to cover and to rescue the people and goods. Due to relentless efforts of the fire brigade, no loss of human life was reported, although two members of the rescue team sustained minor injuries.
- Remarks of the Municipal Authorities: During the course of events, authorities of Kolkata Municipal Authority informed that the market authority was warned about the improper installation of fire safety measures and necessary precautions to be taken to prevent fire.

- The matter was subsequently taken by Kolkata Police for necessary investigation from video footage of CCTV cameras and other possible clues.
- Regarding the limitation of fire control measures, it was revealed that all the 24 overhead water tanks had been empty for the last 6 months. Hence, the fire brigade team was unable to promptly handle the situation due to lack of water.
- Finally it can be concluded from the case study that Inadequate number of fire extinguishers, non-availability of water, absence of pumping stations, narrow and extremely congested road and passage for the fire brigade machine, unplanned construction of building, improper and poor maintenance of the building, obsolete electrical wiring due to old and partly demolishing structures of the building, unauthorized hooking for electricity for the footpath dwellers and businessmen, and last but not the least the lack of environmental awareness among the local people and inhabitants and carelessness of the businessmen about the possibility of fire should be taken into consideration for future management and planning of fire.

6 Analysis and Discussion: Geospatial Modeling for Quick Response to Urban Fire Hazards (Application of Game Theory)

Game theory is "the study of mathematical models to solve various problems related to decision-making." Accordingly, this technique is used in geography and particularly in recommending any plan for regional and sustainable development. For example, an optimal planning for ideal location of minimum number of points or centers with respect to the Shortest Path Analysis (SPA) can be obtained by the application of game theory. Originally, it addressed zero-sum games in which any matrix can be prepared with some "constraint" value, i.e., any value $\leq$ to the constraint value is treated as "0" and inserted in the matrix. Today, game theory applies to a wide range of behavioral relations, and it is now an umbrella term for the science of logical decision-making in humans, animals, and computers. Modern game theory began with the idea regarding the existence of mixed-strategy equilibrium in two-person zero-sum games and its proof by John von Neumann. Von Neumann's original proof used Brouwer fixed-point theorem on continuous mappings into compact convex sets, which became a standard method in game theory and mathematical economics.

6.1 Analysis of Shortest Path in Management of Fire Hazards

The connection between two positions on a vector network can be graphically analyzed with distance path. The problem of finding shortest paths in a graph has a surprising variety of applications.

The main features of Shortest Path Analysis (SPA) can be summed up as follows:

- Finding the best route to drive between *x* and *y* or figuring how to direct packets to a destination across a network.
- The shortest path from top to bottom defines the best separation between left and right.
- Finding the center point requires knowing the distance (i.e., shortest path) between all pairs of vertices.

6.2 Determination of Shortest and Alternative Routes

For any efficient fire management and decision-making as well as to reach at the fire victim's locations as quick as possible, the fire engine should definitely follow the shortest path or route between the fire station to the fire victim's locations (wards). For the present purpose, both shortest and alternative paths are identified so as to make a comprehensive management plan to control the fire hazard as shown by the following (Table 9.2).

6.3 Steps in the Solution of Shortest Path

- The center populations are shown by a map with all the intervening distance (in km).
- In order to prepare a shortest path distance matrix.
- The distance constraint is taken as 8 km.
- The cells with distance less than or equal to 8 km are marked in the distance matrix.
- Column reduction is done in two ways: If they are similar, only one column is retained and others are eliminated. If they are exclusive, the larger column is eliminated and others columns are retained.
- Row reduction is done in two ways: If they are similar, only one row is retained and others are eliminated. If they are inclusive, the smaller row is eliminated and other rows are retained.
- Repeat the process of column reduction and row reduction until an irreducible minimum matrix is obtained. This matrix indicates the shortest path on the basis of given matrix.

Table 9.2 Preparation of the shortest and alternative route

Fire station	Ward no.	Shortest path name	Shortest length	Alternative path name	Alternative length
Cossipore Fire Station (F1)	6	Kossipore rd	0.63 km	Ram Gopal rd–Dilarjung rd	0.85 km
Garden reach Fire Station (F2)	76	Garden reach rd–Shastitala rd–Kabitirtha sarani	1.78 km	Ram Surat singh rd–Karl Max sarani–Kabitirtha sarani	3.13 km
Gariahat Fire Station (F3)	68	Suniti Chaterjee rd–Gariahat rd–Frem rd–Gariahat rd	0.39 km	Suniti Chaterjee rd–Gariahat rd–Ballygunge garden rd–Ferm to Gariahat rd	0.51 km
Fire Brigade HQ (F4)	60		2.68 km		2.37 km
	62		0.64 km		2.2 km
Tollygunge Fire Station (F5)	95	Russa rd (south)–Radha govindo nath sarani	0.54 km	Golf club rd–Prince Golam Mohammad Shah rd–Radha govindo nath sarani	1.20 km
Lalbazar Fire Station (F6)	39	Chittaranjan avenue rd–Zakaria st	0.73 km	Chittaranjan Avenue rd–Maulana Saukat rd–Zakaria st	0.97 km
Maniktala Fire Station (F7)	32		0.90 km		4.62 km
	34	Maniktala Fire Station–CIT Road–Hem Chandra Naskar Road	2.59 km	Maniktala Fire Station–CIT road–Narkeldangha road–Kavi Sukanta Sarani–Dr Suresh Chandra Bannerjee road	3.33 km
Nimtala Fire Station (F8)	20	Nimtala Ghat st–Rabindra sarani	1 km	Strand rd–Ahiritola st–R K Pant Avenue rd–Rabindra sarani	1.35 km
	22	Strand rd–Sovaram Basak st	0.98 km	Strand rd–Kali Krishna rd–Kalakar st–Sovaram Basak st	1.28 km
Kalighat Fire Station (F9)	73	HM rd–Kalighat rd	0.54 km	Hazra rd–S P Mukherjee rd–Balaram Basu Ghat rd–Kalighat rd	1.17 km
Alinagar Fire Station (F10)					
Baishnabghata Fire Station (F11)	100	Eastern Metropolitan Bypass–Canal side rd–NRC Bose rd	2.98 km	Eastern Metropolitan Bypass link rd–Raja Subodh Chandra Mullick rd	2.1 km
Behala Fire Station (F12)	126	Diamond Harbour rd–Kalikinkar Ro rd–Narayan Roy rd–K K Roychoedhury rd	1.22 km	Diamond Harbour rd–Ho Chi Minh sarani–Brojoman Dey rd	2.27 km
Central Fire Station (F13)	48	Chittaranjan Avenue rd–Prem Chand Boral st	0.84 km	Chittaranjan Avenue rd–Colootola st–Sashi Bhusan Dey st	1.46 km

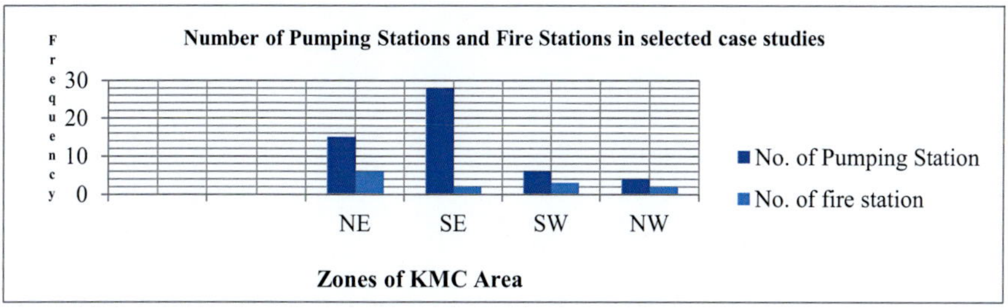

Fig. 9.2 Pumping stations and fire stations in selected case studies

6.4 Fire Engine and Vehicles of West Bengal Fire and Emergency Services

It is also relevant to mention the different types of fire engines and fire vehicles, as the shortest as well as the alternative routes are not uniform in width and the quality of roads is also a concern. Incidentally, more sophisticated and modern engines with a vertical structure, as already been implemented in other developed countries in Asia like Japan, may fit within the existing width condition of roads. Hence, expansion of roadways and free movement of the fire vehicles are essential to reduce the future fire hazards (Fig. 9.2).

6.5 Sources of Water (Availability of Water Pumping Stations with Respect to Fire Stations)

Another essential infrastructural facility for fire management is the availability of water supply and other types of fire extinguishers. Fire history of the KMC area reveals that most of the fires are A, B, and C types require proper water supply and appropriate fire extinguishing systems. Hence, existing water pumping stations as well as nearby water bodies or ponds provide remedial measures for rapid fire control in the fire victim's locations. Hence, water was the main extinguisher of those fires. Therefore, the location map of water points or water sources is very important, as the firemen can locate the nearest water body during the fire at any ward. A survey has been conducted to assess the availability of water to the pumping stations and area-wise and ward-wise pumping stations and fire stations and their ratio have been computed to evaluate the different fire stations with respect to availability of water in different zones of KMC area (Table 9.3 and Fig. 9.3).

From the above graph, it appears that SE sector is the most favorable with respect to fire stations, while SW and NW sectors are in a precarious situation with respect to the availability of water. This should be taken into consideration in the final planning, management, and recommendation (Fig. 9.2).

Table 9.3 Area-wise and ward-wise pumping stations and fire stations

Area	No. of pumping station	No. of wards	Name of wards	No. of fire stations	Name of station	Ratio of pumping station to fire stations
NE	15	11	5, 11, 36, 49, 51, 45, 45, 45, 41, 46, 46, 61, 61, 63, 64	06	F1, F4, F7, F6, F8, F13	2.5
SE	28	13	66, 66, 66, 66, 67, 67, 91, 91, 91, 91, 68, 90, 93, 93, 93, 92, 92, 92, 98, 69, 69, 89, 89, 84, 81, 83, 83, 83	02	F3, F5, F9, F11	14
SW	6	4	128, 74, 74, 73, 73, 82	03	F12	2
NW	4	4	76, 74, 80, 63	02	F2, F10	2
Total	53			13		

Source: Prepared by the researcher

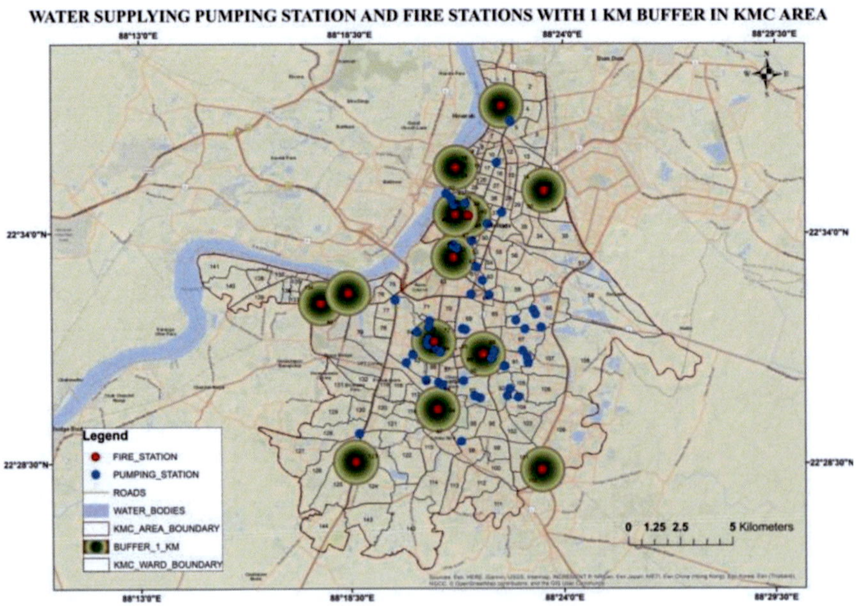

Fig. 9.3 Availability of water pumping stations with respect to fire stations

6.6 *Operational Strategy of 13 Fire Service Stations and Their Constraints for Resolving (Appropriate Fire Management and Planning)*

As per the guidelines of Standing Fire Advisory Council (SFAC), the zone of response of each fire station should be 10 km^2 (One Fire Station per 10 km^2) in urban area. The 13 fire stations under Kolkata Municipal Corporation areas are covering about 65% of the KMC area, and the most vulnerable areas are out of the sphere of influence of fire cover. In fact, most of these areas are highly vulnerable, characterized by dense populations and with a high number of commercial and industrial establishments, as well as several godowns storing combustible materials that are highly concentrated. Hence, all the 13 fire service stations are selected with 2 km buffering in order to examine the area of coverage by each fire service station. Accordingly, mapping is done to show the zone of buffering and their areal coverage of places of fire events (Fig. 9.4 and Table 9.4).

The RS and GIS techniques are applied here to evaluate the locational scenario and spatial pattern of network between fire stations and pumping stations. Here buffering of 1 km and 2 km is made to identify the future planning and implementation of proposed and new fire stations, aiming to achieve a balanced and even distribution of fire stations and pumping stations.

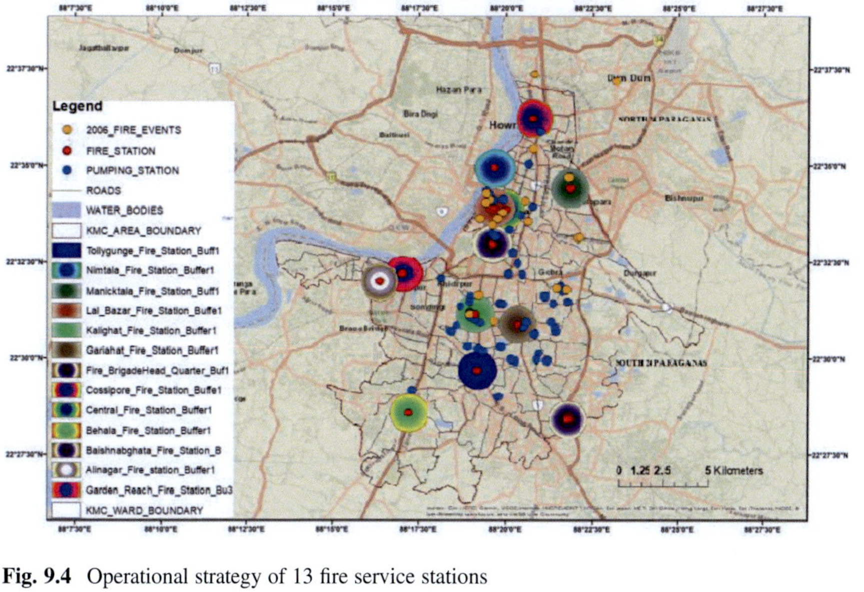

Fig. 9.4 Operational strategy of 13 fire service stations

Table 9.4 Distance and number of fire stations to pumping stations

No. of fire stations	Name of fire stations	Distance from pumping stations		
F1	Cossipore Fire Station	0.75	2.26	–
F2	Garden Reach Fire Station	2.09	2.73	3.49
F3	Gariahat Fire Station	0.56	0.49	1.13
F4	Fire Brigade HQ	0.33	0.56	0.53
F5	Tollygunge Fire Station	0.98	1.35	1.61
F6	Lal Bazar Fire Station	0.49	0.77	0.83
F7	Manicktala Fire Station	2.07	2.5	2.93
F8	Nimta Fire Station	1.14	1.34	1.64
F9	Kalighat Fire Station	0.17	0.42	0.49
F10	Alinagar Fire Station	3.24	4.57	4.45
F11	Baishnabghata Fire Station	3.54	3.75	3.84
F12	Behala Fire Station	1.3	4.69	4.73
F13	Central Fire Statin	0.5	1.03	1.22

Source: Prepared by the researcher

6.7 *Computation of the Value of Response Time Through Calibration*

Finally, the value of response time of fire engine and accessories to the fire victim's locations should be calibrated after Haupt and Haupt (1997) and Banerjee (2018), as expressed by the following equation:

$$T = r0 + K \times r$$

where

T = response time of he fire station.
r = distance (kilometers).
$r0$ (minutes) = operational readiness time (the time taken for the fire engines to leave the fire station upon receiving the call)
K = traffic impedance factor.

For the present purpose, computations of the value of response time through calibration are calculated with respect to three fire events, namely, BAGREE MARKET, AMRI, and STEPHEN COURT.

For example, the above formula can be applied with respect to the incident of Bagree of Fire (September 14, 2018 in ward number 27), where the value of $r = 0.65$ m, $r0 = 7$ m, and $K = 15$ m.

Putting the values, we get $T = (7 + 15) \times 0.65 = 14.3$ m.

- Applying the same equation, AMRI (December 9, 2011, ward No. 93).

  AMRI: $r = 0.74$ km; $r0 = 6$ m; $K = 20$ m; $T = (6 + 20) * 0.74 = 19.24$

- In case of STEPHEN COURT (March 23, 2010, ward no. 60 the value of $r = 0.48$; $r0$ is 5 m and $K = 10$ m and hence $T = (5 + 10) * 0.48 = 7.2$.

Generalizing the above equation, it can be suggested that greater the value of *T* is directly proportional to the extent of damage. The importance of the above equation lies in the fact that *T* factor should be reduced in order to minimize the fire hazard. Therefore, the formula is considered an effective tool for any systematic planning and management of fire hazard.

6.8 Relation Between LULC and Category of Fires

For further assessment of fire hazard, it is now required to find out the degree of association between different sectors and categories of fire as shown by Table 9.5.

Different categories of fire (A, B, C, D, and K) are shown with respect to different sectors (Hospital, Residential, Market, and Industrial) and for numerical and statistical assessment between rows and columns; two-way ANOVA (analysis of variance) is applied. In fact, the basic principle of ANOVA is to test the differences among the means of populations by examining the amount of variation between samples relative to the amount of variation within the samples. Here, two-way ANOVA techniques are used when the data are classified on the basis of two factors. Therefore, we have to calculate the sum of the squares for total variance and for variance between two different categories of one factor as also variance between types of the other factor.

Table 9.5 Relation between LULC and category of fires

	Category of fire (type)					
Sector (LULC)	A (type)	B (type)	C (type)	D (type)	K (type)	
Hospital	2	0	2	0	0	4
Residential	23	10	13	1	2	49
Market	34	20	10	1	0	65
Industrial	33	5	11	5	1	55
	92	35	36	7	3	173

Source: Prepared by the researcher based on Primary Survey (2016 to 2022)

Sum of the squares of deviations between columns (Different Categories of Fire):

$$\text{SS between columns} = \sum \frac{T_j^2}{n_j} - \frac{T^2}{n}$$

Sum of the squares of deviations between rows (Different sectors of LULC):

$$\text{SS between rows} = \sum \frac{T_i^2}{n_i} - \frac{T^2}{n}$$

Calculation of SS between columns and SS between rows is to be determined. Sum of the squares of deviations for residual (SS residual) may be calculated as:

$$\text{SS residual} = \sum {X^2}_{ij} - \frac{T^2}{n} - \left\{ \left(\sum \frac{T_j^2}{n_j} - \frac{T^2}{n} \right) + \sum \frac{T_i^2}{n_i} - \frac{T^2}{n} \right\}$$
$$= \text{SS total} - (\text{SS between columns} + \text{SS between rows})$$

Degrees of Freedom (df) for Total Variance:

$$\left(\sum n_j \times \sum n_i \right) - 1$$

where

n_j and n_i = number of columns and rows respectively

Degrees of Freedom (df) for Variance between columns:

$$\sum n_j - 1$$
$$= (\text{number of columns} - 1)$$

Degrees of Freedom (df) for Variance between rows:

$$\sum n_i - 1$$
$$= (\text{number of rows} - 1)$$

Degrees of Freedom (df) for residual Variance:

$$\left(\sum n_j - 1 \times \right) \left(\sum n_i - 1 \right)$$

where

n_j and n_i = number of columns and rows respectively

Table 9.6 ANOVA (two-way without replication)

ANOVA (two-way without replication, i.e., SRSWOR)						
Source of variation	SS	df	MS	*F*	*P*-value	*F* crit
Rows	436.95	3.00	145.65	3.29	0.06	3.29
Columns	1264.30	4.00	316.08	7.14	0.00	3.26
Error	531.30	12.00	44.27			
Total	2232.55	19				

On the basis of above steps, the following table is prepared:

The computed $F = 3.29$ and is equal to the tabulated value of 3.29; hence, we accept the null hypothesis, i.e., the variation of fire event is not caused by different categories of fire. On the other hand, between rows computed F 7.14 is greater than the tabulated value of 3.26. Thus, we accept the alternative hypothesis, i.e., the variation in fire events is caused by the different sectors of LULC in KMC area (Table 9.6).

Giving due consideration of primary survey and the Geomatics techniques employed here, it can be mentioned that KMC is not in a position of safe zone with respect to fire incident. Hence, proper management and recommendations are needed to ameliorate the fire hazard.

7 Findings: A General Overview of the Vulnerability of Fire Hazards over Space and Time

From the above study and research, a general overview of vulnerability of fire hazard over space and time for KMC area can be drawn as follows:

- Width of the road including road density, road frequency, and road width.
- Availability of water for fire preventive measures.
- Proximity of fire stations to the water pumping stations and other natural and man-made water bodies.
- Determination of shortest path from fire station to fire victim places.
- Determination of alternative route when traffic is very high.
- Fire feeding materials used in building (including slums).
- Short circuit or illegal hooking.
- Illegal storage of gas cylinder, diesel, motor oil, and wooden furniture.
- Unauthorized chemical industries (paper and plastic materials).
- Breaking of the guidelines issued by Fire Safety and Emergency Department.
- Spark from the kitchen open stove used for cooking in improper places.
- Electrical components including loose wiring and computers.
- Environmental awareness and cleanliness to prevent any further fire incidence as far as possible.

8 Conclusions and Outlook

In this research, the historical background of all the fire incidents have been discussed in detail, and here the constraints or the shortcomings are highlighted in order to prepare a comprehensive management-oriented action plans so that the extent of damage can be minimized in any future fire incident in the study area. Such plans can be implemented in other metropolitan or mega-city structures in India as well as in other developing countries of the world. Accordingly, location of fire stations with one km buffer, location of water-pumping stations and other natural water bodies, and location of different fire events (2006–2022) and transport network are mapped with an aim at examining the optimality criteria over space so that final planning of location and management of fire service stations, water pumping stations, and road network be taken and implemented in future. Such planning should come within the purview of the concept of "smart city" since GPS, GIS, and RS (Geomatics system) and radio-coverage mapping can be used as a quick access from fire service centers to places of fire incidents.

References

Anderson, J. R., Hardy, E. E., Roach, J. T., & Witmer, R. E. (1976). A land use and landcover classification system for use with remote sensor data. U.S. GeologicalSurvey, Professional Paper 964.

Banerjee, D. (2018). *Optimal site proposal of fire stations by buffer analysis method: A case of Kolkata, West Bengal* (pp. 186–193).

Basu, B., & Sil, S. (2001). Arsenic mapping for North 24- Parganas District of West Bengal using GIS and remote sensing technology. In *Map India conference 2003*. GIS development.net.

Chakraborty, D., Das, D., Rahaman, M. M., Chowdhur, U. K., & Biswas, B. (2009). *Status of ground water arsenic contamination in W.B., India* (A 20-Year Study Report) (Vol. 53, pp. 542–551). https://doi.org/10.1002/mnfr.200700517

Chatterjee, R. S., Fruneau, B., Rudnat, J. P., Roy, P. S., Frison, P. L., Lakhera, R. C., Dadhwal, V. K., & Saha, R. (2007). Subsidence of Kolkata (Calcutta) City, India during the 1990s as observed from space by differential synthetic aperture radar interferometry (D-InSAR) technique. *Remote Sensing and Environment, 102*, 176–185.

De Vligher, B. M., & Basigos, P. S. (1995). Fire hazard modelling using remote sensing and GIS. A case study of the eparchy of Pylias (Messinia, Grece). *Earsel Advances in Remote Sensing, 4(3)*, 21–29. Retrieved February 12, 2015, from http://www.earsel.org/Advances/4-3-1995/4-3_04_De-Vliegher.pdf

Dhar, S. B. (2013). Recent trends of population growth in Kolkata municipal corporation area. *Geographical Review of India, 75*(2), 181–196.

Federal Emergency Management Agency, the United States Administration. (1999). *Profile of the urban fire problem in the United States*. Retrieved February 17, 2015, from http://www.usfa.fema.gov/downloads/pdf/publications/urban.pdf

Ghosh, S. (2017). *Application of geomatics in fire hazard management of Kolkata Municipal Corporation area, West Bengal (Abstracts) 39th IIG, December (26–28)* (p. 209).

Ghosh, S. (2018). *Application of geomatics in fire hazard vulnerability management of Kolkata Municipal Corporation (KMC) Area, West Bengal* (p. 135). IEELE.

Ghosh Dastidar. (2012). AMRI hospital: Saving lives? *Journal of Case Research, 4*(01), 57–69.

Ghosh, S., & Sen, A. (2021). Planning and management of fire hazards in Kolkata Municipal Corporation Area of West Bengal, India. *Geographical Review of India, 83*(1), 61–80.

Jensen, J. R. (2013). *Remote sensing of the environment: An earth resource perspective* (p. 618). Pearson India Pvt. Ltd.

Haupt, R. L., and Haupt, S. E. (1997) Sue Ellen Haupt.—2nd ed. p. cm. Red. ed. of: Practical genetic algorithms. c1998. "A Wiley-Interscience publication". Includes bibliographical references and index. ISBN 0-471-45565-2 1. Genetic algorithms. I. Haupt, S. E. II. Haupt, Randy L. Practical genetic algorithms. III. Title. QA402.5.H387 2004 519.6¢2—dc22 2004043360 Printed in the Untied States of America.

Lentile, L. B., et al. (2006). *Remote sensing techniques to assess active fire characteristics and post-fire effects*. USDA Forest Service Publication. Retrieved January 20, 2015, from http://digital.commons.unl.edu/usdafsfacpub/194

Melesse, A. M., et al. (2007). *Remote sensing sensors and applications in environmental resources mapping and modelling, sensors*. Retrieved January 15, 2015, from www.mdpi.org/sensors

Mukherjee, S., Bebermeier, W., & Schütt, B. (2018). An overview of the impacts of land use land cover changes (1980–2014) on urban water security of Kolkata. *Land, 7*(2018).

Pal, I. and Ghosh, T. (2014) Fire Incident at AMRI Hospital, Kolkata (India): A Real Time Assessment for Urban Fire, Journal of Business Management & Social Sciences Research (JBM & SSR) Volume 3, No.1, pp. 9–13, retrieved from www.borjournals.com Blue Ocean Research Journals 9 on 4 May, 2018.

Picotte, J. J., & Robertson, K. M. (2010). Accuracy of remote sensing wildland fire-burned area in Southeastern US Coastal Plain Habitats. In *Proceedings of 24th tall timbers fire ecology conference, FL, USA* (pp. 86–93).

Rudra, K. (2014). Changing river courses in the Western Part of the Ganga–Brahmaputra Delta. *Geomorphology, 227*, 87–100.

Sen, R. (2015). Mobility improvement plan for Barasat municipal area. *International Journal of Innovative Research and Development, 4*(2).

Sikdar, P. K., & Sahu, P. (2009). Understanding wetland sub-surface hydrology using geology and isotopic signature. *Hydrology Earth System Sciences, 13*(7), 1313–1329.

Chapter 10
SKIMEQIS: An Effective Information Tool of Seismic Disaster Resilience for Sikkim Himalayan Region

R. Sivakumar, Snehasish Ghosh, and S. S. Sridhar

Abstract Seismic disaster resilience defines the early preparation of communities to resist themselves for future earthquakes taking cues from past earthquakes and their impacts. The United Nations Sustainable Development Goals (SDGs) highlighted the importance of Disaster Risk Reduction (SDG-11) for growing cities and settlements, where a target has been fixed to promote inclusive, safe, resilient, and sustainable growth of the settlements by adjusting to any kind of disasters. The entire Himalayan region is highly prone to seismicity (Zone-IV & V), where the occurrence of earthquakes has increased in the last few decades. The development of unplanned settlements and their extensive growth in this region makes them more vulnerable towards seismically-induced damages. Hence, it is necessary to focus on seismic disaster resilience for the existing settlements in the Himalayan region. Geoinformatics-based customization is capable of developing advanced information

R. Sivakumar (✉)
Earthquake Research Cell, Department of Civil Engineering, College of Engineering and Technology, Faculty of Engineering and Technology, SRM Institute of Science and Technology, Chennai, Tamil Nadu, India

Department of Civil Engineering, College of Engineering and Technology, Faculty of Engineering and Technology, SRM Institute of Science and Technology, Chennai, Tamil Nadu, India
e-mail: sivakumr@srmist.edu.in; eqr.cell.ktr@srmist.edu.in

S. Ghosh
Earthquake Research Cell, Department of Civil Engineering, College of Engineering and Technology, Faculty of Engineering and Technology, SRM Institute of Science and Technology, Chennai, Tamil Nadu, India
e-mail: snehasig@srmist.edu.in

S. S. Sridhar
Earthquake Research Cell, Department of Civil Engineering, College of Engineering and Technology, Faculty of Engineering and Technology, SRM Institute of Science and Technology, Chennai, Tamil Nadu, India

Department of Computing Technologies, College of Engineering and Technology, Faculty of Engineering and Technology, SRM Institute of Science and Technology, Chennai, Tamil Nadu, India
e-mail: sridhars@srmist.edu.in

S. Hazra et al. (eds.), *Geographies of the Indian Subcontinent*,
https://doi.org/10.1007/978-3-031-74813-4_10

tools, which have the ability to visualize integrated geospatial information in a common platform. The present study attempts to develop the Sikkim Earthquake Information System (SKIMEQIS), an effective tool of seismic disaster resilience for the Sikkim Himalayan region using ArcObject programming in VB. NET (Visual Basic. Network Enabled Technologies) framework by the preparation of various graphical user interfaces. To understand the impact of seismological parameters on the existing settlements, this tool visualizes thematic and analytical information together, which depicts that the central part of Sikkim along the Main Central Thrust (MCT) and Ramgarh Thrust (RT) thrust zone coincides with the major earthquake epicenters, where magnitude varies between 3.5 and 5. Major settlements such as Gangtok, Namchi, Ranipool, and Singtam are already developed in this active seismic zone. Suitable decisions for the expansion of these settlements and further development plans can be decided by analyzing the seismological parameters through the SKIMEQIS, which will be helpful as an effective tool for disaster resilience by visualizing the seismic risk factors on settlements.

1 Introduction

Disaster resilience measures the potentiality of revival of individuals and communities from any hazards without interrupting long-term developmental activity (DFID, 2011). According to the Hyogo Framework for Action, disaster resilience defines the capacity of individuals, communities, and public and private organizations to organize themselves for the prevention of future hazards by obtaining knowledge from previous disasters and declining probable risks for the future at international, regional, national and local levels (UNISDR, 2005). The interdisciplinary studies of natural disasters have given rise to the concept of disaster resilience, which integrate risk and resilience (Combaz, 2014). The United Nations Sustainable Development Goals (SDGs) highlighted the importance of Disaster Risk Reduction (SDG-11) for the growing cities and settlements where it emphasizes on significant reduction in affecting people, death and economic loss. Further, a target has been fixed to promote inclusive, safe, resilient, and sustainable growth of the settlements by adjusting to any kind of disaster (SDGs, 2020). The Sendai Framework for Disaster Risk Reduction (SFDRR) during 2015–2030 produces an integrated approach of development and recovery for disaster risk reduction (Maly & Suppasri, 2020).

Disaster management is a cyclic process with three major stages involved, corresponding to the period before, during, and after a disaster strikes. The pre-disaster stage involves preparedness and mitigation efforts. Disaster resilience has become an important component of preparedness due to the increasing frequency of various hazards and their subsequent damages. Seismic disaster resilience is a measure of the degree of early preparation of communities to resist the effects of future earthquakes by developing seismic microzonation, retrofitting building structures, and strengthening the existing infrastructure and geospatial information tool to understand the reduction in disaster risk from future earthquake (Jha, 2020). Communities and societies require accumulating knowledge from past disasters and adopting them to sustain themselves by making necessary adjustments. It is more important that earthquake resilience be focused on disaster risk reduction strategy and retrofitting infrastructures for preventing earthquake damages in the preparedness and mitigation stages rather than emergency relief and recovery during or after the disaster (Ibrion et al., 2015). Even though earthquake is less predictable and its risk cannot be eliminated completely, it is possible to minimize its damages by making the disaster risk strategy through Geo-informatics-based customized tools, which can visualize the necessary information in a common platform.

GIS customization is the process of computer-based programming in the .Net framework to analyze the databases for quick information retrieval and execution (Maguire, 2005). The process of GIS customization defines the development of Graphical User Interface (GUI) by writing suitable programming code. It can be written in Visual Basic, Python, Java, and Visual C^{++}, which helps to produce analytical as well as visual application tools in GIS platform. The customization can be integrated in various GIS software to satisfy the user's requirement. Various customized GIS tools have been proposed for analyzing geospatial data and developing various models. Johnson et al. (2018) explained community vulnerability of Kullu district in Indian Himalayas in the context of flood-induced damages and suggested risk reduction strategies by the adaptation of technological advances (mobile mapping and satellite TV communication techniques), gathering of local knowledge, and accumulation of previous events records and other lines of evidence. Kanji and Agrawal (2019) discussed the crucial social elements and their interdependencies by analyzing interviews and questionnaire surveys to justify the role of social responsibility for disaster resilience and sustainable development. Ekhlaspoor et al. (2022) developed a web-based tool to determine the index value of seismic resilience of building structures in Iran by drawing the functionality and vulnerability curve. Poornima and Murugan (2021) developed the Natural Disaster Resilience Approach (NDRA) for quick and automated retrieval of disaster information to support disaster resilience measurement by the implementation of three-tier preventive checks such as fake user profile, originality of posted authenticity, and faster resilience within the affected region. Sun et al. (2020) studied artificial intelligence for disaster risk reduction and developed an AI-based decision support tool, where application is prepared for all four phases of disaster management. Stevens et al. (2007) produced the ICITY software tool in the GIS platform to help urban planning and decision-making by adopting disaster risk reduction (DRR) strategy. Similarly, the ETECS tool was developed to provide early warning of earthquakes and tsunamis along the seaside cities and probable estimation of damages (Youhai et al., 2006).

The available tools in different GIS software are not always sufficient to analyze all kinds of databases. Hence, it is necessary to design new visual interfaces for analyzing and visualizing the geospatial databases in an easily understandable manner as user's requirements. The Earthquake Information System tool for Sikkim Himalayan region (SKIMEQIS) is proposed as a disaster resilience tool by providing quick information on earthquake hazard and its influencing parameters in a specific settlement area. Construction activity in hilly terrains is always a challenge. It has been observed that various parts of the Himalayan region are affected by different events such as earthquakes, landslides, floods, etc., which induce subsequent damages. The damages can be reduced by providing awareness to the society about the hazard and its probable disaster where geospatial information plays an important role during pre-disaster phase. The Sikkim Himalayan region is part of north eastern Himalaya and seismically more active (Verma, 1991; Nath et al., 2009). It extends from 27°00′ N to 28°07′ N latitudes and 88°00′ E to 88°55′ E longitudes. Topographically, it is a high altitude mountain region with rugged topographic features. Geomorphologically, it covers ridges, peaks, deep valleys, and numerous glaciers, whereas geologically, Gneisses and Schists are dominated in the south, which is thrusted by Main Central Thrust (Sarkar et al., 2012). The major settlements of the study area are Gangtok, Nathang, Ranipool, Tadong, Singtam, etc., all of which are expanding their area to cater to the existing tourist requirements. Such increasing development in such a rugged and seismically active terrain can lead to significant damages during future earthquakes. To prevent such damages, it is necessary to have a Seismic information tool to protect the existing settlements and infrastructures from earthquake-induced damages.

2 Methodology

The methodology proposed for the development of customized tool as shown in Fig. 10.1. In the first stage, Visual Basic (VB) programming language has been selected for writing a suitable programming code. The VB.NET framework has been chosen and interaction has been established with GIS software through ArcObject Software Development Kit. In the second stage, the GUI has been developed to produce the SKIMEQIS tool. In the third stage, the user interface has been integrated with ArcGIS by the use of Component Object Model (COM) technology. In the fourth stage, the SKIMEQIS tool has been implemented in the ArcGIS platform to visualize the seismic hazard and its influencing parameters and seismic hazard impact on settlements. Finally, the proposed tool has been tested in the selected study area to see if it executes the required information.

2.1 *Language Selection and Software Integration*

2.1.1 Interaction Between VB and GIS Software Through SDK

Different GIS software can be selected to implement the information system tool. In the present study, ArcGIS 9.3 has been chosen due to its excellent integration capacity with ArcObjects libraries and SDK (ESRI, 2004, 2005). The ArcGIS

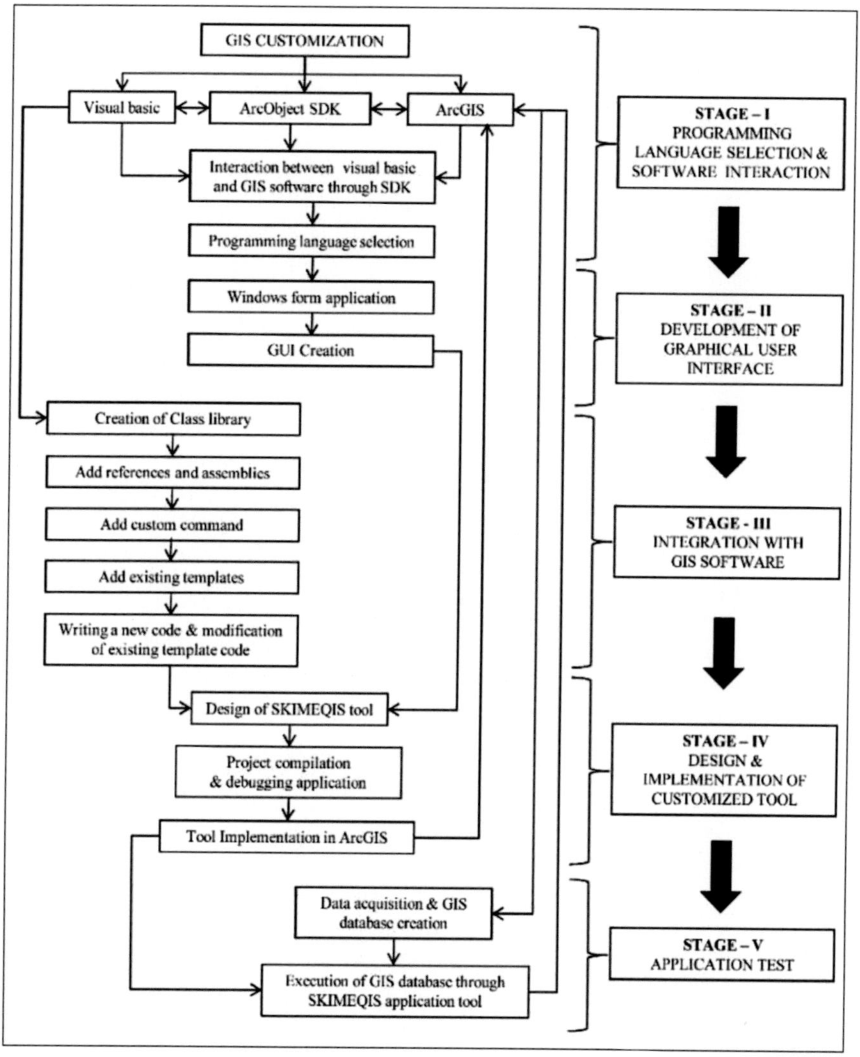

Fig. 10.1 Flow chart of methodology

application package was built on the basis of ArcObjects software libraries. Any kind of available function in ArcGIS can be incorporated through ArcObjects (Stevens et al., 2007). ArcObjects provides Application Programming Interfaces (APIs) for the development of new user interface, which helps in the development of desired tools using the Component Object in Visual Basic.NET framework.

ArcObjects has different Dynamic Link Libraries (DLLs), which contain a group of classes to represent the various components of the ArcGIS package. The geodatabase library includes object classes of different basic components such as feature class, feature, and field (Stevens et al., 2007). The object classes allow the programmer to implement the tool at a higher level. It helps the programmer to search and execute the required feature class rapidly. The search method of the feature class is called by its specified ID or name with the help of the user interface to display the required database (Stevens et al., 2007).

2.1.2 Preferred Programming Language

ArcObjects can be accessed using any COM-compliant programming language. For the development of a new tool for ArcGIS, it is necessary to select one of the versions of the VB software, which can be integrated with ArcGIS and requires an external development environment such as the .NET framework. It is possible to design tools using programming to analyze the data and visualize the information on a GIS platform (Lo & Yeung, 2007). The selection of programming language depends on the ability of the programmer and the suitability of the proposed application.

In order to develop SKIMEQIS tool, in this study, the codes were written in VB and integrated with ArcGIS software through ArcObject SDK to display the graphical user interface. This tool is a customized extension tool for ArcMap Desktop. The available code templates were used and modified according to the needs of the proposed application. The proposed program is helpful in developing an open structure GUI and implementing a suitable format in ArcGIS by creating a customized DLL. An implementation code has been written to create a custom command in a DLL that allows passing an object from the DLL to ArcMap.

2.2 *Development of Graphical User Interface*

The user interface has been proposed in GIS for further enhancement of the efficiency of the data analysis for better interpretation. The GUI interface has great impact on GIS users. The design of GUI interface depends on the requirement of the users in a specific applied field (Wang, 2009). Users normally interact with the GIS software with the help of GUI, which directly visualizes and analyzes the GIS data (Maguire, 2005) named as Windows-Icon-Menu-Pointer (WIMP).

It is necessary to develop a suitable user interface to visualize and analyze spatial data (Dodson, 1993). The user interface is a single part of the system and needs to be improved for better visualization, analysis and to get the results (Frank et al., 1992; Knill, 1993). It is possible to develop an adaptive user interface to implement models and analyze and visualize the information in the GIS platform. The various extended GUI tool for ArcGIS were developed in this environment such as SELES (Fall & Fall, 2001), SpaSim (Moreno et al., 2002), UrbanSim (Waddell, 2002), ICITY (Stevens et al., 2007), SHETRAN (Stephen et al., 2010), CART (Cheng et al., 2010) and EaHaAsTo (Sivakumar & Ghosh, 2017) etc. In this study, the SKIMEQIS

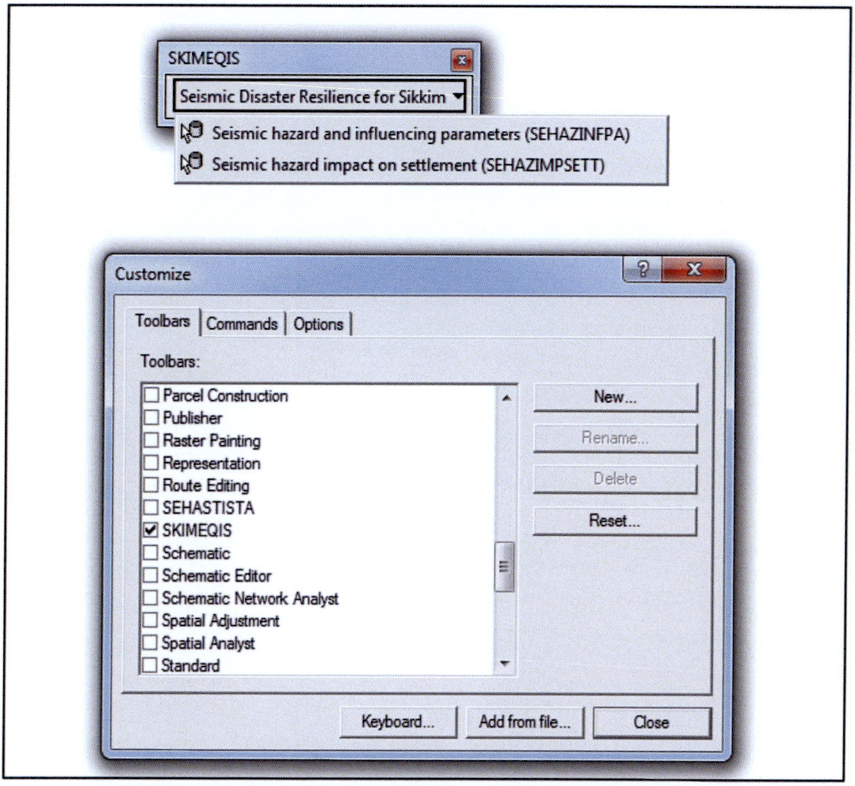

Fig. 10.2 Execution of SKIMEQIS tool in ArcGIS

graphical user interface has been developed using ArcObjects and VB application to visualize the earthquake information in the ArcGIS platform (Fig. 10.2).

2.3 *Integration with ArcGIS*

VB programming was used to develop the proposed application using Automation Technology. The proposed strategy is built based on the COM technology for integrating the selected expert system with ArcGIS. Microsoft-Automation is a COM-based technology that allows an application to programmatically manipulate and execute the set of interfaces. The programming tool accesses the Automation objects and interacts between the user and the system to implement the graphical user interface (GUI) (Sabhan et al., 2003).

To develop the proposed tool and integrate the same with ArcGIS, the new project was created in the VB platform. The new ArcMap class library was created to execute the tool in ArcGIS platform. To create ArcGIS desktop application, the relevant assemblies were added, such as ESRI.ArcGIS.ADF.Local, ESRI.ArcGIS. ArcMapUI, ESRI.ArcGIS.Geodatabase, ESRI.ArcGIS.Carto, ESRI.ArcGIS. DataSourcesFile, ESRI.ArcGIS.Display, ESRI.ArcGIS.Geometry, ESRI.ArcGIS. SystemUI, etc. Further, the reference of the selected assembles, namely, System. Drawing, was also included. The existing ArcMap commands were added for passing the desired CLSID or ProgID. Similarly, the ArcGIS item template, such as base command, base toolbar, and base menu, were introduced to implement its function. The proper code was written and the template modified according to the application's need. The windows form based GUI were incorporated with the SKIMEQIS toolbar by writing a suitable code. The project was built by the compilation of all solution explorer components.

The build operation of .dll file is stored in the Bin\Debug project file. The debug information (.pdb) and a library (.tlb) files are generated in the same directory file by registration tool. Similarly, the obj project directory preserves the temporary files, which are required during project compilation. By running the debug process, the project is integrated with ArcGIS software through ArcMap.exe and plugged in with ArcGIS software.

2.4 *Execution of Tool in ArcGIS*

2.4.1 Design of SKIMEQIS Tool

The proposed resilience tool, SKIMEQIS, was developed through coding in VB software and geodatabase visualization in GIS platform. SKIMEQIS is easy to use as well as tightly fitted with the ArcGIS. This tool facilitates an automated extraction of earthquake information as per users' need. The proposed tool is executed from the customize tool option.

The SKIMEQIS tool is easy to use due to its simple GUIs. The tool is implemented in GIS platform, as shown in the Fig. 10.2. It consists of a main toolbar, namely, SKIMEQIS, which have the content a menu namely Seismic Disaster Resilience for Sikkim. The menu consists of two major tools namely Seismic Hazard Influencing Parameters (SEHAZINFPA) and Seismic Hazard Impact on Settlements (SEHAZIMPSETT). SEHAZINFPA provides the general information on geology, geomorphology, geotectonics, slope, relief, seismotectonics, soil, isoseismal, Peak Ground Acceleration (PGA), Lineament density, structural buffer, earthquake epicenter distribution, and earthquake hazard zone. SEHAZIMPSETT visualizes seismic parameters in Gangtok and Namchi settlement area.

2.4.2 Function of SKIMEQIS Tool

The main function of the proposed tool is to visualize the spatial information and other related graphics on viewers. The users can call the spatial information based on the existing maps/graphics from the respective folders using check box functions. GUI makes a call to interface tools which allow calling the data management functions to display the existing database. The check box allows the user to visualize the earthquake information from the respective folder as they need by calling the layer (.lyr) file from the respective folder. All information is controlled by the check box function. The refresh/remove button is attached with tools to remove or clear the information from the display at a time. The SEHAZINFPA tool helps to compare all the thematic as well as analytical parameters with seismic hazard and epicenter distribution. Similarly, the SEHAZIMPSETT tool compares seismic influencing data and its zonation, in particular Gangtok and Namchi settlement region.

3 Results

The SKIMEQIS tool enables quick extraction of information for a particular area. It is possible to display multiple information through this tool and its various user interfaces to compare with each other. In the present study, the SKIMEQIS tool has been prepared to visualize the Seismic Hazard Influencing Parameters (SEHAZINFPA) and Seismic Hazard Impact on Settlements (SEHAZIMPSETT) information to make suitable decisions for infrastructure development in Sikkim.

A comparison is made between geological formation, slope, relief, structural features with earthquake epicenter distribution and seismic hazard using the SEHAZINFPA tool (Fig. 10.3). The study shows that the epicenters are distributed along the active thrust zone in the southern part of the region. The frequency of earthquakes is more in Gorubathan geological formation. The coarse loamy and loamy skeletal soils of the region are also correlated with earthquake epicenter distribution and the region comprises of low relief (<2000 m) and high degree of slope (>60°). It is recognized that the epicenters are mostly distributed on coarse loamy and fine loamy soils around Gangtok, Namchi, Sigtam, and Ranipool, which spread across the earthquake hazard zones EHZ-IV—EHZ-VI.

The SEHAZINFPA tool is also used to visualize analytical parameters such as PGA, lineament density with past seismicity and seismic hazard. The study finds that frequent, shallow earthquakes with magnitudes ranging from 5.0 to 5.5 have been noticed in areas of high lineament density (>3.0 km/sq. km) and active tectonic zones around Namchi, Nathang, Nayabazar, and Gangtok region, where the central thrust is dominant, including earthquake hazard zone EHZ-IV—EHZ-VI (Fig. 10.3).

The SEHAZIMPSETT tool visualizes the specific settlement locations of Gangtok and Namchi (Fig. 10.4). The observation shows that the Gangtok area comes under EHZ—IV-VI (high to severe seismic hazard zones), where thrust faults are

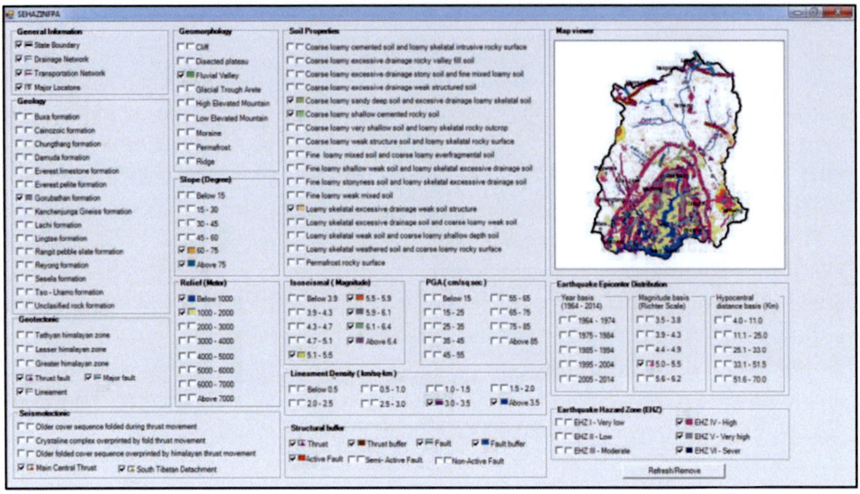

Fig. 10.3 Thematic, analytical and earthquake information using SEHAZINFPA tool

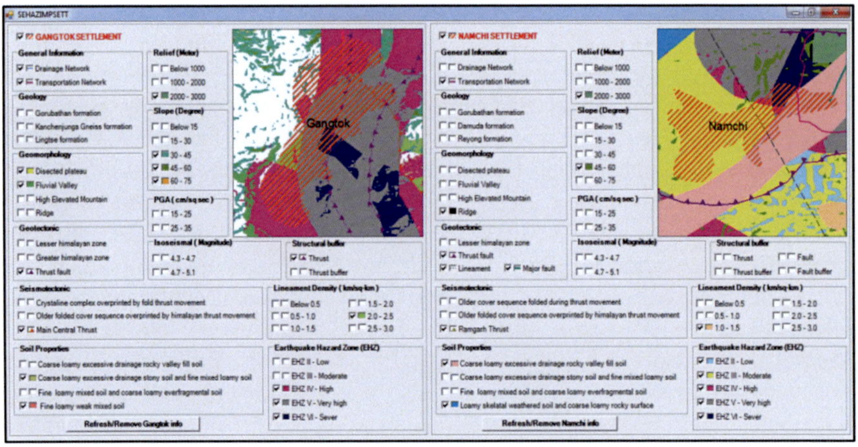

Fig. 10.4 Settlement in Sikkim region observed

dominant. Geomorphologically, it is part of a dissected mountain and fluvial valley region, with an elevation of 2000–3000 m, and a higher slope (>30°). The region covers higher lineament density (2.0–2.5 km/sq.km.) with significant PGA (15–35 cm/sq.-s) and the isoseismal zone ranges between 4.3 and 5.1 in this region. Similarly, it is recognized that Namchi is the part of moderate EHZ –II where Ramgarh thrust and Tista lineament are dominant. The lineament density varies 1.0–1.5 km/sq.km. with higher degree of slope (>45°). The region comes under moderate relief (2000–3000 m) with significant PGA (15–35 cm/sq.-s), and the isoseismal zone ranges between 4.3 and 5.1.

4 Discussions

The visualization tool depicts (Fig. 10.3) that the central part of Sikkim, in between MCT and RT zone, is underlain by Gorubathan, Damuda, Buxa, and Chungthang formations, all of which are seismically active. The major Tista lineament also crosses this region; thus, these are indicators of active sesmotectonic activity in the region (Mukul et al., 2014). Further, it is also observed that the major unit of the Sikkim window is active Ramgarh Thrust (Srivastava & Mitra, 1994; Mitra et al., 2010), which is located in this region. Therefore, complex geological formation and structural features with multiple thrusting are recognized in this region (Mitra et al., 2010). The earthquake epicenters are correlated with structural parameters and analysis suggests that with moderate depth earthquakes occur more frequently in this region. The major settlements such as Gangtok, Namchi, Ranipool, and Singtam are developed within this zone. Gangtok town is located within the Rani Khola river basin, which is adjacent to MCT boundary and forms an active displacement zone, where the magnitude of earthquakes ranges between 4.7 and 5.1. Active tectonic phenomenon in this region is also supported by the presence of Gangtok lineament surrounding Gangtok, Ranipool and Tadong, where clustered buildings have been built (Mukul et al., 2014). The town of Namchi is developed in the Kamrang Khola river basin region, which is influenced by the Ramgarh thrust and various structural faults, where repetition of geological formations leads to variations of earthquake magnitude between 4.3 and 4.7, thereby indicating the influence of the recent tectonic movement. By establishing the communication system between settlements, various infrastructures are developed in this active seismotectonic zone, such as road networks, residential houses, tourist lodges, power supply system, bridges, etc., which are very vulnerable during earthquakes. The proposed tool is capable of visualizing seismic hazard and its influencing parameters for the Sikkim region. To minimize the earthquake-induced damage, it is necessary to control the growth of settlement specifically in zones of high seismicity and active tectonics or develop better seismically resistant structures. The tool described in this chapter will be helpful to identify such vulnerability and inform taking suitable decisions before any construction work.

5 Conclusions

The present study has developed the SKIMEQIS tool, which visualizes Seismic Hazard Influencing Parameters (SEHAZINFPA) and Seismic Hazard Impact on Settlements (SEHAZIMPSETT) information for Sikkim region. It visualizes different seismological parameters such as geology, soil, relief, structural features, lineament density, peak ground acceleration, etc. and depicts the greater influence of the fault displacement as well as subsequent earthquakes, which pose significant risks for the existing settlements and infrastructures specifically in expanding towns such as Gangtok, Ranipool, Tadong, Nayabazar, Namchi, Rangpo, and Singtam. Development is necessary to maintain economic growth and to retain the tourist industry. However, suitable decision for the expansion of various settlement and further development in this region can be decided by analyzing the seismological parameters through SKIMEQIS disaster resilience tool, which will help in minimizing the earthquake induced damages in the study region. The study infers that the SKIMEQIS will be an effective geospatial information tool of seismic disaster resilience in Sikkim region by visualizing the seismic risk factors on settlements.

Acknowledgements The authors are thankful to SRM Institute of Science and Technology, Kattankulathur, for providing all necessary facilities and constant encouragement for doing this research work.

References

Cheng, W., Wang, K., & Zhang, X. (2010). Implementation of a COM-based decision-tree model with VBA in ArcGIS. *Expert Systems with Applications, 37*, 12–17.

Combaz, E. (2014). *Disaster resilience: Topic guide*. GSDRC, University of Birmingham. https://research.manchester.ac.uk/en/publications/disaster-resilience-topic-guide

DFID. (2011). *Defining disaster resilience: A DFID approach paper*. DFID.

Dodson, R. D. (1993). Advances in hydrological computation. In D. Maidment (Ed.), *Handbook of hydrology*. McGraw Hill.

Ekhlaspoor, A., Dehkordi, M. R., Eghbali, M., & Samadian, D. (2022). Pre-event assessment of seismic resilience index for typical Iranian buildings via a web-based tool. *International Journal of Civil Engineering, 20*, 257–272.

ESRI. (2004). *ArcGIS desktop developer guide ArcGIS 9*. ESRI Press.

ESRI. (2005). *ArcObjects technology*. http://www.esri.com/getting_started/developers/arcobjects.html. Accessed 25 Apr 2005.

Fall, A., & Fall, J. (2001). A domain-specific language for models of landscape dynamics. *Ecological Modelling, 141*, 1e18.

Frank, A. U., Kuhn, W., Willauer, L., & Mark, D. M. (1992). *User interfaces for GIS: Discussion at the specialist meeting* (Report on the Specialist Meeting, NCGIA Research Initiative 13, 22–26 June. NCGIA Report 92/3). National Center for Geographic Information and Analysis, University of California at Santa Barbara. http://www.dpi.inpe.br/gilberto/references/userinterf_report.pdf

Ibrion, M., Mokhtari, M., & Nadim, F. (2015). Earthquake disaster risk reduction in Iran: Lessons and "lessons learned" from three large earthquake disasters – Tabas 1978, Rudbar 1990, and bam 2003. *International Journal of Disaster Risk Science, 6*, 415–427.

Jha, R. (2020). *SDGs, Indian cities and seismic sustainability*. 371.

Johnson, R. M., Edwards, E., Gardner, J. S., & Diduck, A. P. (2018). Community vulnerability and resilience in disaster risk reduction: An example from Phojal Nalla, Himachal Pradesh, India. *Regional Environmental Change, 18*, 2073–2087.

Kanji, R., & Agrawal, R. (2019). Building a society conducive to the use of corporate social responsibility as a tool to develop disaster resilience with sustainable development as the goal: An interpretive structural modelling approach in the Indian context. *Asian Journal of Sustainability and Social Responsibility, 4*, 5.

Knill, J. (1993). Geographical information systems: The environmental view. In P. Mather (Ed.), *Geographical information handling-research and applications* (pp. 7–15). Wiley.

Lo, C. P., & Yeung, A. K. W. (2007). *Concepts and techniques of geographic information sysytem*. Prentice Hall.

Maguire, D. J. (2005). GIS customization. In P. A. Longley, M. F. Goodchild, D. J. Maguire, & D. W. Rhind (Eds.), *Geographical information system: Principles, techniques, management and applications (abridges edition)*. Wiley.

Maly, E., & Suppasri, A. (2020). The Sendai framework for disaster risk reduction at five: Lessons from the 2011 great East Japan earthquake and tsunami. *International Journal of Disaster Risk Science, 11*, 167–178.

Mitra, G., Bhattacharyya, K., & Mukul, M. (2010). The lesser Himalayan Duplex in Sikkim: Implications for variations in Himalayan shortening. *Journal Geological Society of India, 75*, 289–301.

Moreno, N., Ablan, M., & Tonella, G. (2002). SpaSim: A software to simulate cellular automata models. In *Proceedings of integrated assessment and decision support, first biennial meeting of the international environmental modelling and software society, Lugano, Switzerland, June 24–27*.

Mukul, M., Jade, S., Ansari, K., & Matin, A. (2014). Seismotectonic implications of strike–slip earthquakes in the Darjiling–Sikkim Himalaya. *Current Science, 106*, 198–210.

Nath, S. K., Singh, K. K., & Abhishek Raj, T. (2009). Earthquake hazard in Northeast India – A seismic microzonation approach with typical case studies from Sikkim Himalaya and Guwahati city. *Journal of Earth System Science, 117*(S2), 809–831. https://doi.org/10.1007/s12040-008-0070-6

Poornima, N., & Murugan, M. (2021). Natural disaster resilience approach (NDRA) to online social networks. *Journal of Ambient Intelligence and Humanized Computing, 12*, 5651–5678.

Sabhan, W. A., Mulligan, M., & Blackburn, G. A. (2003). A real-time hydrological model for flood prediction using GIS and the WWW. *Computers, Environment and Urban Systems, 27*, 9–32.

Sarkar, S. S., Ali, M. A., & Bhattacharya, G. (2012). *Geology and mineral resources of Sikkim* (Report). Geological Survey of India.

SDGs. (2020). *The sustainable development goals report 2020* (p. 68). United Nation.

Sivakumar, R., & Ghosh, S. (2017). Earthquake hazard assessment through geospatial model and development of EaHaAsTo tool for visualization: An integrated geological and geoinformatics approach. *Environmental Earth Sciences, 76*, 442.

Srivastava, P., & Mitra, G. (1994). Thrust geometries and deep structure of the Outer and Lesser Himalaya, Kumaon and Garhwal (India): Implications for evolution of the Himalayan fold-and-thrust belt. *Tectonics, 13*, 89–109.

Stephen, J., Birkinshaw, P., & Ewen, J. (2010). Graphical user interface for rapid set-up of SHETRAN physically-based river catchment model. *Environmental Modelling & Software, 25*, 609–610.

Stevens, D., Dragicevic, S., & Rothley, K. (2007). iCity: A GIS-CA modelling tool for urban planning and decision making. *Environmental Modelling & Software, 22*(2007), 761–773.

Sun, W., Bocchini, P., & Davison, B. D. (2020). Applications of artificial intelligence for disaster management. *Natural Hazards, 103*, 2631–2689.

UNISDR. (2005). *A guide for implementing the Hyogo framework for action by local stakeholders*. UNISDR.

Verma, R. K. (1991). Seismicity of the Himalaya and the Northeast India, and nature of continent-continent collision. In Sharma, K. K. (Ed.). *Geology and geodynamic evolution of the Himalayan Collision Zone, Part 2. Phys Chem Earth, 18*(1–2), 405–429.

Waddell, P. (2002). UrbanSim: Modeling urban development for land use, transportation, and environmental planning. *Journal of the American Planning Association, 68*, 297–315.

Wang, H. (2009). Influence and impact relationship between GIS users and GIS interfaces. *Human Centered Design HCII LNCS, 5619*, 815–824.

Youhai, G., Qimin, F., & Jing, J. A. (2006). GIS-based earthquake and tsunami emergency command system for seaside cities. *Journal of Ocean University of China, 5*(2), 181–186.

Part III
Statistical Modelling

Chapter 11
Forecasting of Onset of Southwest Monsoon with Preferred Number of Pre-monsoon Cyclonic Disturbances Over Bay of Bengal

Sayantika Mukherjee

Abstract The study is aimed to commence the concept of soft computing technique in the form of rough set theory to view the onset of monsoon over Kolkata (22° 32′, 88° 20′), which is situated over Gangetic West Bengal (GWB), India. The concept of rough set theory is primarily based on the condition–decision rule. The method is thus suitable for the present study, where the consideration is to observe the number of cyclonic disturbances during pre-monsoon months as condition and the onset of monsoon whether it would be right time, advance or delay, that is considered as decision. The strength, certainty and coverage factors for the decision algorithms are computed with 19 years' (1990–2008) data. The result of the study reveals that the number of pre-monsoon cyclonic disturbances including depression, deep depression, cyclonic storm, severe cyclonic storm, very severe cyclonic storm and super cyclonic storm has a immense impact on predicting the onset of southwest monsoon over Kolkata. The validation has been done with data from 2009, 2010 and 2011.

1 Introduction

A large percentage of weather-related tragedies are caused by cyclonic disturbances over oceans. Disruptions caused by cyclones that start with a low-pressure system thereafter transform into depression, deep depression, cyclonic storm, severe cyclonic storm, very severe cyclonic storm, and super cyclonic storm (Table 11.1) (Bhardwaj & Singh, 2020). There were several cyclonic disturbances across the Bay of Bengal during the pre-monsoon months. There are several phases to the advent and retreat of the broadscale Asian monsoon, which signify important changes in the large-scale atmospheric and ocean circulations in the Indo-Pacific area (Fasullo & Webster, 2003; Rao, 1976; Murakami & Nakazawa, 1985; Lau et al., 1998; Hsu et al., 1999; Wu & Zhang, 1998). With a standard deviation of roughly 8 days, the monsoon arrives in Kerala around June 1 (Pai & Nair, 2009).

S. Mukherjee (✉)
Amity University Kolkata, Kolkata, India
e-mail: sgmukherjee@kol.amity.edu

S. Hazra et al. (eds.), *Geographies of the Indian Subcontinent*,
https://doi.org/10.1007/978-3-031-74813-4_11

Table 11.1 Classification of cyclonic disturbances (Bhardwaj & Singh, 2020)

System	Pressure deficient (hPa)	Associated wind speed Knot (kmph)
Low pressure area	1.0	<17 (<32)
Depression	1.0–3.0	17–27 (32–50)
Deep depression (DD)	3.0–4.5	28–33 (51–59)
Cyclonic storm (CS)	4.5–8.5	34–47 (60–90)
Severe cyclonic storm (SCS)	8.5–15.5	48–63 (90–119)
Very severe cyclonic storm (VSCS)	15.5–65.6	64–119 (119–220)
Super cyclonic storm	>65.6	>119 (>220)

The monsoon's northward movement is indicative of a deep convection shift that is occurring on a massive scale from equatorial to continental regions (Rao, 1976; Sikka & Gadgil, 1980; Webster et al., 1998). The displacement of westerly troughs in the circumpolar westerlies and the shift of the subtropical jet (STJ) to the north of the Himalayan periphery were factors that Yin (1949) was the first to link to the process of monsoon commencement. According to Murakami and Ding (1982), the onset is connected to the diabatic heating that warms the Eurasian region. Ananthakrisha et al. took into consideration synoptic characteristics linked to the arrival of the southwest monsoon over Kerala (Ananthakrishnan et al., 1968). In order to assess the dynamic foundation for anticipating the date of the commencement of monsoon rains in India, researchers implemented a general circulation model that possesses a particular characteristic (Goswami & Gouda, 2010). Variability of onset of SW monsoon was studied with different aspects (Ananthakrishna & Soman, 1988, 1991; Anathakrishnan et al., 1983; Gadgil, 2003; Bhullar, 1952; Flatau et al., 2003). Dynamical aspects of monsoon were observed (Flatau et al., 2001). Not only that, convective organization during monsoon is analyzed by Goswami and Patra (2004). Interannual variability of the onset of monsoon and a new attempt has done to predict the onset of monsoon (Joseph et al., 1994; Pai & Nair, 2009). Onset vortex of the summer monsoon is a significant aspect of monsoon (Krishnamurti & Ramnathan, 1982; Krishnamurti et al., 1981). Periodically, cyclonic disturbances coincide with the southwest monsoon and extend into Gangetic West Bengal, accompanied by the Bay branch of the monsoon. Occasionally, meteorologists would see the emergence of a low pressure area over the Bay of Bengal and forecast that the monsoon would arrive that year several days earlier than expected. However, when the system continued to intensify and ultimately became a cyclone, the situation ultimately shifted drastically. The monsoon in the Gangetic West Bengal region is delayed as a result. Monsoon season can arrive early or on schedule when there are no cyclonic disruptions.

The study's objective is to use the rough set theory to estimate the number of cyclonic disturbances in order to predict when the southwest monsoon would arrive over Kolkata.

2 Data

The data used for this present study are taken from the website of IMD, *www.imd.gov.in* during the pre-monsoon months from1990 to 2011. Nineteen years of data are used for the training set and 3 years are granted for the validation set. The records of the dates of the onset of southwest monsoon over Kolkata are collected from the Regional Meteorological Centre (RMC), Kolkata for the years 1990 to 2011.

3 Methodology and Its Implementation

When the number of cyclonic disturbances is nil, then it is granted as <1 category and another category is ≥ 1 as identified since the threshold ranges of occurrences of cyclonic depressions during the pre-monsoon months over Kolkata for the study period. The methodology adopted in this study is the rough set theory (RST), an important component of the soft computing technique (Zadeh, 1965). This chapter presents rough set rule induction as one way to deal with data uncertainty while creating predictive "if–then" rules that generalize data values to the entire region (Chaudhuri, 2006, 2008). The condition decision support system of rough set theory is implemented to estimate the certainty of the condition that is the number of cyclonic disturbances during the pre-monsoon months as the input parameter and the onset of monsoon on the decision (Pawlak & Skowron, 2007; Chaudhuri, 2008; Zhao et al., 2008; Chaudhuri et al., 2012) (Annexure 1).

Annexure 1

The basic concept of the RST is the concept of approximation space, which is an ordered pair, Information system IS = (U, A).

U is the universe (a finite set of objects, $U = \{x1, x2, \ldots, xm\}$.
A is the set of attributes (features, variables), non-empty set.

An equivalence relation can be related with every subset of attributes as

$$I(B) = \{(x, y) \in U\text{: for every } a \in B,\ a(x) = a(y)\}$$

where

$a\ (x) \rightarrow$ values of attribute 'a' for every element x
$a\ (y) \rightarrow$ values of attribute 'a' for every element y
$U/I\ (B) \rightarrow$ the family of equivalence classes of $I\ (B)$

The information system is termed as the decision table when the set of attributes A is divided into two disjoint classes C and D representing condition and decision attributes respectively and is given by

$$S = (U,\ C,\ D)$$

where

$C = \{c_1(x), c_2(x), \ldots, c_n(x)\} \rightarrow$ Set of condition attributes.
$D = \{d_1(x), d_2(x), \ldots, d_m(x)\} \rightarrow$ Set of decision attributes

Thus, support of the decision rule is

$$\text{Supp}_x(C, D) = |C(x) \cap D(x)|$$

Strength of the decision rule is

$$\sigma_x(C, D) = \frac{\text{Supp}_x(C, D)}{|U|}$$

Certainty factor is

$$\text{Cer}_x(C, D) = \frac{|C(x) \cap D(x)|}{|C(x)|} = \frac{\text{Supp}_x(C, D)}{|C(x)|}$$

The coverage factor of the decision rule is

$$\text{Cov}_x(C, D) = \frac{|C(x) \cap D(x)|}{|D(x)|}$$

Certainty factor and coverage factor follow

$$\sum_{x \in C} \text{Cer}_x(C, D) = 1$$

$$\sum_{x \in D} \text{Cov}_x(C, D) = 1$$

In this present study, the decision algorithms or the approximation of sets are framed as:

$$\text{If (Number of CD} < 1\text{), then (right time onset of monsoon)} \quad (1)$$

$$\text{If (Number of CD} < 1\text{), then (advance onset of monsoon)} \quad (2)$$

$$\text{If (Number of CD} < 1\text{), then (delay onset of monsoon)} \tag{3}$$

$$\text{If (Number of CD} \geq 1\text{), then (right time of monsoon)} \tag{4}$$

$$\text{If (Number of CD} \geq 1\text{), then (advance of monsoon)} \tag{5}$$

$$\text{If (Number of CD} \geq 1\text{), then (delay of monsoon)} \tag{6}$$

Number of CD → Number of cyclonic disturbances [depression (D), deep depression (DD), cyclonic storm (CS), severe cyclonic storm (SCS), very severe cyclonic storm (VSCS), super cyclonic storm (SuCS)].

4 Results and Discussions

In the present study the method of the rough set theory is adopted to identify the number of cyclonic disturbances, including depression, deep depression, cyclonic storm, severe cyclonic storm, very severe cyclonic storm and super cyclonic storm, during the pre-monsoon months for the right time, delay and advance onset of southwest monsoon over Kolkata. Six decision algorithms with different conditions are framed with the available dataset. Figure 11.1 shows that for the occurrence of cyclonic disturbances over Kolkata during the pre-monsoon season, the lower value of the number of depressions gives a high value of coverage factor both for delay and advance onset of SW monsoon, and the value is 1.0. This means that 100% chances of delay and advance onset of monsoon over Kolkata satisfy the condition and the decision rules, decision algorithm 1. This also shows the higher value of strength and

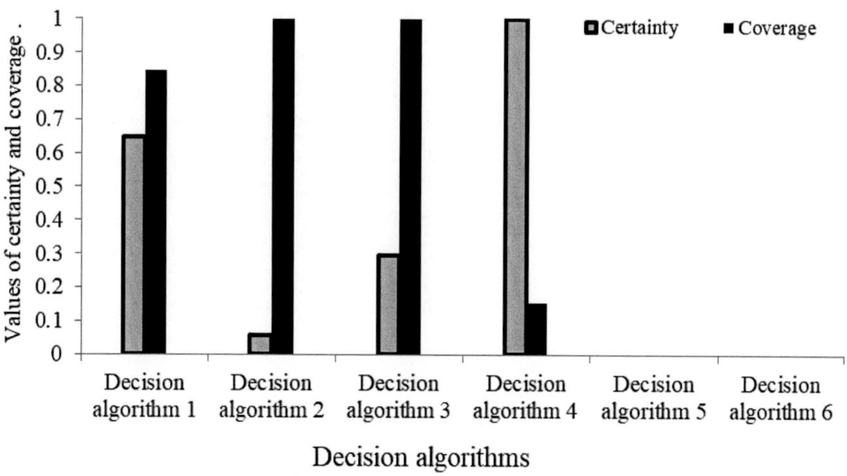

Fig. 11.1 The diagram shows variation in the certainty and coverage of the six decision algorithms for different numbers of depression during the pre-monsoon months over the Bay of Bengal in evaluating the onset of monsoon

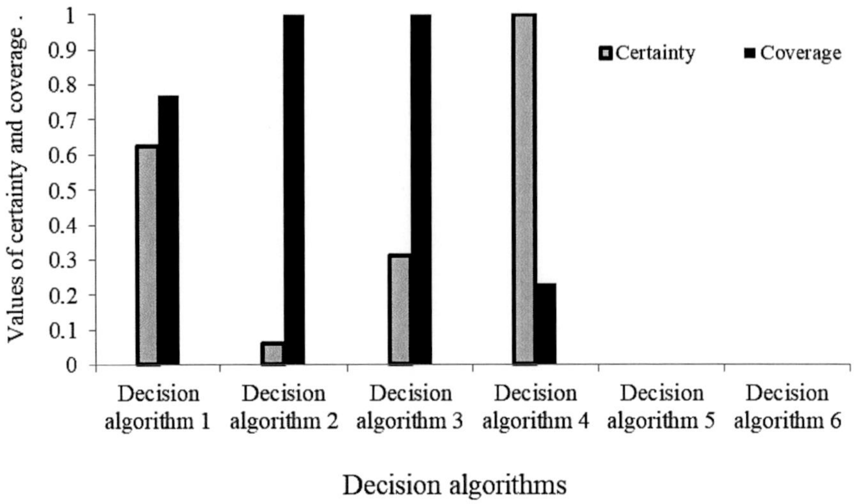

Fig. 11.2 The diagram shows variation in the certainty and coverage of the six decision algorithms for different numbers of deep depression during the pre-monsoon months over the Bay of Bengal in evaluating the onset of monsoon

certainty factor is for the low number of depressions for delay of onset of monsoon than advance of monsoon. The figure further depicts that for the onset of monsoon, the higher value of the certainty factor is for the right time onset of monsoon, and the value is 1. This means that 100% chances of right time onset of monsoon while number of depressions is high.

Figure 11.2 shows that for the onset of monsoon, the higher value of the coverage factor is both for the advance and delay of the SW monsoon over Kolkata while the number of deep depressions is low and the value is 1.0. This means that 100% chances of advance and delay of onset of monsoon which satisfy the condition rule which also satisfy the decision rule and leads to good prediction. The figure also shows that the value of certainty is 0.31 for delay of onset of monsoon. The schematic depicts that for the right time onset of monsoon over Kolkata, the strength, certainty and coverage factor are quite high. The values are 0.52, 0.62 and 0.76, respectively, while the number of deep depressions is low. This figure further shows that the value of coverage is 1.0 for right time onset of SW monsoon while the number of deep depressions is high, which means that 100% chances of right time onset of monsoon.

Figure 11.3 shows that for advance onset of SW monsoon over Kolkata, the higher value of the coverage factor is for the lower number of cyclonic storm during the pre-monsoon season and value is 1.0. This means that 100% of low numbers of cyclonic storms satisfy both the condition and the decision rules, while 60% leads to a good prediction for the onset of monsoon to be delayed over Kolkata. The figure further shows that for right time onset of monsoon, the value of the certainty factor is

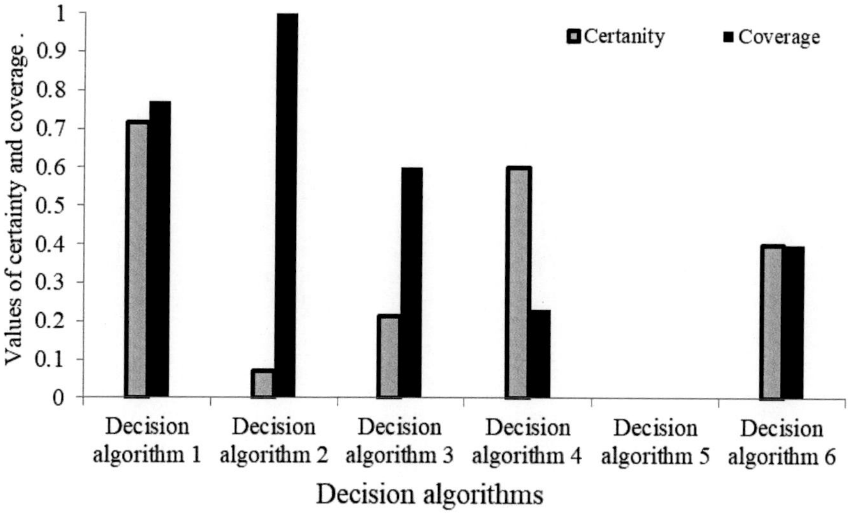

Fig. 11.3 The diagram shows variation in the certainty and coverage of the six decision algorithms for different numbers of cyclonic storm during the pre-monsoon months over the Bay of Bengal in evaluating the onset of monsoon

0.71, the coverage factor is 0.76 and strength is 0.52. It also shows that for a higher number of pre-monsoon cyclonic storms, the value of certainty factor is 0.6, which means 60% chance of right time onset of monsoon. In addition to that, the value of certainty and coverage factor is 0.4; that means a 40% possibility of delayed onset of SW monsoon over Kolkata while number of Cyclonic storms is high.

Figure 11.4 shows that for low number of SCS during the pre-monsoon months over Kolkata, the value of the coverage factor is maximum for the advance and delay of onset of monsoon and the value is 1.0. The figure further shows that for the right time onset of monsoon, the maximum value of the coverage factor is 0.92 and certainty factor is 0.6. This is also shown from the figure that when the number of SCS is high, the certainty factor gives highest value and the value is 1.0.

Figure 11.5 shows that for the right time onset of SW monsoon over Kolkata during the pre-monsoon season, the value of the certainty and coverage factor is 0.69, while the number of VSCS is low. This means that 69% of the total number of VSCS during the pre-monsoon months over Kolkata under study satisfy both the condition and the decision rules and can lead to a good prediction. The figure also shows that the highest value of coverage factor is identified for the advance onset of SW monsoon while the number of VSCS is low. The value of coverage factor is 0.6 for the delay of onset of monsoon. In addition to that, when the number of VSCS is high, the value of coverage factor for the right time onset of monsoon is 0.6, which means 60% chances of right time onset of monsoon may be possible. For delayed onset of monsoon, the value of coverage factor is 0.4, while the number of VSCS is high.

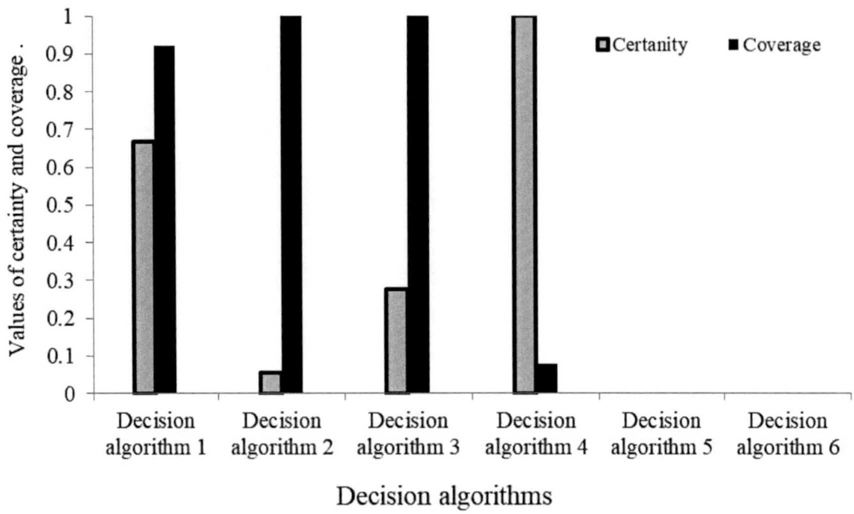

Fig. 11.4 The diagram shows the variation in the certainty and coverage of the six decision algorithms for different numbers of severe cyclonic storm during the pre-monsoon months over the Bay of Bengal in evaluating the onset of monsoon

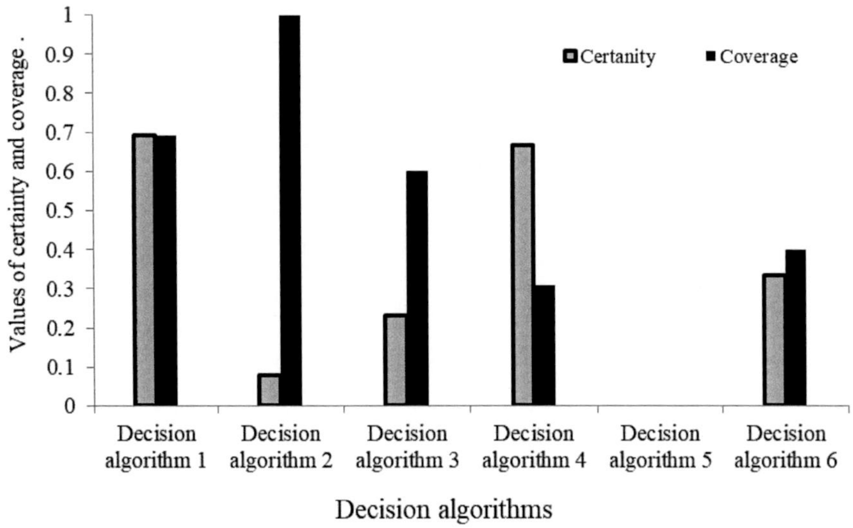

Fig. 11.5 The diagram shows the variation in the certainty and coverage of the six decision algorithms for different numbers of very severe cyclonic storm during the pre-monsoon months over the Bay of Bengal in evaluating the onset of monsoon

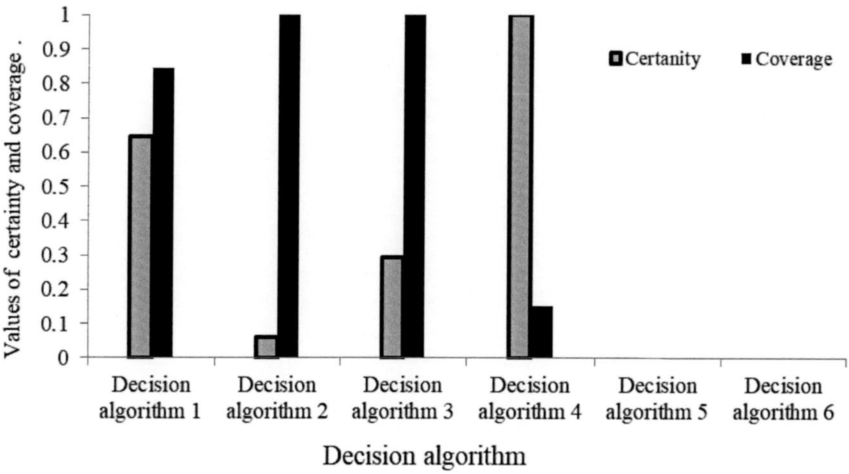

Fig. 11.6 The diagram shows variation in the certainty and coverage of the six decision algorithms for different numbers of super cyclonic storm during the pre-monsoon months over thevBay of Bengal in evaluating the onset of monsoon

Figure 11.6 depicts that while the number of pre-monsoon SuCS is low, the highest value of coverage factor is found for both advance and delay of SW monsoon over Kolkata and the value is 1.0. For the right time onset of monsoon, the values of strength, certainty and coverage factor are 0.5, 0.6 and 0.8, respectively, while the number of SuCS is low. The highest value of coverage factor is found for the right time onset of monsoon over Kolkata, when the number of SuCS is high. The value is 1.0. It can be stated that there is a specific range of number of pre monsoon cyclonic disturbances which satisfies the right time delay and advance onset of monsoon over Kolkata.

5 Validation

The validation set consists of 3 years, 2009, 2010 and 2011. It can be stated from Fig. 11.7 that the value of coverage factor is the highest in 2009 and 2010 for the low number of depression during the pre-monsoon months over Kolkata and the value is 1.0. It is confirmed that the onset of monsoon is delay for both years while the number of depression is low. For the year 2011, the value of coverage is 0.84; this means 84% chance of onset of monsoon at the right time.

Figure 11.8 depicts that for the low number of deep depression, the value of coverage factor is maximum. This also shows 100% possibilities of delay of monsoon during 2009 and 2010. In addition to that, 2011 shows similar chances of right time onset of SW monsoon over Kolkata. Considering the number of cyclonic storm, the years 2009 and 2010 shows 40% and 60% probability of delay of onset of monsoon, respectively.

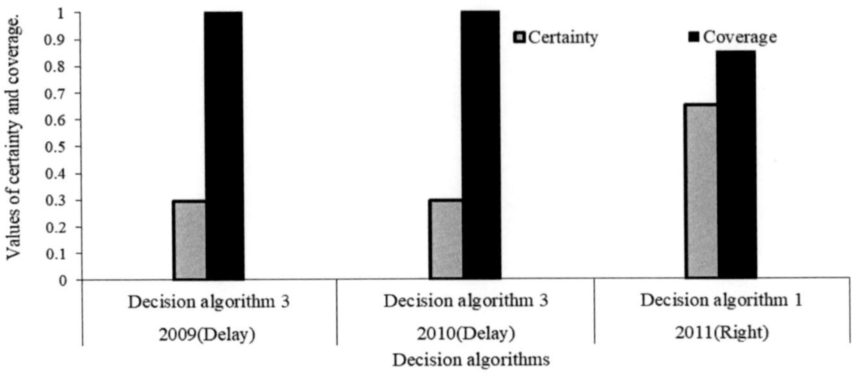

Fig. 11.7 Validation diagram showing the values of certainty and coverage during 2009, 2010 and 2011 for different decision algorithms regarding number of depressions over the Bay of Bengal

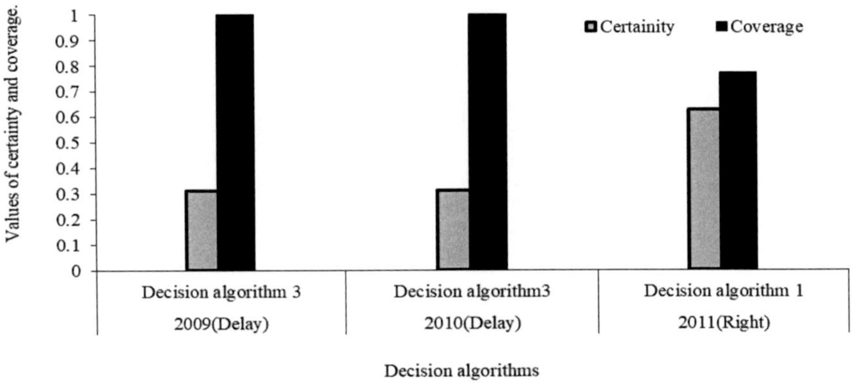

Fig. 11.8 Validation diagram showing the values of certainty and coverage during 2009, 2010 and 2011 for different decision algorithms regarding the number of deep depressions over the Bay of Bengal

On the other hand, the values of certainty and coverage are 0.71 and 0.76, respectively, indicating higher possibility of right time onset of monsoon over Kolkata (Fig. 11.9).

Figure 11.10 shows the regarding number of Severe cyclonic storm there is 92% chance of onset of monsoon at the right time in the year 2011. Both the values of certainty and coverage are high enough. But 2009 and 2010 give false alarm.

Figure 11.11 depicts the value of coverage factor is 0.6 for both the years 2009 and 2010 considering the number of very severe cyclonic storms. But for the year 2011, the values of certainty and coverage are both 0.7, which means 70% chance of onset of SW monsoon at the right time.

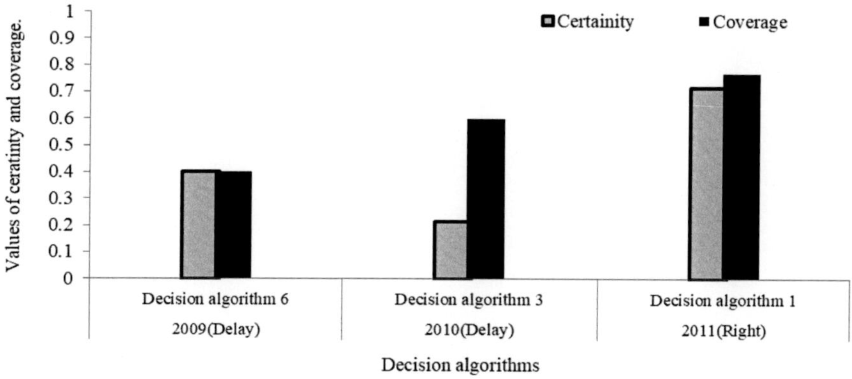

Fig. 11.9 Validation diagram showing the values of certainty and coverage during 2009, 2010 and 2011 for different decision algorithms regarding the number of cyclonic storms over the Bay of Bengal

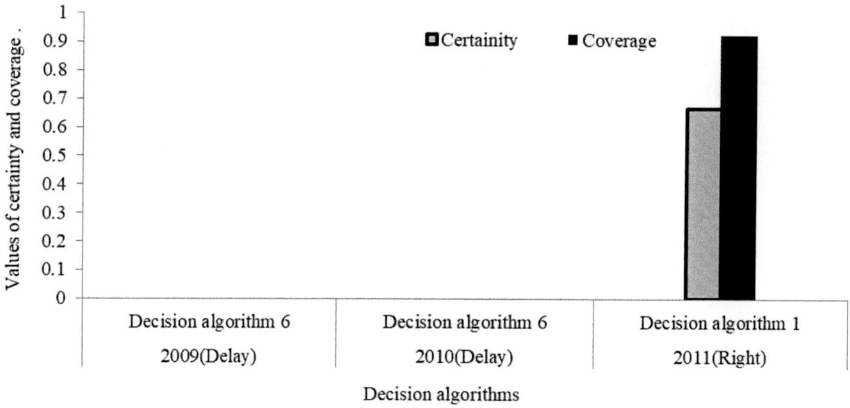

Fig. 11.10 Validation diagram showing the values of certainty and coverage during 2009, 2010 and 2011 for different decision algorithms regarding the number of severe cyclonic storms over the Bay of Bengal

It is revealed from Fig. 11.12 that, the value of coverage factor is 1.0. This means the highest possibility of delayed onset of SW monsoon over Kolkata. Similarly, for the year 2011, the values of certainty and coverage are 0.64 and 0.84, respectively, considering the number of super cyclonic storm. This also reflects a strong possibility of onset of monsoon at the right time.

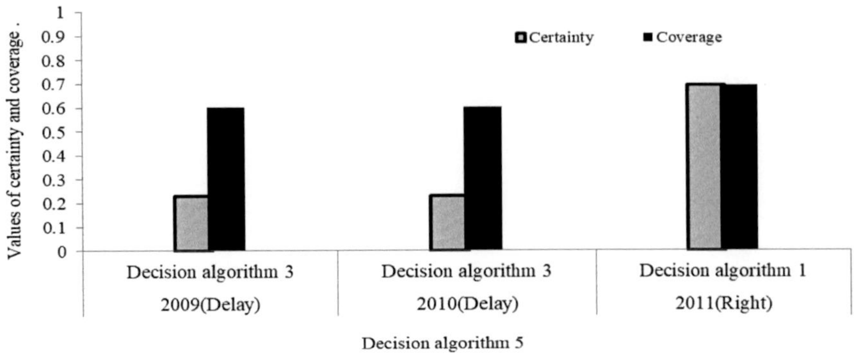

Fig. 11.11 Validation diagram showing the values of certainty and coverage during 2009, 2010 and 2011 for different decision algorithms regarding number of very severe cyclonic storms over the Bay of Bengal

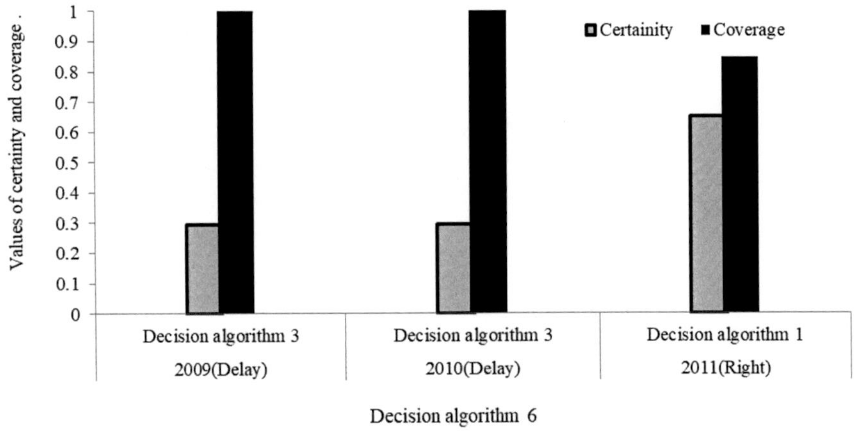

Fig. 11.12 Validation diagram showing the values of certainty and coverage during 2009, 2010 and 2011 for different decision algorithms regarding number of super cyclonic storm over the Bay of Bengal

6 Conclusion

The study leads to conclude that there exists a relationship between the number of pre monsoon cyclonic disturbances and the onset of southwest monsoon. It is depicted that the number of depression <1 leads to the onset of southwest monsoon at the right time (8th June, give or take 7 days) with 85% possibility, where as within the same range delay and advance of in the onset of southwest monsoon is 100% for both. But when the number of depression ≥ 1, there is 100% chance of right time onset of monsoon, which could be a good prediction. Similar result is obtained while considering the number of deep depression for the onset of right time, delay and advance onset of monsoon over Kolkata. It also reveals that the number of Cyclonic storm <1, it leads to the onset of southwest monsoon at advance with 100% possibility where as within the same range of cyclonic storm leads to the delay with 60% accuracy in the onset of southwest monsoon and leads to advance onset of monsoon over Kolkata with 75% chances. The number of Severe Cyclonic storm <1 ranges leads to the onset of southwest monsoon at right time, delay and advance onset of monsoon over Kolkata with high accuracy. Addition to that, while SCS ≥ 1, the onset of monsoon would be predicted with right time only. Similarly, a preferable range of the number of very severe cyclonic storms and super cyclonic storms leads to predicting the onset of southwest monsoon at the right time and delay and advance with moderately good results. The strength, certainty and coverage of decision support algorithm through the rough set theory endorse the above findings. It can be thus stated that the onset of SW monsoon is predicted with the help of a number of cyclonic disturbances.

References

Ananthakrishna, R., & Soman, M. K. (1988). The onset of the south west monsoon over Kerala 1901–1980. *Journal of Climatology, 8*, 282–296.

Ananthakrishna, R., & Soman, M. K. (1991). The onset of south west monsoon in 1990. *Current Science, 61*, 447–453.

Ananthakrishnan, R. V., Ramakrishnan, A. R., & Jambunathan, R. (1968). *Synoptic features associated with onset of southwest monsoon over Kerala* (IMD Forecasting Manual Rep, IV-IS.2).

Anathakrishnan, R., Pathan, J. M., & Aralikatti, S. S. (1983). The onset phase of the southwest monsoon. *Current Science, 52*(16), 755–764.

Bhardwaj, P., & Singh, O. (2020). Climatological characteristics of Bay of Bengal tropical cyclones: 1972–2017. *Theoretical and Applied Climatology, 139*, 615–629. https://doi.org/10.1007/s00704-019-02989-4

Bhullar, G. S. (1952). Onset of monsoon over Delhi. *Indian Journal of Meteorology and Geophysics, 3*, 25–30.

Chaudhuri, S. (2006). A hybrid model to estimate the depth of potential convective instability during severe thunderstorms. *Soft Computing – A Fusion of Foundations, Methodologies and Applications, 10*(8), 643–648.

Chaudhuri, S. (2008). Preferred type of cloud in the genesis of severe thunderstorms – A soft computing approach. *Atmospheric Research, 88*(2), 149–156.

Chaudhuri, S., Middey, A., Goswami, S., & Banerjee, S. (2012). Appraisal of the prevalence of severe tropical storms over Indian Ocean by screening the features of tropical depressions. *Natural Hazards*. https://doi.org/10.1007/s11069-011-0068-2

Fasullo, J., & Webster, P. J. (2003). A hydrological definition of Indian monsoon onset and withdrawal. *Journal of Climate, 16*, 3200–3211.

Flatau, M. K., Flatau, P. J., & Rudnick, D. (2001). The dynamics of double monsoon onsets. *Journal of Climate, 14*, 4130–4146.

Flatau, M. K., Flatau, P. J., Schmidt, J., & Kiladis, G. N. (2003). Delayed onset of the 2002 Indian monsoon. *Geophysical Research Letters, 30*, 1768.

Gadgil, S. (2003). The Indian monsoon and its variability. *Annual Review of Earth and Planetary Sciences, 31*, 429–467.

Goswami, I. P., & Gouda, K. C. (2010). Evaluation of a dynamical Bais for advance forecasting of the date of onset of monsoon rainfall over India. *Monthly Weather Review, 138*, 3120–3141.
Goswami, P., & Patra, G. K. (2004). Characteristics scale of convective organization and monsoon intensity. *Geophysical Research Letters., 31*, L24109. https://doi.org/10.1029/2004GL020268
Hsu, H. H., Terng, C. T., & Chen, C. T. (1999). Evolution of largescale circulation and heating during the first transition of Asian summer monsoon. *Journal of Climate, 12*, 793–810.
Joseph, P. V., Eisheid, J., & Pyle, R. J. (1994). Interannual variability of the onset of the Indian summer monsoon and its association with atmospheric features, EINino, and SST anomalies. *Journal of Climate, 7*, 81–105.
Krishnamurti, T. N., & Ramnathan, Y. (1982). Sensitivity of the monsoon onset to differential heating. *Journal of the Atmospheric Sciences, 39*, 1290–1306.
Krishnamurti, T. N., Ardanuy, P., Ramanathan, Y., & Pasch, R. (1981). On the onset vortex of the summer monsoon. *Monthly Weather Review, 109*, 344–363.
Lau, K. M., Wu, H. T., & Yang, S. (1998). Hydrologic processes associated with the first transition of the Asian summer monsoon: A pilot satellite study. *Bulletin of the American Meteorological Society, 79*, 1871–1882.
Murakami, T., & Ding, Y. (1982). Wind and temperature changes over Eurasia during the early summer of 1979. *Journal of the Meteorological Society of Japan, 60*, 183–196.
Murakami, T., & Nakazawa, T. (1985). Transition from the Southern to Northern Hemisphere summer monsoon. *Monthly Weather Review, 113*, 1470–1486.
Pai, D. S., & Nair, R. M. (2009). South west monsoon onset over Kerala: New definition and prediction. *Journal of Earth System Science, 118*(2), 123–135.
Pawlak, Z., & Skowron. (2007). Rough set and Boolean reasoning. *Information Science, 177*(1), 41–73.
Rao, Y. P. (1976). *Southwest monsoon: Synoptic meteorology* (Meteorological Monographs, No. 1/1976) (p. 367). India Meteorological Department.
Sikka, D. R., & Gadgil, S. (1980). On the maximum cloud zone and the ITCZ over Indian longitudes during the southwest monsoon. *Monthly Weather Review, 108*, 1840–1853.
Webster, P. J., Magana, V. O., Palmer, T. N., Shukla, J., Tomas, R. A., Yanai, M., & Yasunari, T. (1998). Monsoons: Processes, predictability, and the prospects for prediction. *Journal of Geophysical Research, 103*, 14,451–14,510.
Wu, G. X., & Zhang, Y. S. (1998). Tibetan Plateau forcing and the timing of the monsoon onset over South Asia and the South China Sea. *Monthly Weather Review, 126*, 913–927.
Yin, M. T. (1949). A synoptic-aerologic study of the onset of the summer monsoon over India and Burma. *Journal of Meteorology, 6*, 364–400.
Zadeh, L. (1965). Fuzzy sets. *Information and Control, 8*, 338–353.
Zhao, Q., Cook, J., & Harasti, P. R. (2008). Improving short term predictions by assimilating both radar radial-wind and reflectivity observations. *Weather Forecasting, 23*(3), 373–391.

Chapter 12
Natural Disaster Recovery for Flood-Affected Areas Using YOLOv7 Object Detection and Instance Segmentation

Shrimanta Satpati and Jayati Bhadra

Abstract Natural disasters are uncalled for and timely response is necessary to alleviate large damages. There are now exponentially large multimodal datasets (such as images, videos, blogs, and tweets) available in the public domain on social networking sites and online content-sharing platforms for data usability. By including local groups in data gathering and information exchange during natural disasters, it may be possible to establish data-driven approaches to damage assessment and recovery planning as well as a better understanding of flood vulnerabilities, capacities, and risks. We used state-of-the-art real-time YOLOv7 object detection for instantaneous human (and vehicles) recognition and tracking based on video content analysis of feeds. The YOLOv7 object detection pretrained model, efficient for localization and recognition in the COCO Human dataset (for training) was tested. We also custom-trained the YOLOv7 instance segmentation model in segmenting flooded areas for testing damage assessment in real time. It was trained on the River Flooding Detection System dataset, part of a flooding alert system research published in 2022, supported by the São Paulo Research Foundation (FAPESP). With the dataset, the YOLOv7 algorithm was trained and tested with a transfer learning approach on Google Colab's virtual machine. Experimental results revealed that the YOLOv7 model achieved the Precision, Recall, and mAP@0.50 values for the bounding boxes and masks of vehicles as 0.9985, 1, and 0.995, respectively.

S. Satpati (✉)
M.Sc. Big Data Analytics, St. Joseph's University, Bangalore, India

J. Bhadra
Department of Advanced Computing, St. Joseph's University, Bangalore, India
e-mail: jayatibhadra@sjc.ac.in

S. Hazra et al. (eds.), *Geographies of the Indian Subcontinent*,
https://doi.org/10.1007/978-3-031-74813-4_12

1 Introduction

Natural disasters pose a significant hazard, yet we are able to develop answers to a wide range of challenges in this new period where technical advancements are occurring quickly. One of the main catastrophes that can occur anywhere in the world is flooding. Recent years have seen a relative increase in rainfall that is both heavy and quick.

Due to a strong downpour, a rise in water levels causes land regions, particularly low-lying ones, to be submerged. Losses of life and property result from it. The government has contributed to the flood victims' housing needs. Additionally, researchers have been looking for efficient ways to repair the harm. Real-time human identification can help assess the disaster areas for disaster management dissemination. Rescue efforts and basic assistance for victims must first reach the highly waterlogged area. Here is where techniques for identifying flood-affected areas become crucial. Immediate segmentation of flooded areas in real-time can save lives and reduce damage.

Real-time object detection is a crucial task in computer vision and is frequently a crucial part of computer vision systems. Video analytics, robotics, autonomous vehicles, multi-object tracking and object counting, medical image analysis, and other applications use real-time object detection models. When performing image recognition tasks, an object detector algorithm uses an input image to forecast bounding boxes and class probabilities for each object in the image. The majority of methods extract features from the image to estimate the likelihood of the learnt classes using a convolutional neural network (CNN).

YOLO in computer vision stands for "You Only Look Once"; it is a popular family of real-time object detection algorithms (Fig. 12.1). In 2016, the initial version of the YOLO object detector was made available. Santosh Divvala, Ali Farhadi, and Joseph Redmon were the architects of it. This architecture was state-of-the-art for real-time computer vision applications when it was released because it was faster than other object detectors. Since then, other YOLO versions and

Fig. 12.1 YOLOv7 object detection in a dense scene

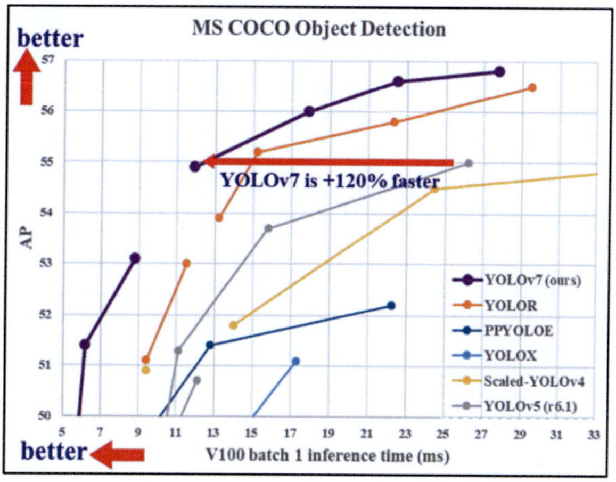

Fig. 12.2 Compared to comparable real-time object detection models, the evaluation of YOLOv7 models shows faster inferences (x-axis) and greater accuracy (y-axis)

variations have been put forth, each of which offers a notable improvement in effectiveness and performance (Fig. 12.2). YOLOv1 through the well-known YOLOv3 were developed by Joseph Redmon, a doctoral student at the time, and his advisor Ali Farhadi. Alexey Bochkovskiy, who carried on Redmon's legacy when he halted his computer vision research owing to ethical considerations, introduced YOLOv4. The most recent version of the YOLO architecture, designated as YOLOv7, was developed by its original creators. A popular example is YOLOv5 which was created by the company Ultralytics for instance segmentation. It is comparable to YOLOv4, however PyTorch is used in place of DarkNet.

There are five sections in this chapter. In Sect. 2, a few current and related research articles are discussed. The methodology was explained in Sect. 3. The experimental data and their analysis are explained in Sect. 4. Finally, Sect. 5 has the conclusion.

2 Literature Reviews

Deep learning-based approaches are used for instance segmentation on flood images. Here, a few recent studies that are pertinent are briefly discussed.

Chatterjee et al. (2022) investigated variants of Detectron2 instance segmentation architectures to detect a water body and compute the changes in its area from the time-lapsed images of water bodies taken by the Sentinel-2 Satellite and compared based on the average precision (AP) for AdaBelief optimizer, 99.95 and 94.51 at AP50 and AP75 metrics, respectively. In addition, they developed an ensemble approach for the efficient identification of shrinkage or expansion of water bodies. Goodchild and Glennon (2010) investigated crowdsourcing geographic disaster information collection to the public in order to facilitate the information flow. Recent studies have focused on the difficulty of identifying ground objects in overhead drone imagery. Guirado et al. (2018) employed Faster R-CNN to track and count the number of whales from satellite images. Radovic et al. (2017) used transfer learning based on the you-only-look-once (YOLO) algorithm to detect airplanes on the ground from aerial views. Han et al. (2012) conducted real-time object detection in drone imagery using region-based CNN (R-CNN) and kernelized correlation filter (KCF). Yildirim et al. (2022) did automated vehicle detection and instance segmentation from high-resolution UAV imagery using YOLOv7 model. This methodology was used in our research. Geetha et al. (2017) used crowd-sourced images to detect the extent of flood areas using a color-based segmentation threshold. Rahnemoonfar et al. (2018) used an integration of densely connected CNN and RNN networks, which is able to accurately segment out semantically meaningful object boundaries with end-to-end learning. The proposed network was applied on UAV aerial images of flooded areas in Houston, TX. We achieved 96% accuracy in detecting flooded areas on a large UAV dataset. Rahnemoonfar et al. (2020) presented a high-resolution UAV imagery dataset, FloodNet, captured after the hurricane Harvey. In this chapter, they compared and contrasted the performances of baseline methods for image classification, semantic segmentation, and visual question answering on their dataset. Salluri et al. (2020) used the pretrained CNN models like ResNet50, VGG-16 and VGG-19 on the DISASTER dataset for object detection of flood and earthquake images. VGG-19 has the highest accuracy of the three pre-trained models, at 94.22%.

This chapter implements both instance segmentation of static flood images and video flood segmentation on test data. *There is a fundamental difference between semantic and instance segmentation*. Suppose, an image containing three cats in it. The semantic segmentation segments the image into two—one segment containing all the cats, and the remaining portion has been segmented as background. On the other hand, the instance segmentation without the background separates the three cats into their own segments.

The novelty of this chapter is to segment and detected the flooded area (single object) captured from the satellite/aerial high-resolution color images or high-resolution videos with high accuracy using the latest state-of-the-art most advanced real-time object identification model YOLOv7.

3 Proposed Methodology

The most recent member of the YOLO (You Only Look Once) family of models is the v7 version. YOLO models are single-stage object detectors. In a YOLO model, visual frames are enhanced by a backbone. YOLO first divides the image into *N* grids before doing object detection in a single step (Fig. 12.3). These grids are of the same **S × S** size. These characteristics are merged and mixed in the neck, after which they are transmitted to the network's head. It is utilized to find and locate any objects (different types too) that might be present in any of these areas. Bounding

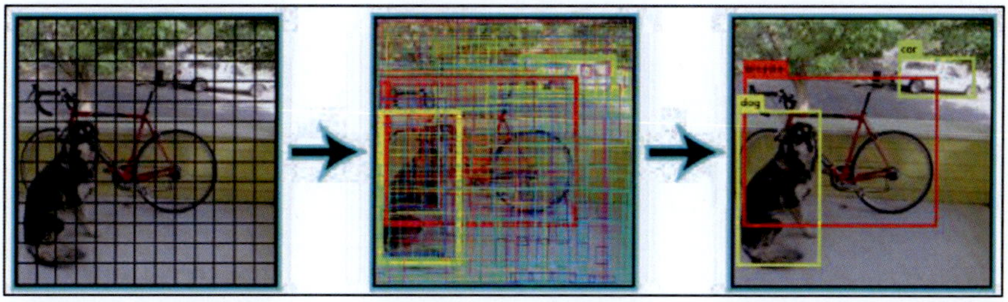

Fig. 12.3 YOLO object detection model

box coordinates, B, for any prospective objects are predicted for each grid along with their object labels and a likelihood rating for their presence.

This results in a considerable overlap of expected objects from the grids' cumulative predictions, as one may have guessed. The YOLO algorithm employs postprocessing using Non-Maximal Suppression to suppress all the bounding boxes with significantly lower probability scores in order to handle this redundancy and narrow the predicted objects down to those of interest.

To do this, YOLO examines the probability scores attached to each decision and selects the one with the highest score. The bounding boxes having the biggest Intersection over Union with the selected high probability bounding box are then removed. Then, this process is repeated until the only bounding boxes left are the ones needed.

For YOLOv7, some new adjustments were introduced. These modifications demonstrate how they contributed to the dramatic gain in performance in YOLOv7 over earlier iterations.

A number of architectural improvements were made in YOLOv7, the most recent detector in the YOLO series, to increase the model's speed and detection accuracy. On its backbone, an expanded efficient layer aggregation network (E-ELAN) based on expanding, shuffling, and merging cardinality was proposed. While maintaining the initial gradient path, this network continuously enhances its learning capabilities. Additionally, a compound model scaling strategy was used in order to preserve the properties the model had in its original design and maintain an ideal structure. Additionally, the reparametrized convolution was suggested in YOLOv7, called the Trainable bag of freebies. Soft labels and label designers were introduced to enhance training. This method is important because it allows for the dynamic adjustment of fine and coarse labels during training. It also has a coarse for the auxiliary heads and a fine for the lead loss head. In the YOLOv7 architecture, the head responsible for the final output is called the lead head, and the head used to assist in training is called the auxiliary head. To create coarse-to-fine hierarchical labels for auxiliary head and lead head learning, respectively, YOLOv7 employs the lead head prediction as guidance (Fig. 12.4).

In addition to the aforementioned, YOLO now has brand-new features that weren't present in earlier iterations (Fig. 12.5). The first one allows YOLO to

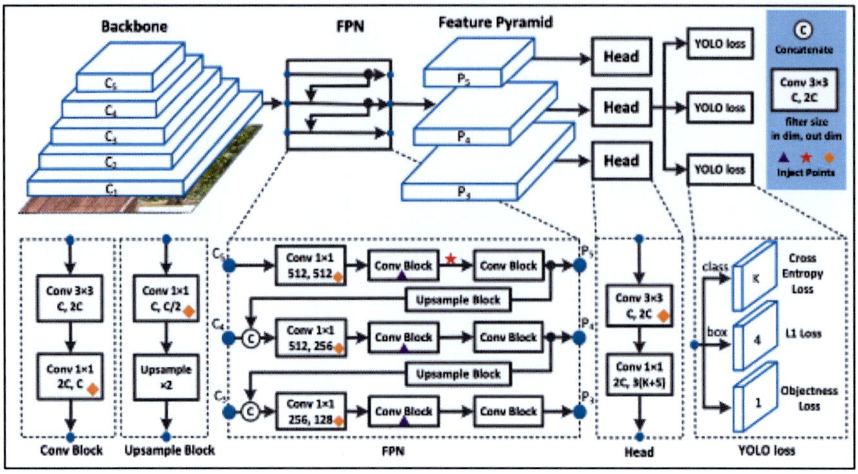

Fig. 12.4 YOLO network architecture as depicted in PP-YOLO

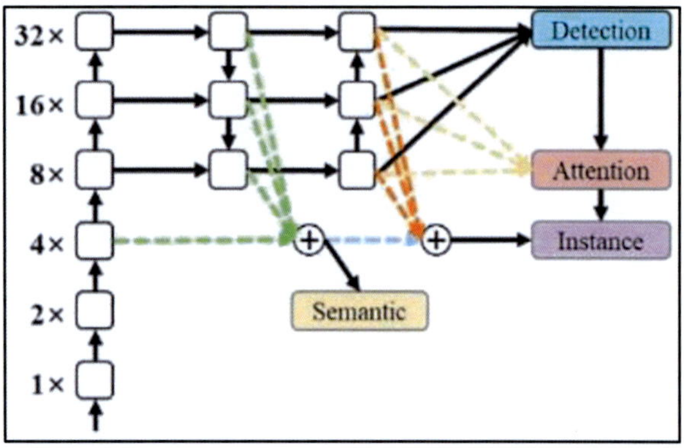

Fig. 12.5 The architecture for image segmentation with YOLOv7-mask

segment objects pixel-by-pixel and is called instance segmentation. Pose estimate, which is the second technique, is useful for comprehending bodily motion.

3.1 Objects Detection Using YOLOv7 Pretrained Model

In the range of 5 FPS to 160 FPS, YOLOv7 outperforms all other known object detectors in terms of speed and accuracy, and on GPU V100, it has the greatest accuracy of 56.8% AP of all real-time object detectors with 30 FPS or more. YOLOv7 outdoes: YOLOR, YOLOX, Scaled-YOLOv4, YOLOv5, DETR, Deformable DETR, DINO-5scale-R50, ViT-Adapter-B, and many other object detectors in speed and accuracy (Fig. 12.6). We use the pretrained YOLOv7 model trained on MS COCO dataset from scratch without using any other datasets or pre-trained

MS COCO

Model	Test Size	AP^{test}	$AP_{50}{}^{test}$	$AP_{75}{}^{test}$	batch 1 fps	batch 32 average time
YOLOv7	640	51.4%	69.7%	55.9%	161 *fps*	2.8 *ms*
YOLOv7-X	640	53.1%	71.2%	57.8%	114 *fps*	4.3 *ms*
YOLOv7-W6	1280	54.9%	72.6%	60.1%	84 *fps*	7.6 *ms*
YOLOv7-E6	1280	56.0%	73.5%	61.2%	56 *fps*	12.3 *ms*
YOLOv7-D6	1280	56.6%	74.0%	61.8%	44 *fps*	15.0 *ms*
YOLOv7-E6E	1280	56.8%	74.4%	62.1%	36 *fps*	18.7 *ms*

Fig. 12.6 Performance of YOLOv7 on MS COCO dataset for object detection

Fig. 12.7 Objects detection in drone footage of floods in Zhengzhou (tested on YouTube video)

weights. This was done to identify the humans and vehicles (in the coco.yaml file) filtered by the specific classes for tracking and counting in images or each frame of a video. This is useful for analyzingUAV feeds in real-time (Fig. 12.7).

3.2 Instance Segmentation Using YOLOv7 on Custom Data

3.2.1 Dataset and Implementation

This implementation is based on YOLOv5, but it uses the **YOLOv7-seg.pt, YOLOv7x-seg.pt, YOLOv7-seg-dev.pt** pretrained weights files, generated after training on MS COCO dataset for instance segmentation. All of the installation, data preparation, and usage are as same as YOLOv5. It was trained on the **Floodwall dataset [1][2 – see sample images]** (a River Flooding Detection System dataset), part of a flooding alert system research published in 2022, supported by **São Paulo**

Research Foundation (FAPESP). The dataset was labelled by the researchers of Floodwall using LabelMe annotation software.

4 Experimental Results and Its Analysis

The YOLOv7 instance segmentation model was custom trained on the Floodwall dataset which contains 999 annotated images along with their corresponding masks (the annotations are in COCO json format for YOLOv7 instance segmentation). The dataset was uploaded in Roboflow to get the zipped folder of training set (904 images), validation set (75 images), and testing set (20 images) folders for training. The YOLOv7 instance segmentation model was trained and tested with a transfer learning approach in the cloud-based Google Colaboratory on Python 3 Google Compute Engine backend (GPU), the NVIDIA Tesla T4 GPU for faster training and less time for running epochs. In the study, the transfer learning approach was adopted because a relatively medium data set was used. That is, instead of training the model end-to-end from scratch, the training process of the YOLOv7 model was initialized from the pretrained weights obtained on the COCO dataset. The hyperparameters utilized in training phase of the model are given in Table 12.1.

4.1 *YOLOv7-seg.pt Pretrained Weights*

YOLOv7-seg.pt. as the pre-trained weights file was used for our custom instance segmentation training.

The command we use:

```
!python /content/YOLOv7-segmentation/segment/train.py --data /
content/YOLOv7-segmentation/data/Flooded-1/data.yaml --batch 15 --
```

Table 12.1 Experimental hyperparameter configuration

Hyperparameters	Values	Meanings
lr0	0.01	Initial learning rate (SGD = 1E-2, Adam = 1E-3)
lrf	0.1	Final OneCycleLR learning rate (lr0 * lrf)
Momentum	0.937	SGD momentum/Adam beta1
Weight_decay	0.0005	Optimizer weight decay 5e-4
Box	0.05	Box loss gain
cls	0.3	cls loss gain
obj	0.7	obj loss gain (scale with pixels)
Image size	640 × 640	The dimensions of the images in the dataset
Epochs	40	The number of iterations for training the batches
Batch size	15	The number of images trained in each batch

```
weights /content/YOLOv7-segmentation/YOLOv7-seg.pt --cfg /content/
YOLOv7-segmentation/models/segment/YOLOv7-seg.yaml --epochs 40 --
name YOLOv7-seg --img 640 --hyp /content/YOLOv7-segmentation/data/
hyps/hyp.scratch-high.yaml
```

The data.yaml has paths to training, validation, and testing data along with the class name "Flooded area," which were instance segmenting in the image. The dataset is divided into 15 batches for faster training, with 40 epochs for better feature learning. The image dimensions are 640 × 640. The **hyperparameters for high-augmentation COCO training from scratch are initialized in** hyp.scratch-high. yaml. The oldest optimizer is used, the stochastic gradient descent (SGD) optimizer, which does not use momentum while determining the update weights instead of the Adam optimizer. SGD (lr = 0.01) is used.

4.2 Results and Inferences

The YOLOv7 training and validation approach generated four different types of losses: the classification loss, bounding box loss, segmentation loss, and objectiveness loss. Figure 12.8 demonstrates that the model did not show any overfitting during training and that every loss value had a decreasing trend. After 40 training rounds, the minimum value in the training and validation loss curves was reached.

YOLOv7 loss are: The predicted objectness probability and the complete intersection over union (CIoU) with the matched target as ground truth are used to calculate the objectness loss score. The value is 0 if there are no matches. The box (or regression) loss, is calculated as the **mean (1 – CIoU)** between all candidate anchor boxes and their matched target. The classification loss, using Binary Cross Entropy Loss between the predicted class probabilities for each anchor box and a one-hot encoded vector of the true class of the matched target. The segmentation loss is given by train/seg_loss. Even with just training in the 40 epochs completed in 0.641 h, the model demonstrated higher performance in detection and segmentation tests. The validation losses are: val/box_loss = **0.0024499**, val/seg_loss = **0.0011653**, val/obj_loss = **0.0015382**, val/cls_loss = **0**.

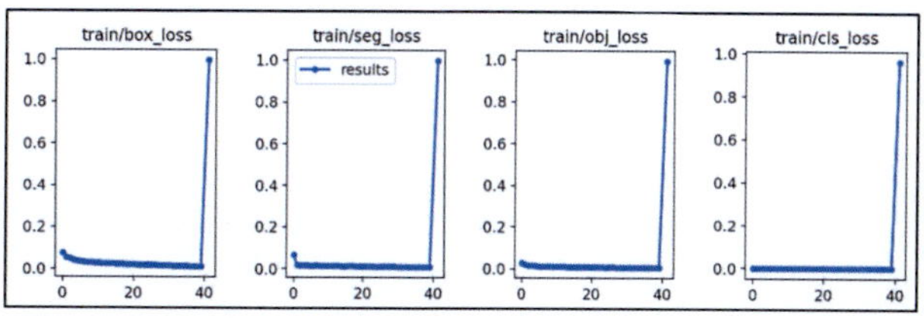

YOLOv7 losses after custom training	**Converged (at 40th epoch)**
train/box_loss	0.012134
train/seg_loss	0.0091952
train/obj_loss	0.0061019
train/cls_loss	0

Fig. 12.8 Training loss graphs and the YOLOv7 loss values table

Obtaining the average precision (mAP) across several classes is done by finding the average precision (AP) for each class. False positives (FP) and false negatives (FN) are both taken into account by the mAP, which takes into account the trade-off between precision and recall (FN). The majority of detecting applications can use mAP because of this characteristic.

Formulas

$$\text{Intersection over Union(IoU)} = \frac{\text{Intersection Area}}{\text{Union Area}}$$

$$\text{Recall} = \frac{\text{TP}}{\text{TP} + \text{FN}}$$

$$\text{Precision} = \frac{\text{TP}}{\text{TP} + \text{FP}}$$

$$\text{mAP} = \frac{1}{N}\sum\nolimits_{i=1}^{N} \text{APi}$$

Precision, recall, and mean average precision (mAP), measured at the intersection over union (IoU) threshold value of 0.50, were used as accuracy measures or metric values to evaluate the performance of the trained model.

The model's precision, recall, and mAP@0.50 values for the bounding boxes and masks of the flooded area in the **Floodwall dataset** were, respectively, **0.9985, 1, and 0.995** (Fig. 12.9).

To more thoroughly examine the model's performance, Fig. 12.10 includes the visual detection and segmentation results of the model on the testing dataset in addition to the accuracy assessment and computation time evaluation. It was found that the YOLOv7 model accurately suited the object boundaries and correctly forecasted the bounding boxes and segmentation masks of the flooded areas. The model also did well at segmenting and recognizing flooded areas with various backgrounds and other objects. Additionally, it was noted that the model performed remarkably well in identifying flooded areas of various sizes and orientations. With a confidence score of nearly 90%, the model correctly identified every location of flooding in the images in the testing data of Floodwall dataset (Fig. 12.10).

5 Conclusion

In this study, the cutting-edge deep learning-based YOLOv7 was used to automatically recognize flooded areas and segment instances on the **Floodwall dataset**. Additionally, using YOLOv7's instance segmentation capability, pixel-wise masks of each detected flooded region were obtained. The experimental findings revealed

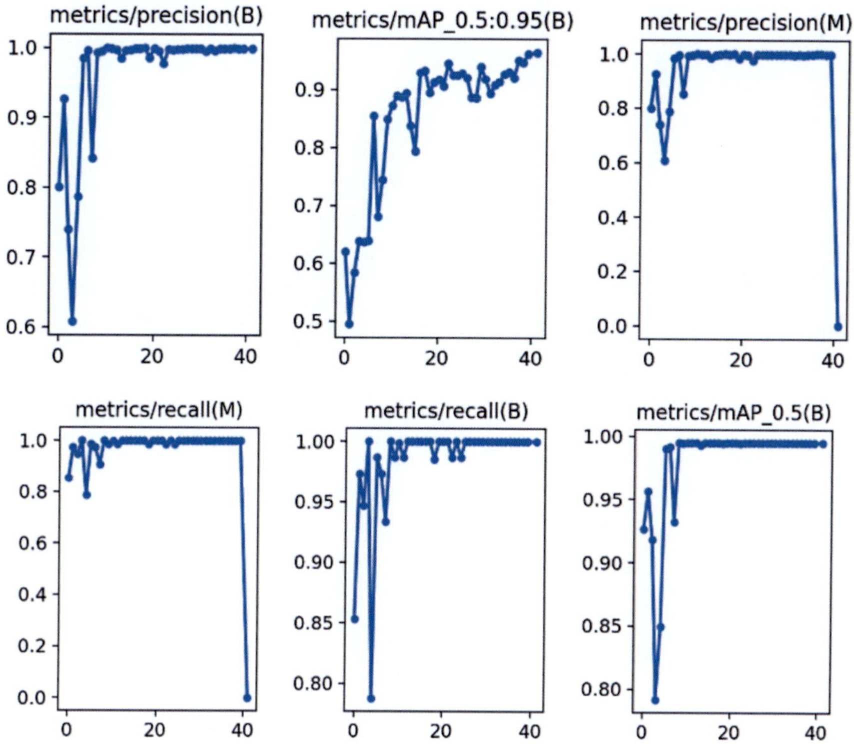

Fig. 12.9 Performance evaluation of the YOLOv7 model

Fig. 12.10 Inferencing on testing data of the Floodwall dataset

that the model had a mAP of roughly 99% for properly identifying flooded areas on the test data of Floodwall dataset. According to the study's findings, YOLOv7 successfully identified flooded areas from test drone footage with a good level of accuracy and speed (Fig. 12.11). On the new data taken from the internet, the trained YOLOv7 is working fine with a good probability of the bounding boxes of instance segmentation. More images for training and running for more epochs for training can

Fig. 12.11 Testing on new data from the internet

give better results than what is got in this research. The following model can be deployed on Streamlit or other Python-based frameworks to give real-time flooded area instance segmentation results.

References

Chatterjee, R., Chatterjee, A., & Islam, S. K. (2022). Deep learning techniques for observing the impact of the global warming from satellite images of water-bodies. *Multimedia Tools and Applications, 81*, 6115. https://doi.org/10.1007/s11042-021-11811-1

Goodchild, M. F., & Glennon, J. A. (2010). Crowdsourcing geographic information for disaster response: A research frontier. *International Journal of Digital Earth, 3*(3), 231–241.

Guirado, E., Tabik, S., Rivas, M. L., Alcaraz-Segura, D., & Herrera, F. (2018). Automatic whale counting in satellite images with deep learning. *bioRxiv*, 443671.

Geetha, M., et al. (2017). *Detection and estimation of the extent of flood from crowd sourced images*. International Conference on Communication and Signal Processing (ICCSP), IEEE.

Radovic, M., Adarkwa, O., & Wang, Q. (2017). Object recognition in aerial images using convolutional neural networks. *Journal of Imaging, 3*(2), 21.

Han, S., Shen, W., & Liu, Z. (2012). *Deep drone: Object detection and tracking for smart drones on embedded system*. Stanford University.

Yildirim, E., Sefercik, U. G., & Kavzoglu, T. (2022). *Automated vehicle detection and instance segmentation from high-resolution UAV imagery using YOLOv7 model*. 5th Intercontinental Geoinformation Days (IGD), Netra.

Salluri, D. K., Bade, K., & Madala, G. (2020). Object detection using convolutional neural networks for natural disaster recovery. *International Journal of Safety and Security Engineering, 10*(2), 285–291.

Rahnemoonfar, M., Murphy, R., Miquel, M. V., Dobbs, D., & Adams, A. (2018). *Flooded area detection from UAV images based on densely connected recurrent neural networks*. Proceedings of the IEEE International Geoscience and Remote Sensing Symposium (IGARSS), pp. 1788–1791.

Rahnemoonfar, M., Maryam, et al. (2020). FloodNet: A high-resolution aerial imagery dataset for post flood scene understanding. *IEEE Access, 9*, 89644–89654.

Chapter 13
Rainfall Variability and Irrigation Scheduling for Paddy Cultivation in the Kankai Irrigation System, Eastern Nepal

Chandrayee Chakraborty and Sukla Hazra

Abstract Rainfall variability has a great impact on agricultural production and irrigation water requirements in the monsoon countries. For agricultural development as well as for planned river water management, an irrigation system was developed in the Eastern Terai of Nepal known as Kankai Irrigation System (KIS) in the lower part of the river Kankai. In the study, irrigation water requirement in different growing seasons for main crop paddy is calculated with the help of the Cropwat (8.0) model using climate data, crop coefficient and soil characteristics. The FAO (Food and Agricultural Organisation) CROPWAT 8.0 Simulation software and the CLIMWAT 2.0 tool are used to calculate the crop water requirement, irrigation water requirement and irrigation schedules for the crop paddy for the monsoon and spring season. Canal water demand and supply, irrigation schedules and planning depend on the inter-seasonal and intra-seasonal rainfall variability. The Penman-Monteith method is used to estimate the reference evapotranspiration and crop evapotranspiration (ET_0) and the US Department of Agriculture (USDA) soil conservation (SC) method is used to find out the effective rainfall. The study results show that the ET_0 varied from 3.12 to 7.99 mm (average 4.9 mm) and the effective rainfall varied from 3.6 to 167.8 mm. The net irrigation requirements are 276.2 and 980.5 mm for paddy in monsoon and spring rice, respectively. There is a higher demand for irrigation in winter (Rabi) and spring season (Zaid) and lower demand during the wet season (Kharif). Correlation between rainfall variability and irrigation requirement is highly positive. The value for correlation coefficient (r^2) is 0.98, slope 7.93 and the intercept 15.

C. Chakraborty (✉)
Department of Geography, East Calcutta Girls' College, West Bengal State University, Barasat, India

S. Hazra
East Calcutta Girls' College, West Bengal State University, Barasat, India

S. Hazra et al. (eds.), *Geographies of the Indian Subcontinent*,
https://doi.org/10.1007/978-3-031-74813-4_13

1 Introduction

Irrigation systems are a major component of modern agricultural land management, helping to boost crop output and quality (Chougule & Mashalkar, 2022). However, irrigation is no longer only used to manage agricultural water; it is also used most successfully to manage water resources in a range of geomorphological situations worldwide (Smith et al., 2009). Irrigation systems, except rainfed crops, basically include intentionally supplying water to encourage crop development (both in terms of quality and quantity) (Haseeb et al., 2023). The rapid increase in water consumption puts stress on the water resources.

To examine extreme events, agricultural systems and effective water management, particularly for any river basin, a region's rainfall and temperature, or its meteorological characteristics, are required. In the rainfed farming system, availability of precipitation data is crucial for planting, cultivation, watering, and other agricultural activities. Good and effective rainfall is crucial for hydrological analysis in many anthropogenic systems and the management of water resources (Jamaludin & Suhaimi, 2013). In this context, meteorological study is highly needed using different climatic indices. Rainfall variability (coefficient of variation) for the years 1981–2021 is calculated and the Cropwat 8.0 model is used here to determine the crop water and irrigation requirements. Cropwat is a decision support programming system developed by the Land and Water Development Division of FAO (1990) for irrigation management. Finally, a correlation is done between the variability of rainfall and irrigation requirement as irrigation scheduling depends on both of these to increase the irrigation efficiency.

In summer, the weather and climate of the Indian sub-continent are significantly influenced by the monsoon, and agriculture is governed by monsoon rainfall. The variability of monsoon rainfall is highly unpredictable. During the lean season, intra-seasonal and inter-seasonal variability has an impact on the growth of the irrigation system and water management. Irrigation scheduling of any region is controlled by climatic condition. Rainfall distribution and irrigation supply both are necessary for crop yield. Therefore, climatic parameters like rainfall temperature are needed to be measured for this study. To study rainfall, Nery et al. (2017) examined the concentration indicators. The concentration index and the proportion of precipitation on rainy days were associated with 10%, 20%, 30%, and 50% of each rainfall station. Ahmad et al. (2018) used Sen's slope estimator, the Mann-Kendall test, and Innovative Trend Analysis (ITA) to examine the seasonal and yearly precipitation variability at the upper Indus Basin. Based on the percentile scale, they determined the seasonal range of precipitation intensities, ranging from mild to heavy, and they examined the flood situation. In Chile, rainfall concentration and its fluctuation were analyzed both spatially and temporally between 1970 and 2016, according to Sangüesa et al. (2018). These indicators were computed for two distinct climatic types: humid and sub-humid and dry and semi-arid. The Gini Index and PCI were computed, and the Mann-Kendall test was used to analyze trends at a 5% significance level. The daily, monthly and maximum amounts of precipitation in a given day were considered. The mean yearly rainfall, PCI and mean Gini index coefficient were all compared.

Irrigation is regarded as the fundamental infrastructure and the most important input for agricultural productivity because it acts as a substitute for rainfall. Evapotranspiration makes up the bulk of the water lost during irrigation, which utilizes massive volumes of water in agriculture (Barrow, 2016). Strategic water management and the application of suitable irrigation technologies can improve agriculture (Adeoti & Sinh, 2009). Enhancing the efficiency of water usage on irrigated land leads to a better utilization of many agricultural inputs, including manpower, agricultural machinery, high-yielding crops, fertilizers, and tillage (Sharma et al., 2015).

Previous studies show the importance of the calculation of the crop water demand for agricultural development in different fields. The CROPWAT is a computer program used to estimate the crop water demand and the irrigation requirement using the soil, climate data, and crop data of an area to assess the water use efficiency (Paredes et al., 2017). In the paper Comparative Study of Monsoon Rainfall Variability over India and the Odisha State, Gouda et al. (2017) discussed the monsoon rainfall variability using the daily gridded rainfall data from India Meteorological Department (IMD). In the article "Monsoon variability, crop water requirement, and crop planning for Kharif rice in Sagar Island, India," Mandal et al. (2015) worked on crop water requirement that depends on rainfall variability. Kuo et al. (2001) worked on CWR in Taiwan. The FAO Penman-Monteith method was preferred for the calculation in the determination of the reference evapotranspiration (ET_0) as it is very useful for the calculation of actual crop water use data globally (Pereira et al., 2015). Kuo and Tan (2016) have done a regression relationship between irrigation water requirements and crop yield for paddy fields. Surendran et al. (2015) used the model for different agroecological units to find out the CWR and seasonal climatic water balance, water resource assessment, and irrigation scheduling. The authors worked on crop water requirements of different crops in different agroecological units. They calculated two things; one is climatic water balance using evapotranspiration and effective rainfall of each AEU and the other is water balance analysis for future demand for domestic, agricultural, and industrial purposes for sustainable water resource management. Vashisht and Satpute (2015) worked on CWR and irrigation scheduling for rice and wheat. Kuo and Tan (2016) assessed the capacity building of irrigation associations and irrigation water requirements for paddy in Taiwan. Mehanuddin et al. (2018) calculated CWR and effective rainfall for perennial crops. Raut et al. (2010) worked on a river basin for monthly irrigation requirements of different seasons and corresponding irrigated areas, i.e., net irrigation area. Ewaid et al. (2019) worked on irrigation schedules of different cropped areas in Southern Iraq. The researchers worked in the climate of Iraq which is of the subtropical semi-dry type. The Food and Agriculture Organization (FAO) CROPWAT 8.0 simulation software and the CLIMWAT 2.0 tool together were used in this study for Dhi-Qar Province in southern Iraq to find the crop water requirements (CWRs) and irrigation schedules for some major crops. They analyzed the TAM and RAM for the irrigation schedule. Implications of climate change on

CWR were done by Manasa and Shivapur (2016). Onyancha et al. (2017) in the article *FAO-Cropwat model-based estimation of the crop water requirement of major crops in Mwala, Machakos country* described it is very important to assess the crop water requirements of various crops grown under irrigation, especially in arid and semi-arid regions. That helps for efficient use of water and better irrigation practices.

2 Study Area

The Kankai Irrigation System is located in Jhapa district of Mechi zone in Eastern Nepal now in Province 1 (Koshi Province). It is located between 26° N and 27° N and 87° E and 88° E. The elevation of the area varies from 120 m (north) to 75 m (south) above mean sea level, with a gradual slope from north to south. The project covers the people of seven VDCs (Village Development Committees): Satashidham, Shivagunj, Panchgachhi, Mahabhara, Dharampur, Baigundhura, and Topgachhi. Now the VDCs are transformed into Nagarpalika and Gaonpalika. The water source of the Kankai Irrigation System (KIS) is the Kankai River, an Indo-Nepal transboundary perennial river having a total catchment area of 1190 km^2. It originates from the Mahabharat Mountain Range and flows down through Terai plain and crosses the Indo-Nepal Boarder and merges into River Ganges in India. Nepal and India both are river-based countries. The physical condition of Nepal and India is same in some places. The Terai of Nepal and India belongs to the same physiographic condition in terms of relief, soil, climate, etc. Hence, the nature of river flow is also same in both areas. Some rivers are originated in Nepal Himalaya and meet the other rivers in India, i.e., they are Indo-Nepal transboundary river. The river flows in the Terai region are properly managed in some cases to develop a good irrigation system or to control flood or to maintain the proper river health, ecosystem, and river basin management. Kankai is such a transboundary river and hence it is selected for the study (Fig. 13.1).

3 Data Used

The study calculates the different values using three types of inputs, namely, climatic data, i.e., hydro meteorological data, soil data, and crop data.

3.1 Meteorological Data

Average climatic data for 40 years (1981–2021) of the Sanischare Meteorological Station, Damak, and Ilam Tea Estate were obtained from the NASA power data office and from the software CLIMWAT 2.0 which is a climatic database to be used

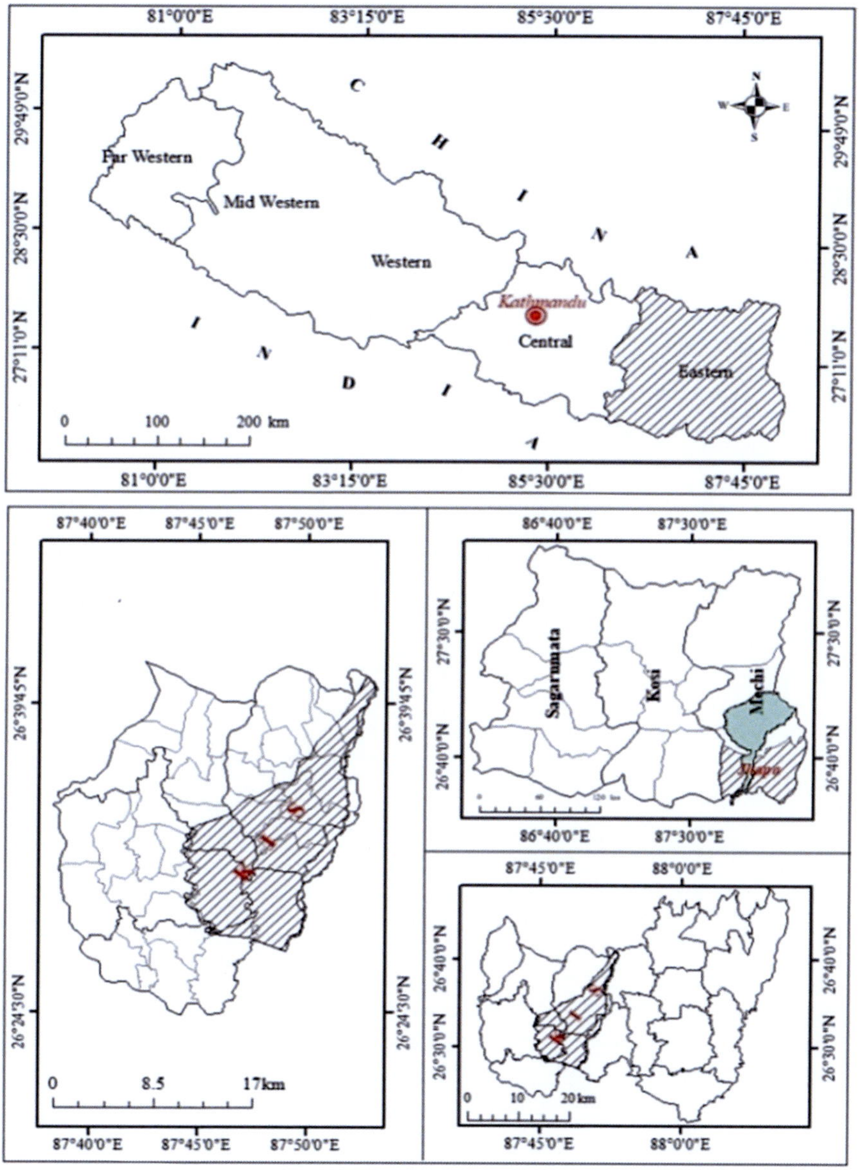

Fig. 13.1 Location Map of Kankai Basin and Kankai Irrigation System (KIS)

in association with the CROPWAT 8.0 program. CLIMWAT and CROPWAT show the monthly climatic parameters, i.e., monthly maximum and minimum temperature (°C), wind speed (km/h), mean relative humidity (%), sunshine hours (h), rainfall data (mm), and effective rainfall (mm). For this analysis, total annual rainfall, total seasonal rainfall, monthly mean rainfall of four decades (1981–1991, 1991–2001,

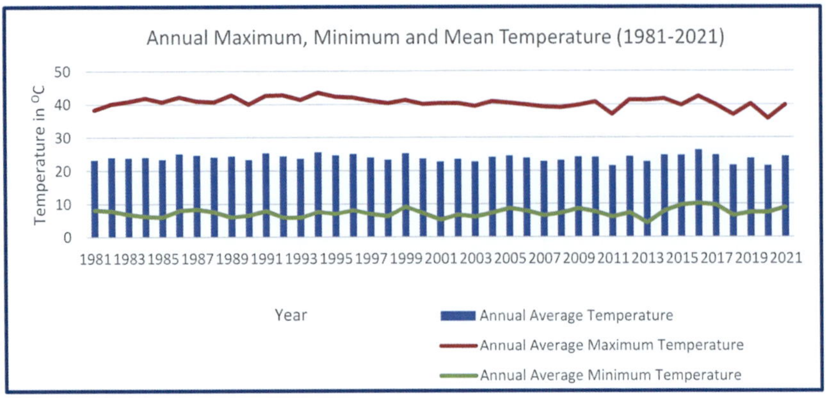

Fig. 13.2 Graphical representation of Total Annual Rainfall of three stations in Kankai Basin (1981–2021)

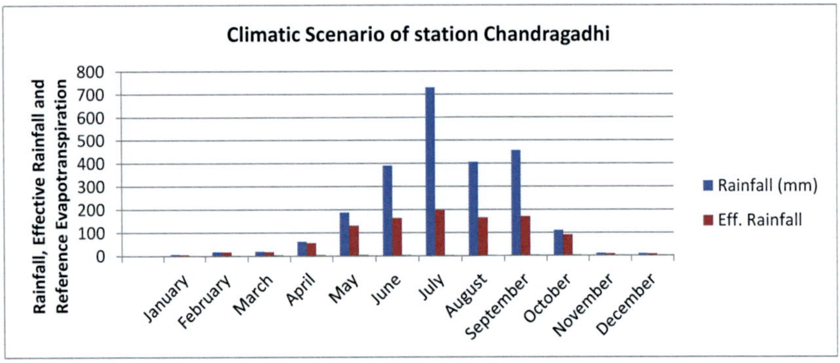

Fig. 13.3 Graphical representation of Total Annual Rainfall, Effective Rainfall, and Reference Evapotranspiration of station Chandragadhi(1981–2021)

2001–2011, and 2011–2021), and seasonal mean rainfall (pre-monsoon, monsoon, and post-monsoon) were considered. Rainfall and temperature conditions of the three stations are graphically represented (Figs. 13.2 and 13.3; Table 13.1).

3.2 *Crop Data*

The major crops of the study area are paddy, wheat, maize, potato, mustard, vegetables, pulses, spring maize, and spring rice. The data of paddy crop for two seasons (monsoon and spring) were collected from FAO manual 56 and subsequently integrated to the CROPWAT software for analysis. The data gives the rooting depth, crop coefficient, critical depletion, yield response factor, and length of plant growth stages. Planting dates are given according to the crop calendar of KIS.

Table 13.1 Effective Rainfall and Reference Evapotranspiration of station Chandragadhi(1981–2021)

Month	Jan	Feb	Mar	Apr	May	June	July	Aug	Sept	Oct	Nov	Dec
Eff. Rainfall (mm)	3.9	8.9	18.7	49.1	102.6	148.9	167.8	156.2	146.8	60.8	4.2	3.6
ET_{ref} mm/day	3.58	5.15	7.28	7.99	7.11	5.74	4.09	3.87	3.78	3.71	3.35	3.12

3.3 Soil Data

The soil parameters are obtained from the FAO CROPWAT 8.0 model which includes detailed information on the soil near the climatic station, such as total available moisture content, initial moisture depletion, maximum rain infiltration rate, and maximum rooting depth (Table 13.2).

4 Methodology

The variability of rainfall is a natural phenomenon in the monsoon countries. The variability of rainfall is of two types: areal and temporal. Here, temporal rainfall variability is considered. Rainfall variability indicates the value to what degree the range of rainfall amount varies. Rainfall variability of three seasons is calculated here (1981–2021): spring season or pre-monsoon (Feb–May), Kharif season or monsoon (June–October), and Rabi season or post-monsoon (Nov–Jan). Here, the agricultural production depends on rainfall in the Kharif season (monsoon paddy) and irrigation in the post-monsoon (winter crops and vegetables) and pre-monsoon (spring rice and spring maize). Rainfall variability has been computed using the Mann-Kandell test and Sen's slope. Rainfall trends for each month and three seasons (pre-monsoon, monsoon, and post-monsoon) are calculated here. Mean, standard deviation, and coefficient of variation are the outputs of this method.

Mann-Kendall's test was used to compute the analysis of monthly and seasonal rainfall trends. Mann-Kendall's study examined the pattern of rainfall and its seasonal variations throughout the study region. This test was developed and applied by Mann (1945), Kendall (1975), and Gilbert (1987) to find patterns in the precipitation data. This test was evaluated for the following hypotheses: "One should reject the null hypothesis (H0) and accept the alternative hypothesis (Ha)" or "One cannot reject the null hypothesis Ho:"

where Ho shows that the computed time series has no trend.
Ha shows that the time series under investigation has a pattern.

The significance level for the test's *p*-value is set at $p = 0.05$ (95%). The null hypothesis was taken into consideration when the estimated *p* value exceeded the significance level. Sen's slope estimator was employed to assess the trend's size.

Table 13.2 Soil data (FAO)

Soil Name	Medium (Loam)
Total Available Moisture (FC-WP)	290 mm/meter
Maximum rain infiltration rate	40 mm/day
Maximum rooting depth	900 cm
Initial soil Moisture depletion (as %TAM)	0%
Initial Available soil Moisture	290 mm/meter

The expression for the Mann-Kendall test is (Feidas et al., 2007):

$$S = \sum_{k=1}^{n-1} \sum_{kj=k+1}^{n} \operatorname{sgn}(x_j - x_k) \tag{13.1}$$

where $\operatorname{sgn}(x_j - x_k)$ is defined as follows: where n is the time series's size, x_j and x_k are the data values in the time series *j* and *k*, respectively

$$\operatorname{Sgn}(x) = \begin{cases} +1; \text{if } x > 0 \\ 0; \text{if } x = 0 \\ -1; \text{if } x < 0 \end{cases} \tag{13.2}$$

In this case, a big positive value for *S* represents a substantially growing trend, whereas a large negative value for *S* shows a strongly falling trend, where *n* is the sample size. The *Z* test (Table 13.3) can be used to assess if a trend is significant or not when the MK test is run on a time series that contains a lot of values. For rainfall series, a positive slope value shows an upward or increasing trend, whereas a negative value indicates a downward or decreasing trend of rainfall. This method was applied for the monthly and seasonal (pre-monsoon, monsoon, and post-monsoon) rainfall data.

Joss Swennenhuis created CROPWAT 8.0 specifically for the FAO's Development and Management Service, which deals with water resources and irrigation. It was created by the Institute of Irrigation and Development Studies and the University of Southmpton. A computer software called CROPWAT may be used to estimate crop evapotranspiration (ET_c), irrigation scheduling, and reference crop evapotranspiration (ET_o) (FAO, 1992).

Table 13.3 Monthly trend analysis values of rainfall of Chandragadhi station (1981–2021)

Months	$Y = a + bx$	R^2
January	$y = 0.0292x - 54.849$	0.0036
February	$y = 0.0252x - 41.688$	0.0005
March	$y = 0.6817x - 1345.2$	0.0476
April	$y = 0.8309x - 1610.6$	0.0741
May	$y = 3.9563x - 7793$	0.4402
June	$y = 3.4085x - 6576.3$	0.2039
July	$y = 3.5364x - 6655.8$	0.0655
August	$y = 6.5353x - 12772$	0.2562
September	$y = 4.4617x - 8693.6$	0.1691
October	$y = 1.9649x - 3863$	0.0935
November	$y = -0.0495x + 103.4$	0.0023
December	$y = -0.0746x + 152.52$	0.0256
Pre-monsoon	$y = 1.3735x - 2697.6$	0.3946
Monsoon	$y = 4.4855x - 8674.3$	0.3458
Post-monsoon	$y = 17.004x - 32971$	0.3528

5 Reference Evapotranspiration (ET_0)

The Windows CROPWAT model uses the FAO Penman–Monteith equation for the computation of the ET_0 where most of the parameters are taken from the climatic data. The *Penman–Monteith equation* is as follows:

$$\lambda \mathrm{ET} = \frac{\Delta(\mathrm{Rn} - G) + \mathrm{Pa}\ \mathrm{Cp}((\mathrm{es} - \mathrm{ea})/\mathrm{ra})}{\Delta + \gamma(1 + \mathrm{rs}/\mathrm{ra})} \tag{13.3}$$

where γ is the psychometric constant, rs and ra are the surface and aerodynamic resistances, respectively, and Pa is the mean of air density at constant pressure. Additionally, Δ is the slope of the relationship between saturation vapor pressure and air temperature, and G is the soil heat flux. Finally, Cp is the specific heat of the air (Mehanuddin et al., 2018).

6 Effective Rainfall (Eff. Rain)

In Cropwat, there are four popular empirical techniques for determining effective rainfalls: fixed proportion of rainfall, consistent rainfall, empirical formula, and USDA soil conservation service method. This research employed the USDA soil conservation service methodology. These four empirical equations were given by Smith (1991):

USDA soil conservation service method:

$$\mathrm{PE}_{\mathrm{eff}} = P_{\mathrm{tot}}{}^{*} \frac{125 - 0.2P_{\mathrm{tot}}}{125} \quad \text{for } P_{\mathrm{tot}} < 250^{\mathrm{mm}}$$

$$\mathrm{PE}_{\mathrm{eff}} = 125 + 0.1P_{\mathrm{tot}} \quad \text{for } P_{\mathrm{tot}} > 250^{\mathrm{mm}} \tag{13.4}$$

where $\mathrm{PE}_{\mathrm{eff}}$ is the effective rainfall and P_{tot} is the total rainfall (Ewaid et al., 2019).

7 Crop Water Requirement (CWR) and Crop Evapotranspiration (ET_c)

Crop evapotranspiration is the term used to describe the quantity of water lost from a cropped field due to evapotranspiration. This amount is measured in millimeters per day (mm/day). Crop evapotranspiration (ET_c), which may be calculated using the following formula, is used to estimate CWR (Gangwar et al., 2017).

$$\mathrm{ET}_c = K_c \times \mathrm{ET}_0 \tag{13.5}$$

In this case, K_c is the crop coefficient. It is the crop's ET_c to ET_0 ratio. Because of variations in ET during the growth phases, the crop's K_c will change during the growing period, which may be separated into four different stages: early, mid-season, late-season, and crop development. Hence,

$$ET_0 \text{ in different stages} = ET_c / K_c \tag{13.6}$$

The amount of water (or depth) required for crops is determined by how much water is lost to evaporation (ET). Water requirements for different crops often vary based on the location, climate, type of soil, cultivation technique, amount of effective rain, etc., and also rely on different phases of the crop's life. Here, the crop water requirements (CWR) for the spring rice and monsoon rice are 980.5 and 276.2 respectively in this study area.

Irrigation Schedule The right amount of water to irrigate with and the right time to water are determined by irrigation scheduling. The CROPWAT model calculates the ET_0, CWR, and IRs to develop the irrigation schedules under different water supply conditions.

8 Analysis

Pre-monsoon and monsoon seasons have shown a favorable tendency over the past 40 years; however, there has been no change in rainfall trends during the post-monsoon season, according to trend analysis. So, a link is established between the crops' need for irrigation and the coefficient of variance. The computed data and the aforementioned figures make it evident that variability is maximum throughout the progressive monsoon, pre-monsoon, and post-monsoon periods. Using the Mann-Kendall test (a non-parametric test), the monthly trend analysis of rainfall was conducted, and the coefficient of variation was also calculated. We looked at the coefficient of variation both seasonally and monthly. The coefficient of variation (C.V.) value varies from 55 to 152 during the pre-monsoon, 35 to 53 during the monsoon season, and 107 to 290 during the post-monsoon when rainfall variability is at its highest. In the Terai portion of the basin, more irrigation is required for crops when the variation is considerable between the post-monsoon and pre-monsoon. The almost identical pattern was seen at three locations. Between 1981 and 2021, there was no trend in the variance of rainfall in January, February, and March (pre-monsoon) and October, November, and December (post-monsoon). For all three sites, the other month's rainfall readings show an increasing trend in precipitation. In this test, the null hypothesis was taken into consideration when the estimated p-value exceeded the significance level. Pre-monsoon, monsoon, and post-monsoon rainfall patterns were calculated using the same methodology. Rainfall during the monsoon and pre-monsoon seasons increased; however, it did not during the post-monsoon period. Sen's slope was computed in order to assess the trend's degree. The monsoon season has a larger Kendall's Tau value, a higher Sen's slope value, and a relatively low p value. These show how consistently the rain fell during the monsoon season.

A trend's significance is measured using a *p*-value of 0.05. A nonparametric method used to assess trends in rainfall data is Sen's Slope estimator of rainfall trend. There is no trend when the Sen's slope value is 0. Sen's slope values that are both positive and negative represent rising and falling trends in the time series, respectively. If the *p*-value is less than 0.157, the trend is positive, and if it is more than 0.157, the trend is zero. *P*-values are larger than 0.157 from October to March, indicating no trend during this time, and a positive trend from April to September. The null hypothesis, which states that there is no trend in the series, is accepted since the estimated *p*-value alpha is 0.05. When it comes to seasonal precipitation data, which is collected for the pre-monsoon, monsoon, and post-monsoon seasons, there is a positive and negative tendency. Rainfall variability was quite high from October to March and relatively low during the monsoon. It showed that the pre- and post-monsoon periods had little rainfall and significant fluctuation, which reflected the primary features of the monsoon climate.

The ET_0 values obtained from the CROPWAT software for different months are shown in Table 13.1. The highest ET_o is found in April where the lowest is found in December. The lowest value is in winter and it gradually increases with summer; this depicts positive relation between the ET_o and solar radiation. The most important meteorological factors affecting ET_o are temperature and solar radiation.

Effective rainfall is the part of rainfall which is effectively used by the crop after losses by surface runoff and deep infiltration; it is used to evaluate the CWR. Knowledge of rainfall amount, frequency, and intensity is essential for planning its full use for agriculture as these vary temporarily and spatially. Average effective rainfall for the 40 years is the highest (167.8 mm) in July and the lowest (3.6 mm) in December. When the effective rainfall is high in monsoon, it requires less or no irrigation, but when it is low in winter and spring season it requires higher irrigation.

The net irrigation requirement (NIR) values of two paddy crops in different stages are shown in Figs. 13.4 and 13.5 (Table 13.4). The irrigation amounts were obtained by decade (the 10-day average irrigation requirements (IR)). High values of total NIR of the crops were observed during the Zaid period when there is no rain in spring season (Table 13.5). According to Shouqin et al. (2014), when temperature is high, evaporation rates are high in association with soil moisture depletion and it needs a high irrigation water requirement.

During the monsoon, paddy requires lots of water which is naturally supplied by rainwater. Hence, net irrigation requirement is very low from June second week to October when rainfall is high. In the nursery or planting stage the demand of water is high. Hence the requirement of irrigation is high as rainfall is low. Actual irrigation requirement of monsoon paddy is zero (Table 13.3) (Fig. 13.6). Soil moisture depletion is low in this season (Fig. 13.7) and 92% of the KIS area is under the monsoon paddy. During the spring time, the total requirement of water is based on the canal irrigation system. In different stages, different amount of irrigation water is

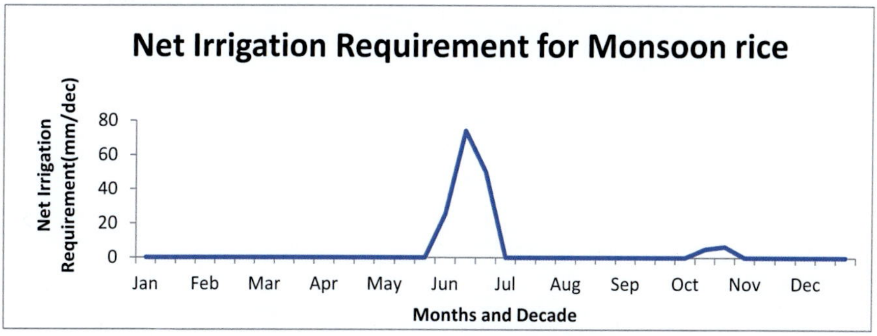

Fig. 13.4 Net Irrigation Requirement for Monsoon Rice (1981–2021)

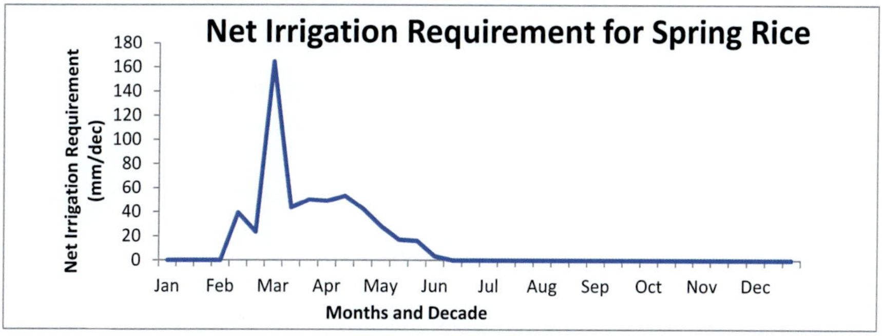

Fig. 13.5 Net Irrigation Requirement for Spring Rice (1981–2021)

Table 13.4 Different parameters calculated by CROPWAT 8.0 Model

Calculated parameters of paddy	Monsoon rice	Spring paddy
Crop Evapotranspiration (mm)	592.8	531.5
Net Irrigation Requirement (mm)	161.0	530.9
Actual Irrigation requirement (mm)	−670.9	155.5
Total Rainfall (mm)	2023.5	505.6
Effective Rainfall (mm)	1159.4	374.0
Total Rain loss (mm)	864.1	131.6
Total Gross Irrigation (mm)	142.5	402.6
Total Net Irrigation (mm)	99.8	281.8
Actual water use by crop (mm)	488.5	529.5
Moist deficit at harvest (mm)	0.0	4.0

needed. Actual irrigation requirement is 155.5 mm. In maximum stages (Figs. 13.8 and 13.9), the range of irrigation requirement varies from 20 to 50 mm/dec. But maximum requirement of irrigation is found in the first decade of March, almost 189 mm/dec which is the last nursing stage. This is successfully supplied by the canal irrigation system.

Table 13.5 Net irrigation requirement of Monsoon and Spring rice

Months	Decade	Stages		Net irrigation requirement (mm/day)	
		Monsoon rice	Spring maize	Monsoon rice	Spring rice
January	1				
	2				
	3				
February	1				
	2		Initial stage		2.3
	3		Initial stage		2.1
March	1		Development stage		7.9
	2		Development stage		19.1
	3		Development stage		34
April	1		Development stage		45.9
	2		Mid		55.4
	3		Mid		43.5
May	1		Mid		28.9
	2		Mid		17.6
	3	Nursery stage	Late		7.5
June	1	Nursery stage/Land Preparation stage	Late	25.4	0
	2	Nursery stage/Land Preparation stage	Late	74	0
	3	Initial stage		50	
July	1	Initial stage		0	
	2	Development stage		0	
	3	Development stage		0	
August	1	Development stage		0	
	2	Mid		0	
	3	Mid		0	
September	1	Mid		0	
	2	Mid		0	
	3	Late		0	
October	1	Late		0	
	2	Late		5	
	3	Late		6.5	

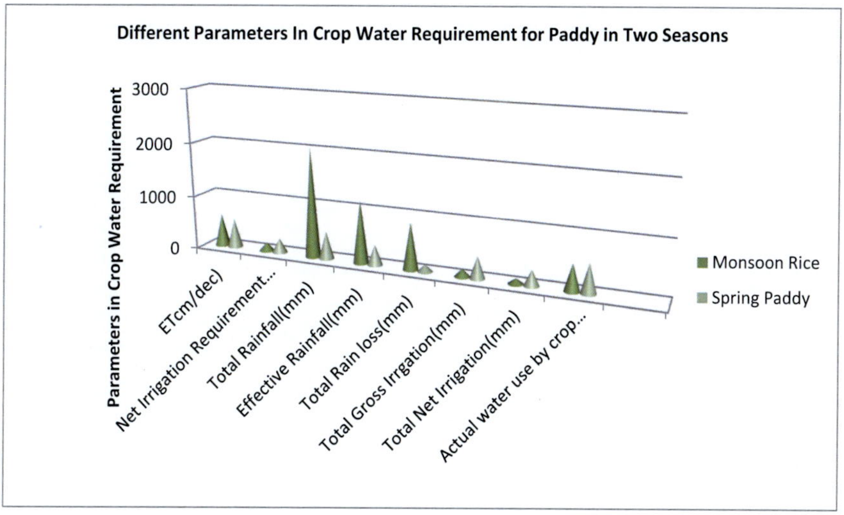

Fig. 13.6 Actual Irrigation Requirement for paddy in two seasons (1981–2021)

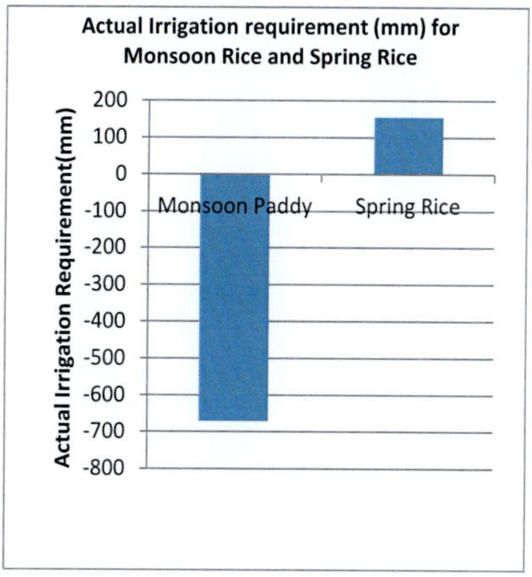

Fig. 13.7 Graph of Irrigation Scheduling for Monsoon Rice (1981–2021)

9 Rainfall Variability and Irrigation Scheduling

The amount and scheduling of precipitation strongly influence the irrigation water requirements (Smith et al., 1998). Precipitation trend analysis shows a positive trend of rainfall for the two seasons pre-monsoon and monsoon for the past 40 years and no trend of rainfall changes during the post-monsoon season (Fig. 13.10). It shows the monthly and seasonal variability. Inter-seasonal variability is also calculated (month-wise). Hence, a correlation is done between the coefficient of variation and irrigation requirement for the crops. Computed data and above charts reveal that rainfall variability is highest in post monsoon season followed by gradual decrease

Fig. 13.8 Crop Evapotranspiration, Effective Rainfall, and Irrigation Requirement in different stages of Paddy in Monsoon Season (1981–2021)

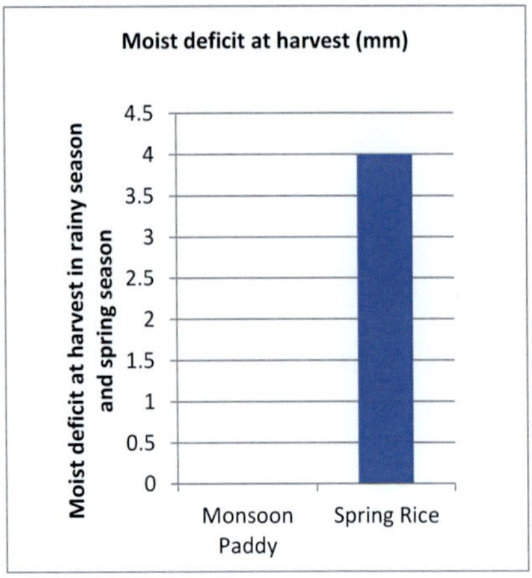

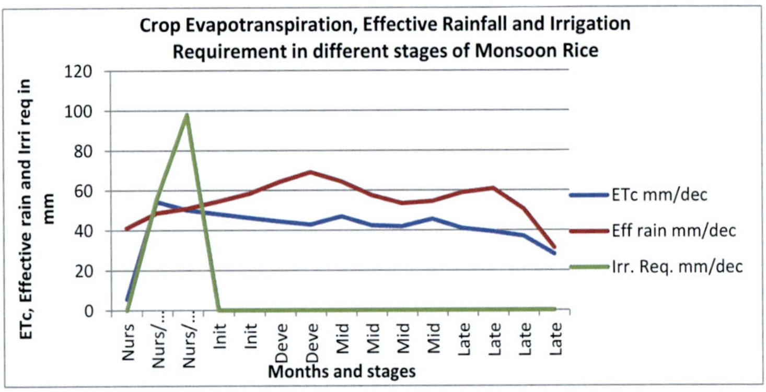

Fig. 13.9 Crop Evapotranspiration, Effective Rainfall and Irrigation Requirement in different stages of Paddy in spring season (1981–2021)

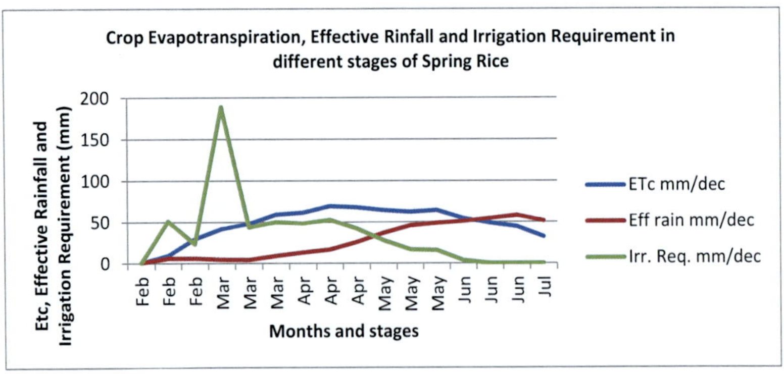

Fig. 13.10 Seasonal Rainfall trend analysis (1981–2021) of station Chandragadhi

during pre-monsoon season and lowest in monsoon season. It means the irrigation requirements is maximum during the post-monsoon as the rainfall consistency is low. The need for irrigation is low in the monsoon season where rainfall is almost consistent and high. The rain water is sufficient for the agricultural activity.

Irrigation scheduling is very much important and depends on this climatic scenario. The graphs show the different parameters of the canal irrigation scenario and the irrigation requirements for the monsoon rice and spring rice (Figs. 13.11 and 13.12). The two most dominant factors in irrigation scheduling, planning, design, and operation are the available water supply and the water demand (Sakthivadivel et al., 1993). Water requirement and canal water supply in different seasons are considered here.

During the spring time, the total requirement of water is based on the canal irrigation system. The irrigation capacity of each canals (S0–S21) are different. In different stages, different amount of irrigation water is needed. Actual irrigation requirement is 155.5 mm. In maximum stages, the range of irrigation requirement varies from 16 to 80 mm/dec. But maximum requirement is found in the first decade of March, almost 236.2 mm/dec which is the last nursing stage. This is successfully supplied by the canal irrigation system.

Analysis of the demand and supply of the canal water during June to October shows that the rain water passes through at high pace, but the demand of irrigation water is almost zero. In the spring time, the supply is also highly sufficient from the canal for irrigation to meet the demand (Figs. 13.13 and 13.14). The spring rice concentration is high in the eastern part of the canal system as the canals of the eastern side are highly developed and irrigation capacity is high.

Analysis of correlation between rainfall variability of different seasons and irrigation requirement (Fig. 13.15) shows that monsoon season with low variability recorded less irrigation requirement, and in the Rabi and the Zaid seasons when the variability is high, the irrigation requirement is high too.

In irrigation scheduling, total available moisture (TAM) in the soil and the readily available water (RAM) are considered. Field capacity is the water content at which drainage becomes very much negligible on a free draining soil. In the minimum soil-water content when the plants permanently wilt, it is called the permanent wilting point. The soil water which is stored between field capacity and the permanent wilting point is called the total available moisture or available water capacity (AWC). The prepuddling stage of monsoon rice is from 17 May to 31 May, the puddling stage is from 1 June to 5 June, the initial stage from 6 June to 25 June, developmental stage from 26 June to 25July, mid stage up to 3 September, and the end stage up to 3 October. During the prepuddling stage the soil moisture depletion ranges from 8mm to 10mm. However in the puddling stage soil moisture depletion drops to zero due to the onset of monsoon (Figs. 13.7 and 13.16). Rain efficiency is 80.8%. At first, the depletion of saturation water is high near 40–50 mm and gradually the value decreases. Percolation is 3.4 mm and the total available moisture in the rootzone is low during the preparation time. During the monsoon period, TAM increases in the soil and almost reaches 120 mm; however, it was zero at the starting stage. RAM also increases but it ranges from 20 to 25 mm, less than TAM. Spring

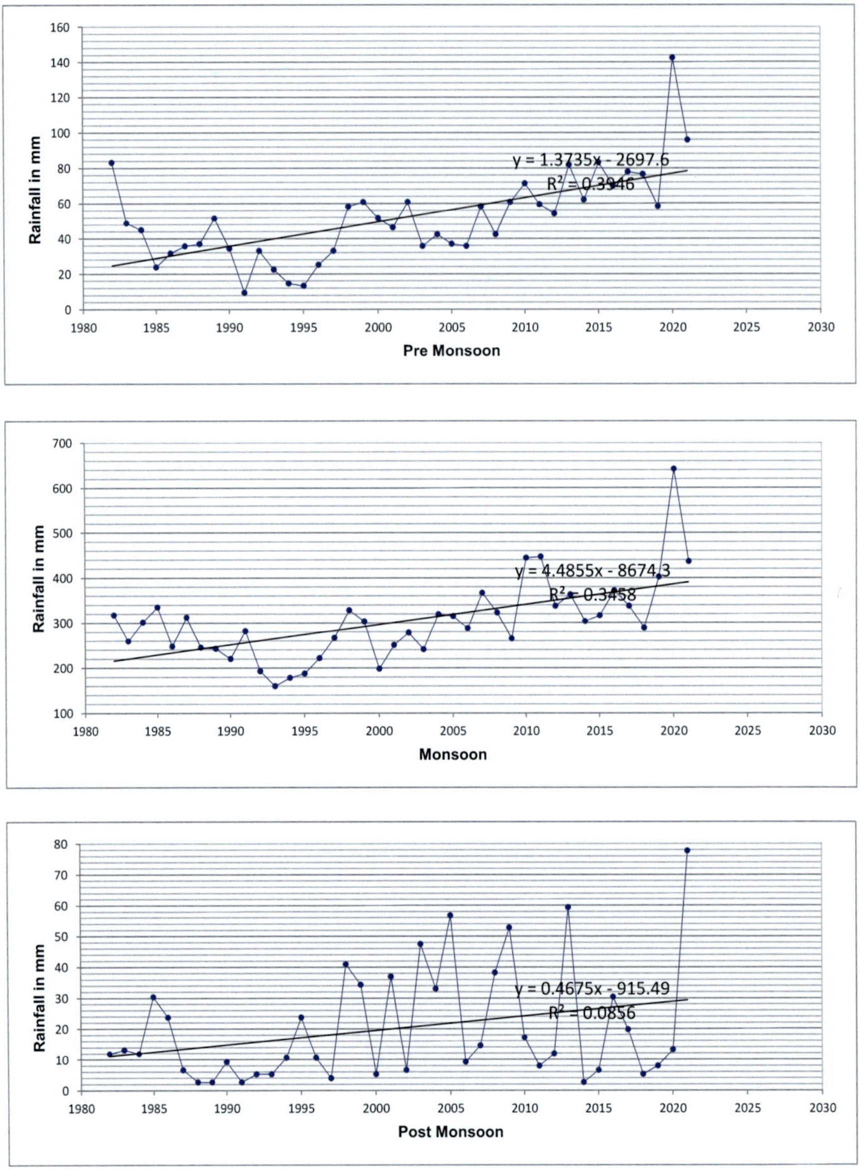

Fig. 13.11 Different parameters in Crop water requirement for paddy in two seasons (1981–2021)

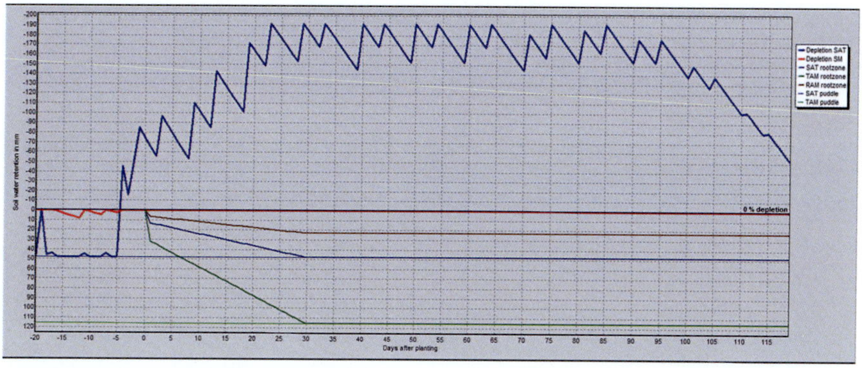

Fig. 13.12 Moist deficit at harvest for paddy in two seasons (1981–2021)

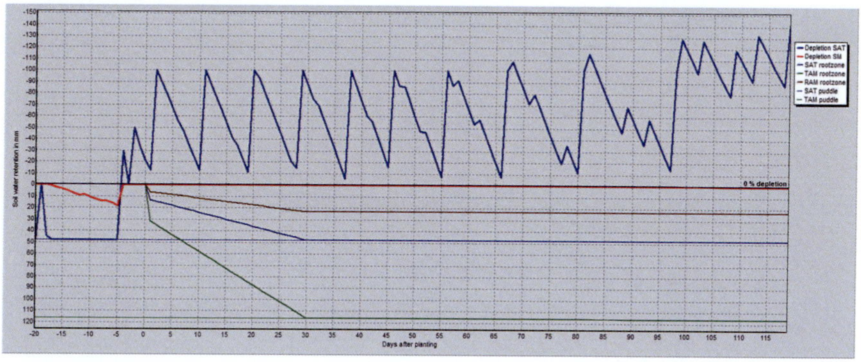

Fig. 13.13 Monthly Irrigation Requirement and main canal water supply for Paddy in Monsoon season (1981–2021)

Fig. 13.14 Monthly Irrigation Requirement and main canal water supply for Paddy in Spring season (1981–2021)

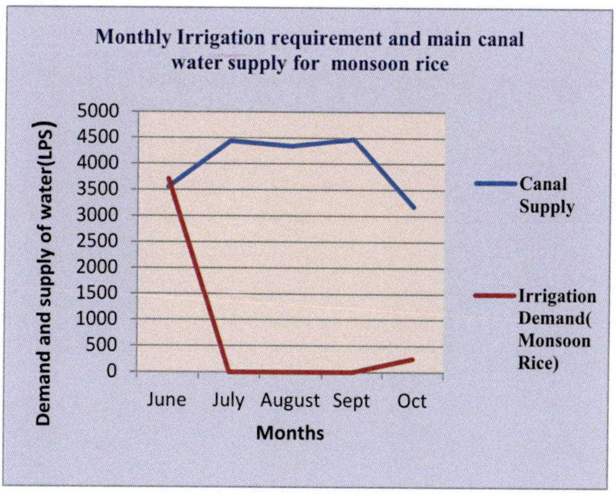

Fig. 13.15 Correlation between Rainfall Variability and Irrigation Requirement for Monsoon and Spring Paddy

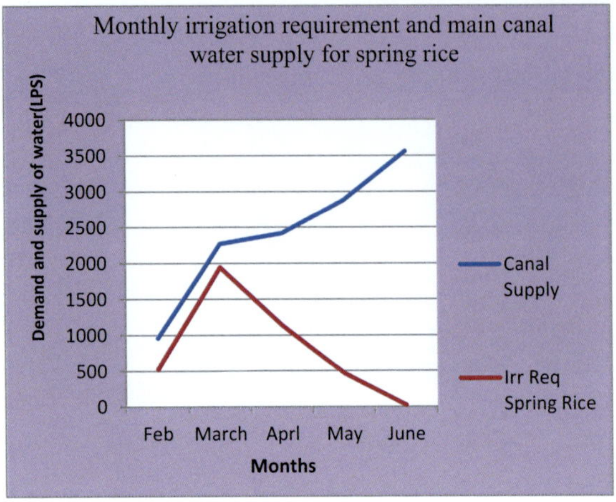

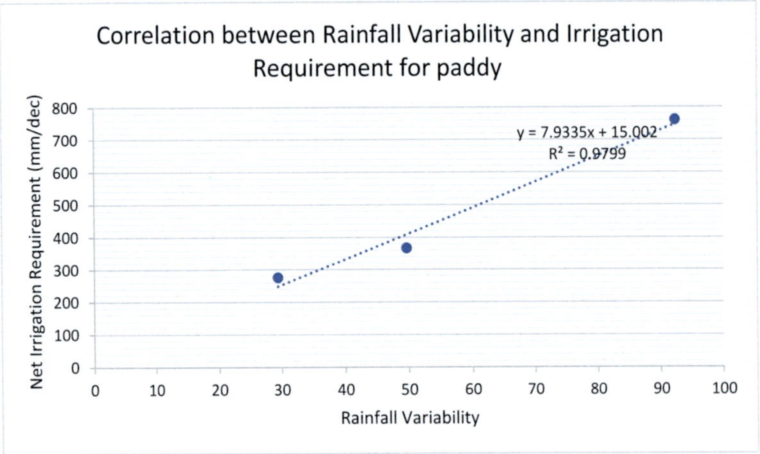

Fig. 13.16 Graph of Irrigation Scheduling for Spring Rice (1981–2021)

season experiences high soil moisture depletion primarily attributed to rainfall deficiencies. TAM and RAM are low at the planting time and it continues up to 55 days after planting. The TAM and RAM slowly gradually increases.

$$\text{TAM} = (\text{Field Capacity} - \text{Wilting Point}) \times \text{current rooting depth of the crop}$$

$$\text{RAM} = \text{TAM} \times P, \text{where } P \text{ is the depletion fraction in the crop coefficient } (K_c) \text{ file.}$$

When the calculated soil moisture deficit falls below the readily available moisture, it results in crop stress. During the winter (post-monsoon) and spring time (pre-monsoon), the depletion level is high as the rainfall variability is high and it

Table 13.6 Irrigation scheduling of Monsoon Rice

Date	Day	Stage	Rain	Ks	Eta	Puddl	Percol.	Soil Moisture Depletion	Net Gift	Loss	Soil Moisture Depletion
			Mm	fract.	%	state	mm	Mm	mm	mm	Mm
5 Jun	– 19	Pre-puddling	170.0	1.0	100	Prep	0.0	1	25.8	0.0	24.0
20 Jun	−4	Puddling	206.0	1.0	100	Prep	0.2	1	74.0	0.0	24.0
22 Oct	End	End	90.0	1.0	0	OK	0.0	0			

Table 13.7 Irrigation scheduling of Spring Rice

Date	Day	Stage	Rain	Ks	Eta	Puddl	Percol	Depl. SM	Net Gift	Loss	Depl. SAT
			mm	fract.	%	%	mm	mm	mm	mm	Mm
26-Jan	– 19	PrePu	0.0	1.0	100	Prep	0.0	1	24.9	0.0	24.00
10-Feb	−4	Puddl	0.0	1.0	100	Prep	0.0	5	74.0	0.0	24.00
12-Feb	−2	Puddl	0.0	1.0	100	Ok	15.0	0	50.0	0.0	0.8
19-Feb	5	Init	0.0	1.0	100	Ok	3.4	0	98.1	0.0	−1. 9
06-Mar	20	Init	0.0	1.0	100	Ok	3.4	0	102.1	0.0	2.1
19-Mar	33	Dev	0.0	1.0	100	Ok	3.4	0	99.4	0.0	−0. 6
21-Feb	45	Dev	0.0	1.0	100	Ok	3.4	0	98.5	0.0	−1. 5
11-Apr	56	Mid	0.0	1.0	100	Ok	3.4	0	97.3	0.0	−2. 7
22-Apr	67	Mid	0.0	1.0	100	Ok	3.4	0	100.3	0.0	0.3

requires high irrigation demand. Throughout the monsoon rice season, soil saturation level remains consistently high due to sustained rainfall (Tables 13.6 and 13.7). Hence, the soil moisture depletion is low. It requires less irrigation water. Rainfall and effective rainfall are high during May to September, and agriculture is rain dependent at that season.

The objective of the irrigation project was to enhance agricultural productivity, local economic development and implementation of sustainable river management of the eastern terai of Nepal. Here the main crops are the monsoon paddy (Kharif—

Table 13.8 Agricultural paddy land change (2000–2020)

Crops	2000(area ha)	2010(area ha)	2020(area ha)
Monsoon Rice	5178	5534	6394
Spring Rice	1308	1594	2025

June-September), wheat, maize, potato, mustard, vegetables, pulses (Rabi—October-January), and spring rice or early paddy (February–May). The major important crops are, viz., monsoon rice (covers 91% of the cropped area), wheat (12%), maize (16%), and early paddy (53%). From the secondary data analysis (KIS data) of the year 2000 and 2020, it is observed that monsoon cropped area increased by 23.5% and spring cropped area increased by 54.8% during the last 20 years (Table 13.8). Correlation between rainfall and production of 2000, 2010 and 2016 indicate that rabi season experiences significant irrigation requirement; however the canal discharge during this period is consistantly below optimal level. In the spring season, due to inadequate water supply from Secondary canal S0-S12 and S13-S21 rotational irrigation was implemented on an alternating yearly basis in the previous years. But now a days due to enhancement in river managment, the canal commitee has successfully maintained uninterrupted irrigation services in spring cultivation period, covering the full extension of the canal network.

10 Conclusion

The consistency and increase of production are mainly due to the availability of irrigation water according to the demand through main, secondary, and tertiary canals. Here, irrigation scheduling shows dependency on rainfall variability. During monsoon, canal flows are high enough, soil moisture depletion is minimal and irrigation requirements are consequently very low. Hence, the percolation is low and the excess water passes through the canals to protect from flooded situation. But in the Zaid and Rabi season, the canal flow is high to meet the irrigation requirement. Sufficient canal flow shows the irrigation efficiency during the lean periods. In the Rabi (post-monsoon) season, the rainfall variability is maximum that results in soil water stress. Hence, three is a variability in rainfall and agricultural production depends on the irrigation. This Japan's project has been implemented efficiently and successfully in Kankai region however further improvements are necessary particularly in the western part. It should be note here that the majority of the economically active local population which tantamount to 85% is directly involved in agriculture and 5% is involved in other professions other than agriculture. In spite of being a male-dominated society, it is found that few women are also involved in social activities.

References

Adeoti, J. O., & Sinh, B. T. (2009). Technological constraint and farmers' vulnerability in selected developing countries (Nigeria and Vietnam). In *7th International conference on 'inclusive growth, innovation and technological change: education, social capital and sustainable development'held on* (pp. 6–8).

Ahmad, I., Zhang, F., Tayyab, M., Anjum, M. N., Zaman, M., Liu, J., et al. (2018). Spatiotemporal analysis of precipitation variability in annual, seasonal and extreme values over upper Indus River basin. *Atmospheric Research, 213*, 346–360. https://doi.org/10.1016/j.atmosres.2018.06.019

Barrow, C. J. (2016). *Water resources and agricultural development in the tropics*. Routledge.

Chougule, M. A., & Mashalkar, A. S. (2022). A comprehensive review of agriculture irrigation using artificial intelligence for crop production. *Computational Intelligence in Manufacturing*, 187–200. https://doi.org/10.1016/B978-0-323-91854-1.00002-9

Ewaid, S. H., Abed, S. A., & Al-Ansari, N. (2019). Crop water requirements and irrigation schedules for some major crops in Southern Iraq. *Water, 11*(4), 756. https://doi.org/10.3390/w11040756

Feidas, H., Noulopoulou, C., Makrogiannis, T., & Bora-Senta, E. (2007). Trend analysis of precipitation time series in Greece and their relationship with circulation using surface and satellite data: 1955–2001. *Theoretical and Applied Climatology, 87*, 155–177. https://doi.org/10.1007/s00704-006-0200-5

Gangwar, A., Nayak, T. R., Singh, R. M., & Singh, A. (2017). Estimation of crop water requirement using CROPWAT 8.0 model for Bina command, Madhya Pradesh. *Indian journal of Ecology, 44*(4), 71–76.

Gilbert, M. (1987). Modelling collective belief. Synthese, 73, 185–204.

Gouda, K. C., Sahoo, S. K., Samantray, P., & Shivappa, H. (2017). Comparative study of monsoon rainfall variability over India and the Odisha State. *Climate, 5*(4), 79. https://doi.org/10.3390/cli5040079

Haseeb, M., Farid, H. U., Khan, Z. M., Anjum, M. N., Ahmad, A., & Mubeen, M. (2023). Quantifying irrigation water demand and supply gap using remote sensing and GIS in Multan, Pakistan. *Environmental Monitoring and Assessment, 195*(8), 990. https://doi.org/10.1007/s10661-023-11546-6

Jamaludin, S., & Suhaimi, H. (2013). Spatial interpolation on rainfall data over peninsular Malaysia using ordinary kriging. *Jurnal Teknologi, 63*(2), 51–58. https://doi.org/10.11113/jt.v63.1912

Kendall, K. (1975). Transition between cohesive and interfacial failure in a laminate. Proceedings of the Royal Society of London. A. Mathematical and Physical Sciences, 344(1637), 287–302.

Kuo, S. F., & Tan, C. H. (2016). *Capacity building of irrigation association and irrigation water requirements for paddy in Taiwan*. 2nd World Irrigation Forum 6–8 November 2016, Chiang Mai.

Kuo, S. F., Lin, B. J., & Shieh, H. J. (2001). *CROPWAT model to evaluate crop water requirements in Taiwan*. International Commission on Irrigation and Drainage. International Commission on Irrigation and Drainage, 1st Asian Regional Conference Seoul.

Manasa, H. G., & Shivapur, A. V. (2016). Implications of climate change on crop water requirements in Hukkeri Taluk of Belagavi District, Karnataka, India. *International Journal of Research in Engineering and Technology, 5*(06), 236–241.

Mandal, S., Choudhury, B. U., & Satpati, L. N. (2015). Monsoon variability, crop water requirement, and crop planning for kharif rice in Sagar Island, India. *International Journal of Biometeorology, 59*, 1891–1903. https://doi.org/10.1007/s00484-015-0995-9

Mann, T. (1945). Studies on the metabolism of semen: 1. General aspects. Occurrence and distribution of cytochrome, certain enzymes and coenzymes. Biochemical Journal, 39(5), 451.

Mehanuddin, H., Nikhitha, G. R., Prapthishree, K. S., Praveen, L. B., & Manasa, H. G. (2018). Study on water requirement of selected crops and irrigation scheduling using CROPWAT 8.0. *International Journal of Innovative Research in Science, Engineering and Technology, 7*(4), 3431–3436.

Nery, J. T., Carfan, A. C., & Martin-Vide, J. (2017). Analysis of rain variability using the daily and monthly concentration indexes in southeastern Brazil. *Atmospheric and Climate Sciences., 7*(2), 176–190. https://doi.org/10.4236/acs.2017.72013

Onyancha, D. M., Gachene, C. K. K., & Kironchi, G. (2017). FAO-cropwat model-based estimation of the crop water requirement of major crops in Mwala, Machakos county. *Research Journal's Journal of Ecology, 4*(2) http://www.researchjournali.com/view.php?id=3751

Paredes, P., Pereira, L. S., Rodrigues, G. C., Botelho, N., & Torres, M. O. (2017). Using the FAO dual crop coefficient approach to model water use and productivity of processing pea (Pisum sativum L.) as influenced by irrigation strategies. *Agricultural Water Management, 189*, 5–18. https://doi.org/10.1016/j.agwat.2017.04.010

Pereira, L. S., Allen, R. G., Smith, M., & Raes, D. (2015). Crop evapotranspiration estimation with FAO56: Past and future. Agricultural water management, 147, 4–20.

Raut, S., Sarma, K. S. S., & Das, D. K. (2010). Study of irrigation and crop water requirements and growth of two Rabi crops grown in a semi arid region using agrometeorology and remote sensing. Journal of the Indian Society of Remote Sensing, 38, 321–331.

Sakthivadivel, R., Merrey, D. J., & Fernando, N. (1993). Cumulative relative water supply: A methodology for assessing irrigation system performance. Irrigation and Drainage Systems, 7, 43–67.

Sangüesa, C., Pizarro, R., Ibañez, A., Pino, J., Rivera, D., García-Chevesich, P., & Ingram, B. (2018). Spatial and temporal analysis of rainfall concentration using the Gini index and PCI. *Water, 10*(2), 112. https://doi.org/10.3390/w10020112

Shouqin, Z., Weihua, Z., Jiake, L., & Chaofu, W. (2014). Temporal variation of soil water and its influencing factors in hilly area of Chongqing, China. International Journal of Agricultural and Biological Engineering, 7(4), 47–59.

Smith, V. L. (1991). Rational choice: The contrast between economics and psychology. Journal of Political Economy, 99(4), 877–897.

Smith, M., Allen, R., & Pereira, L. (1998). Revised FAO methodology for crop-water requirements.

Sharma, B., Molden, D., & Cook, S. (2015). *Water use efficiency in agriculture: Measurement, current situation and trends* (No. 612-2016-40604).

Smith, R. J., Raine, S. R., McCarthy, A. C., & Hancock, N. H. (2009). Managing spatial and temporal variability in irrigated agriculture through adaptive control. *Australian Journal of Multi-Disciplinary Engineering, 7*(1), 79–90. https://doi.org/10.1080/14488388.2009.11464801

Surendran, U., Sushanth, C. M., Mammen, G., & Joseph, E. J. (2015). Modelling the crop water requirement using FAO-CROPWAT and assessment of water resources for sustainable water resource management: A case study in Palakkad district of humid tropical Kerala, India. *Aquatic Procedia, 4*, 1211–1219. https://doi.org/10.1016/j.aqpro.2015.02.154

Vashisht, A., & Satpute, S. T. (2015). Water requirements and irrigation scheduling of direct seeded rice-wheat using CROPWAT model. *Indian Journal of Hill Farming, 28*(2).

Chapter 14
A Study on the Application of Different Statistical Tools in the Field of Social Science Paradox

Chinmoy Ghosh

Abstract Statistics is an inclusive matter expedient in practically all restraints, especially in social science studies. Each and all scholar should have some information in Statistics and necessity practice statistical tools in his or her study, one should know nearly the prominence of statistical tools and how to use them in their study. The eminence pledge of the work must be allocated with: the statistical processes essential to regulator and validate the investigative measures as well as the resultant data making errors in methodical work is inevitable. This is the reason why a throng of different statistical tools is, mandatory some of them meek, some complex, and repeatedly very detailed for convinced resolves. In methodical work, the utmost vital common process is the comparison of data, or sets of data, to enumerate precision (prejudice) and accuracy. Providentially, with a few simple suitable statistical tools most of the info needed. In consistent study work can be got: the "*t*-test, the "*F*-test", and regression analysis. Clearly, statistics are a device. In social sciences, first, identify the category of data, parametric or nonparametric, based on the data, then carry on the study. A thing to be noted at the time of collection of data is that it must be a sample of Big Data, because it will generate nearby analysis, which is actually required by the researcher.

1 Introduction

Statistical approaches help to work in social science paradox, particularly in policy documentation and planning. Statistics plays a vital role in the social sciences, applying severe quantitative approaches to comprehend and solve some of the greatest tests affecting human societies. Statistics is particularly serious at a time when the volume and type of obtainable data are transmuting the social sciences. Social statistics and quantitative data analysis are key tools for thoughtful society

C. Ghosh (✉)
Department of Commerce & Management, East Calcutta Girls' College and Visiting Faculty, School of Management Science, Maulana Abul Kalam Azad Uinversity of Technology, Bidhannagar, West Bengal, India

S. Hazra et al. (eds.), *Geographies of the Indian Subcontinent*,
https://doi.org/10.1007/978-3-031-74813-4_14

and social transformation. We can try to capture people's approaches and record information in comportment and situations using numbers and also label how societies and populations modification.

Depending on the purpose of the study, statistical analysis is used to gather information about the population study situation. The first statistical task is to perform descriptive analysis of the variables. The results of the descriptive analysis must be presented in the form of frequencies and percentages for the qualitative variables. After the descriptive analysis, the association between the variables and the predictive analysis using multiple regression models can be performed.

Social science researchers can use software packages such as SPSS (Statistical Package for the Social Sciences), STATA (statistics and data), Minitab, Open Epi, Graph pad, and many more depending on your use and knowledge of the software. Survey analysis refers to the process of measuring public opinion using surveys. Surveys are used by researchers, market researchers, and organizations of all sizes. Survey data can be analyzed using a variety of statistical techniques. The statistical software packages developed for research professionals are known as statistical software packages. Some of the most popular statistical software packages are SAS (SAS), SPSS (SPSS), and STATA (STATA). However, many survey data analysis techniques can be performed using spreadsheet programs like EXCEL (EXCEL is also part of Microsoft's popular office package). Excel and other spreadsheet programs are easy to use and excel in the entry, coding, and storage of survey data.

The distributions of age, gender, race and any other socio-economic characteristics obtained for the study (e.g., income, education level, and availability of medical care) will help inform the analysis regarding possible age-adjustment, weighting, and other analytical tools available in order to address issues related to bias and unrepresentative samples.

The suitable statistical technique to use in social science be contingent on the study areas, the data type and distribution, and the nature of the explanations. In the vibrant world of social science research, the use of statistical study is not just a methodological optimal; it's a critical tool that brings depth, clarity, and reliability to scholarly surveys. This discipline, at the connection of numbers and narratives, allows researchers to expose patterns, test theories, and provide experimental backing to their assumptions.

2 Research Design Procedure in Social Science

The present study aims to outline the path of using simple data analysis tools and its application in modern research on social science domain.

2.1 Part A*: Data Generation Process (Which Has Been Summarized in Fig. 14.1)*

2.1.1 Milieu Chart

The context of the data found is understood through the use of this display method. For the purpose of creating thematic frames, the information contained in each frame needs to be contextually related. After the context chart is finished, the data findings can be subjected to partial analysis, which is frequently used to validate variables or themes, or interim analysis, which is the process of identifying an early direction or theme in the data. The researcher illustrates how the data are related to one another while keeping the research questions in mind by using a context chart (Cherry, 1984)[1].

Fig. 14.1 Model 1

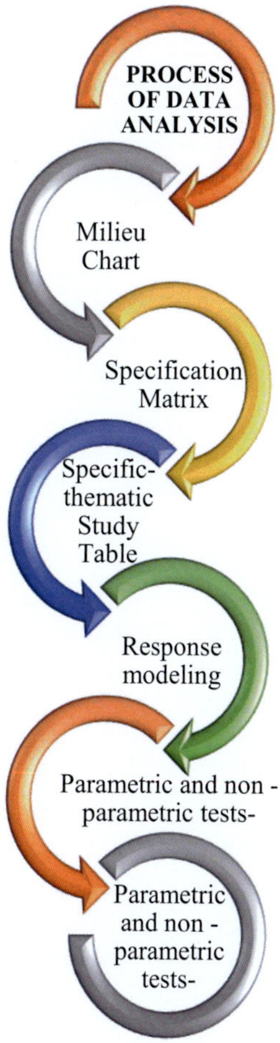

2.1.2 Specification Matrix

By using an object for comparison, like apples and oranges, a variable is used to determine whether or not data is visible or useful in the analysis of qualitative data. This display method will do this. The data is divided into thematic points and arranged in the matrix's labeled columns, rows, and point-guided rubrics (such as strong, sketchy, and adequate). The usefulness of the thematic points as a variable is then assessed in accordance with the point-guided rubric's numerical strength.

2.1.3 Specific Thematic Study Table

This table is made up of columns labeled by coded patterns and rows labeled with themes. An additional method of differentiation for a variable-oriented data analysis is pattern coding. The researcher can quickly conduct a preliminary analysis of the data collected by simply noting which cell the pattern-coded data fills under specific thematic rows in what is commonly referred to as a cross-case analysis table.

2.1.4 Response Modeling

A chart with a single central directive is used in this method. It frequently has a branching tree appearance. For example, the primary decision might be whether to purchase a contract. Two decision boxes are made based on the directive. Following the completion of the survey, the researcher separates the pros and cons boxes into two branches, leaving room for a third branch for the unsure. Due to the subjective and qualitative nature of the data collection, the researcher coded the responses based on context in order to ascertain if they fall under the pros or cons category. In order to analyze the results, the researcher will write those patterned responses in boxes that resemble twigs growing from the relevant branch in this display (Guo, 2013).

2.1.5 Parametric and Non-parametric Tests

It can be challenging to select the appropriate test to compare measurements because there are two test families to consider: parametric and nonparametric. Assuming that the data are sampled from a Gaussian distribution is the foundation for many statistical tests. We call these kinds of tests parametric tests. The table's first column lists popular parametric tests, such as the *t*-test and analysis of variance. Nonparametric tests are those that don't make any assumptions about the probability distribution. The outcome variable is ranked from low to high in all frequently used nonparametric tests, and the ranks are subsequently analyzed. These tests, which include the Gottschalk, L. A. Wilcoxon, and Mann-Whitney tests, are mentioned in the table's second column.

2.2 Part B: *Statistical Parameters for Analysis (Fig. 14.2).*

2.2.1 Mean

The arithmetic mean, additionally referred to as "the average," is computed by dividing the total of a set of numbers by the total number of items in the list. The mean can be used to quickly get a quick overview of your data or to identify the general trend of a data set. The mean also has the benefit of being incredibly simple and quick to calculate.

2.2.2 Standard Deviation

The measure of data spread around the mean is called the standard deviation, which is commonly denoted by the Greek letter sigma. Whereas a low standard deviation indicates that more data are in line with the mean, a high standard deviation indicates that the data are more widely dispersed from the mean. The standard deviation is helpful in a portfolio of data analysis techniques for promptly ascertaining the dispersion of data points.

Fig. 14.2 Model 2

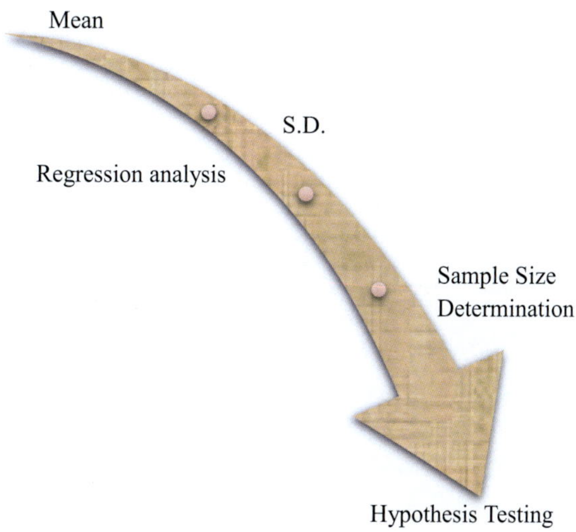

2.2.3 Regression Analysis

Regression analysis uses a scatterplot to show how dependent and explanatory variables relate to one another. The strength or weakness of those relationships is also indicated by the regression line. In statistics classes in high school or college, regression is frequently taught with applications for business or science in identifying long-term trends.

2.2.4 Sample Size Determination

A sample will often suffice when measuring a large data set or population, such as the workforce, rather than interviewing every member of the population. Ensuring the accuracy of a sample requires careful consideration of its size. By applying the proportion and standard deviation techniques, you can precisely ascertain the appropriate sample size required to achieve statistical significance in your data collection.

2.2.5 Hypothesis Testing

Hypothesis testing, sometimes referred to as *t*-testing, determines whether a particular premise is true for your data set or population. A hypothesis test result is deemed statistically significant in data analysis and statistics if it is unlikely to have occurred by chance. In everything from science and research to business and economics, hypothesis tests are utilized (Rockwell, 1975).

Statistical investigation, also known as quantitative study, covenants with hard data and numbers, while qualitative investigation is less perceptible. Statistical analysis can help eliminate abstruseness and reaction from elucidations by decreasing issues to facts and figures. Descriptive statistics, the first level of investigation, can help researchers find outlines and review data.

Case Study: 1

A statistical analysis with a data set and a specific purpose of study can be analyzed as follows:

- 100 respondents provide their views on utilization of mobile application when they are traveling.
- Providing them close-ended questionnaire.
- Data collected from Kolkata south and north.

Basis of that the hypothesis of chi-square as follows:

H0: Using mobile application is not helpful at the time of traveling.
H1: Using mobile application is helpful at the time of traveling.

This has been explained in Table 14.1.

Table 14.1 Mobile application use at the time of traveling

Category	Yes	No
South Kolkata	73	27
North Kolkata	61	39

2.2.6 Analysis (Using Minitab)

Chi-Square Test Mobile application used at the time of traveling

```
Yes No Total
1 73 27 100
67.00 33.00
0.537 1.091

2 61 39 100
67.00 33.00
0.537 1.091

Total 134 66 200

Chi-Sq = 3.256, DF = 1, P-Value = 0.071
```

Here p-value is 0.071 (that is >0.05)

So, the null hypothesis is accepted.

In a Chi-Square test, the null hypothesis is rejected when the test statistic is more than the critical value. This means that there is enough evidence to recommend that the categorical variables are related and that the data are meaningfully dissimilar from what was predictable.

Case Study: 2

A statistical analysis with data set as explained in Table 14.2:

- 15 respondents provide their views on performance analysis parameters in a private bank in Kolkata.
- Providing them close-ended questionnaire (based on a 10-point scale).
- Data collected from Kolkata.
- Principal component analysis.

3 Principal Component Analysis: Clear Vision, Communication, Adaptability, Talent

Table 14.2 Model 2 (Flow chart statistical parameters)

Respondents	Clear vision	Communication	Adaptability	Talent	Strong leadership
1	6	8	7	8	9
2	6	9	8	8	9
3	7	9	6	7	9
4	8	7	9	9	9
5	8	8	8	7	9
6	9	9	7	7	9
7	4	7	9	9	8
8	7	8	9	6	9
9	7	9	8	9	8
10	8	9	8	8	9
11	9	8	8	9	8
12	5	9	9	8	9
13	6	8	7	8	7
14	7	9	9	9	9
15	9	8	7	8	9

```
Eigen analysis of the Correlation Matrix

Eigenvalue 1.8845 1.0884 0.8949 0.7774 0.3549
Proportion 0.377 0.218 0.179 0.155 0.071
Cumulative 0.377 0.595 0.774 0.929 1.000

Variable PC1 PC2 PC3 PC4 PC5
clear vision 0.415 -0.141 -0.782 -0.294 -0.331
communication 0.439 -0.004 0.584 -0.628 -0.268
adaptability -0.423 0.702 -0.069 -0.128 -0.554
talent -0.517 -0.113 -0.173 -0.707 0.436
strong leadership 0.435 0.689 -0.115 -0.050 0.567
```

According to the above analysis, we can say that the two main principal components from the above are clear vision and communication based on Eigen values.

As a social science researcher, it's very easy to analyze the Eigen value, which characterizes the total sum of variance that can be clarified by a given principal component. They can be positive or negative in theory, but in run-through, they explain variance, which is always positive. If Eigen values are greater than zero, then it's a good sign.

Case Study: 3

A statistical analysis with data set-

- 15 respondents provide their views on performance analysis parameters in a private bank in Kolkata.
- Providing them close-oended questionnaire (based on a 10-point scale).

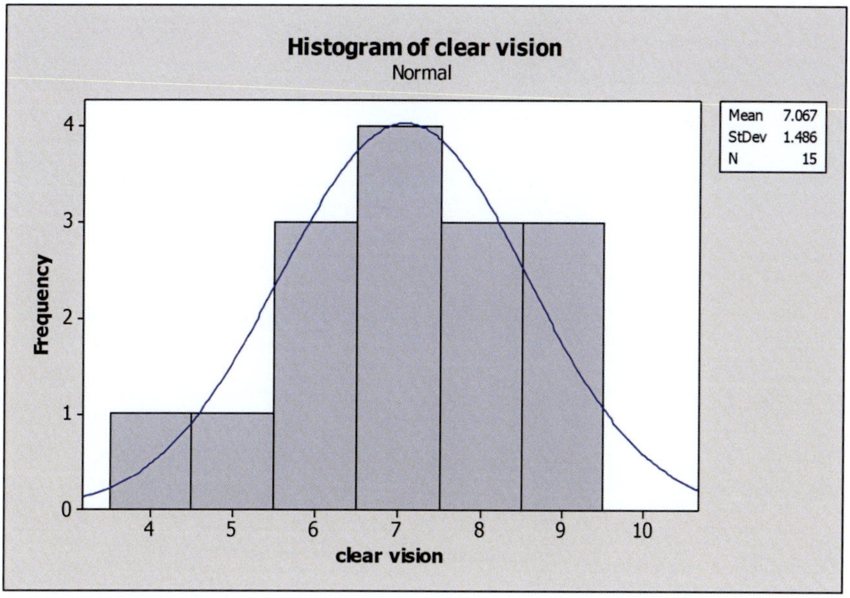

Fig. 14.3 Histogram of clear vision

- Data collected from Kolkata.
- Normality test.

In this case, the researcher can try to analyze normality of the data based in mean and standard deviation (Figs. 14.3, 14.4, and 14.5). The result shows that most data passed the normality supported by mean and S.D. (Figs. 14.6 and 14.7).

4 Use of Quantitative and Qualitative Data in Social Science Research

In advanced coursework, a researcher can employ a mixed methodology, a quantitative approach, or a qualitative one. When choosing a qualitative approach, data analysis options for researchers are numerous. Commonly used tools include matrices, tables, charts, and other visual displays. Through the use of visual aids, researchers can filter through the frequently copious amounts of subjective data collected and identify variables that will be valuable for their qualitative data analysis (Statistical Analysis, 2018).

Making a distinction in their reports between repeatable practices and the nonrepeatable outcomes that resulted from those practices is one way educational researchers attempt to get around the repeatability problem.

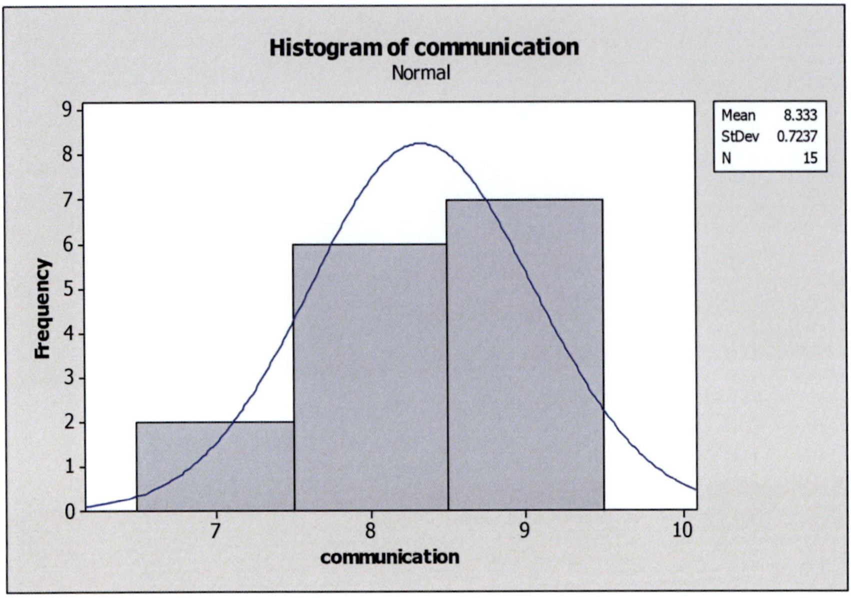

Fig. 14.4 Histogram of communication

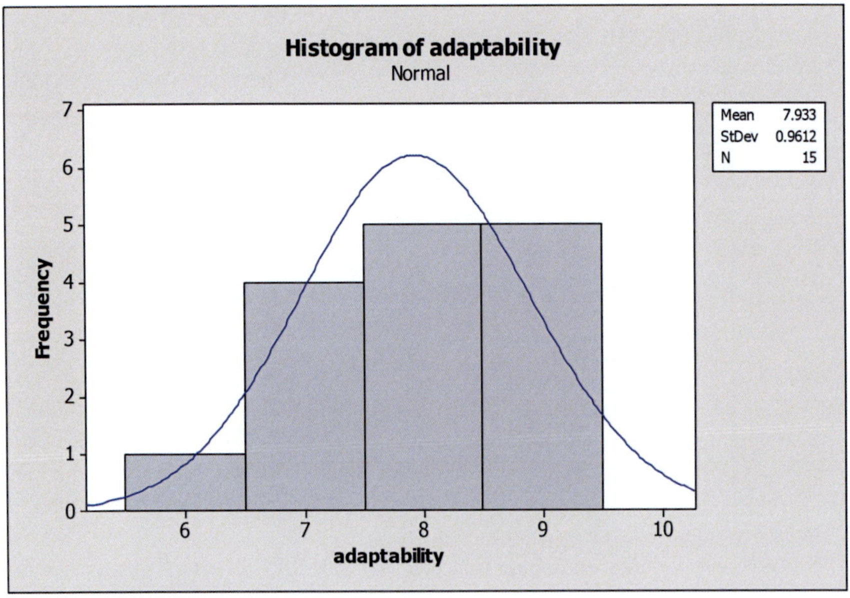

Fig. 14.5 Histogram of adaptability

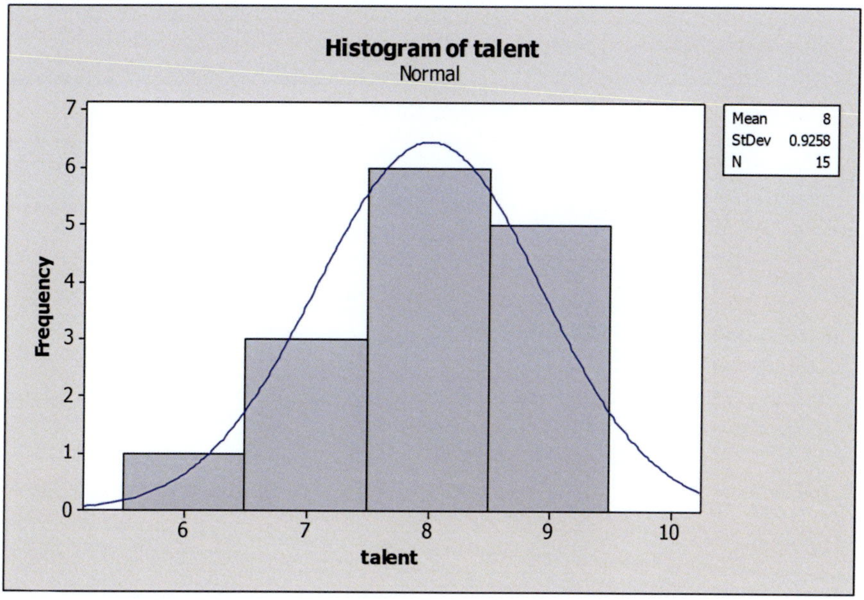

Fig. 14.6 Histogram of talent

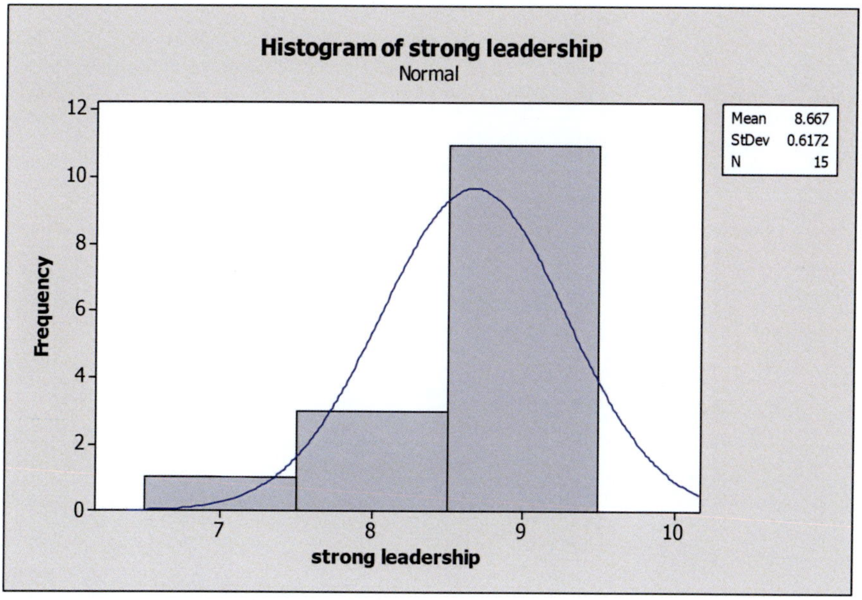

Fig. 14.7 Histogram of strong leadership

The phrase "qualitative research" is broad and covers a variety of research techniques that are frequently employed in the human and social sciences, including sociology, anthropology, ethnography, and other related fields. The main criticism of qualitative research is that the investigator's performance is too crucial to the study's quality. The results are naturally influenced by the researcher because of her personal beliefs because she creates the questions (Schmidt, 2019).

5 Conclusion

Several statistical tool types were explained in this chapter for research and dissertation purposes in various fields. As a result, one should be adept at choosing a statistical tool for their research that produces meaningful results. For their upcoming studies, the researchers can still receive some additional information by purgation of statistical method–based software and their proper utilization.

References

Cherry, R. (1984). Software reviews: StatPac: Statistical analysis package. *Social Science Micro Review, 2*(3), 170–171. https://doi.org/10.1177/089443938600200309

Guo, S. (2013). Advanced statistical analysis. In *Encyclopedia of social work*. Oxford University Press. https://doi.org/10.1093/acrefore/9780199975839.013.840

Rockwell, R. C. (1975). Error-correcting programming in statistical analysis packages. *Social Science Information, 14*(3), 229–233. https://doi.org/10.1177/053901847501400311

Schmidt, T. (2019). Statistical analysis of social network. *Change*. https://doi.org/10.15760/etd.7288

Statistical Analysis. (2018). *Encyclopedia of social network analysis and mining* (p. 2980). Springer. https://doi.org/10.1007/978-1-4939-7131-2_101257

Part IV
Policy Research

Chapter 15
Decomposition Analysis of CO_2 Emission from Electricity Generation in Top Three Emitting Countries, with Emphasis on Indian Scenario

Ujjaini Mukhopadhyay and Ratnakar Pani

Abstract The global surge in CO_2 emission has been instrumental in perpetrating climate change and other environmental jeopardies. With about 40% of CO_2 emission being contributed by the generation of electricity, it becomes imperative to emphasize on curbing electricity-related emission. The study undertakes decomposition analysis of CO_2 emission of the highest three emitting countries, viz., China, the USA and India over the period 1992–2018 using variance analysis method. The major findings indicate that rising population and income are the driving forces of higher emission from electricity generation. However, China and the USA have made technological improvements to neutralize their emission to some extent, but India has hardly been able to perform remarkably in this regard. Unlike the other two countries, both the use of coal in electricity generation and emission intensity of coal in India has increased over time. The study attempts to identify the factors that can play a dominant role in harnessing rising emission in India and suggests policy prescriptions thereof.

1 Introduction

Rampant increase in human activities, accompanied by increased energy combustion, has led to manifold increase in greenhouse gases (GHG), jeopardizing the environment due to global warming and climate change (Pachauri & Mayer, 2014). The major share of greenhouse gas emission stems from burning fossil fuels for electricity, heat, and transportation (Meyerson, 1998). Carbon dioxide is the most important constituent of greenhouse gas, accounting for about 76% of total GHG emissions (Environmental Protection Agency, 2017).

U. Mukhopadhyay (✉)
East Calcutta Girls' College, Lake Town, Kolkata, India

R. Pani
Prafulla Chandra College, Kolkata, India

S. Hazra et al. (eds.), *Geographies of the Indian Subcontinent*,
https://doi.org/10.1007/978-3-031-74813-4_15

The international community, in recognition of the impending catastrophe, has come together for multilateral climate agreements since the Kyoto Protocol in 1997. As part of the Paris Climate Accord, signatory countries have pledged to generate a substantial part of their energy through renewable energy and made a commitment to reduce greenhouse gas emissions. The twenty-second session of the UN climate negotiations Conference of the Parties (COP22) in 2016 in Marrakesh, Morocco, resulted in considerable initiative in reducing CO_2 emissions (UNFCCC, 2016). The 26th session (COP26) held in Glasgow aimed to achieve global net zero by 2050[1] so as to limit the temperature rise to 1.5 °C since the preindustrial levels, requiring countries to declare their 2030 emission reduction targets based on phasing out coal, using electric vehicles and investing in renewables.

Globally, the electricity sector accounts for 31% of GHG emissions. According to the U.S. International Energy Agency data, the contribution of electricity to total CO_2 emission increased from 38% in 1992 to 42% in 2018. The sector is characterized by high carbon footprint due to the substantial use of fossil fuels. According to the World Nuclear Association, over 40% of energy-related carbon dioxide emissions are due to the burning of fossil fuels like coal, oil, and natural gas for electricity generation. Over the past two decades, global electricity generation has expanded approximately 1.6 times, with fossil fuels accounting for the largest share of this growth. According to the World Bank data, overall CO_2 emission from electricity generation has nearly doubled from 7907 million tonnes to 14,182 million tonnes during 1992 to 2018.

There has been a spectacular change in the individual contribution of countries in emission from electricity generation over the last three decades. While in 1992, the highest emitter of CO_2 was the USA, followed by Russia and China, in 2006, China surpassed the USA. Since 2013, China, the USA and India have become the top three highest emitters. Interestingly, India experienced a remarkable growth in electricity-induced emission; while it was in the sixth position in 1992, it emerged as the third largest emitter by 2013. The CO_2 emission in India from the power sector during 2021–2022 was 1002.02 million tonnes (Central Electricity Authority, GOI, 2022). Hence, reduction in carbon dioxide emission from electricity generation becomes a priority in the battle for mitigating greenhouse gas emission.

Decarbonization of the electricity sector warrants focussed policy making due to three reasons: first, the sector provides the means to use non-fossil low-carbon energy like hydro, nuclear, wind and solar; secondly, since electricity is clean at the point of final use, it is effective in improving air quality; and thirdly, regulation of energy-related emission is easier through centralized policies at power stations.

In order to curb carbon dioxide emission, the vital steps involve identification of the prime determining factors of emission, ascertaining their role in driving the changes in emission and formulating appropriate policies in accordance to priorities. The most common tool used for decoupling the contribution of each determining factor of emission is the decomposition analysis.

[1]However, China and India have committed to net zero emission by 2060 and 2070 respectively.

The purpose of this study is to decompose the determining factors of carbon dioxide emission caused due to electricity generation in the three highest emitting countries, viz., China, the USA and India. The objectives of the chapter are twofold: first, to analyze and compare the extent and pattern of emission of the top electricity-induced emitters, with emphasis on India; and secondly, to suggest policy framework and appropriate strategies.

2 Theoretical Framework of Decomposition Analysis

In decomposition analysis, the change in an indicator is distributed into components where each component represents the change in some predetermined factors that drive the change in the indicator. Decomposition is a useful analytical technique aimed to separate the effects that influence an aggregate indicator into disaggregate levels and identify the relationship of the particular effect with the indicator so as to provide information for policy making.

In environmental studies, Myers and Nakamura (1978) were the earliest to use the decomposition technique in order to examine energy use. A number of decomposition methods have been developed and used in studies since then. Ang and Zhang (2000) provide a detailed review of the methods and their variants used in energy and environmental studies. The most extensively used methodological tools to decompose the changes in environmental indicators into changes in the driving factors are Structural Decomposition Analysis (SDA) and Index Decomposition Analysis (IDA). They use data at two particular points of time and determine the contribution of each of the driving forces in altering the indicator. Although both SDA and IDA are aimed at decomposition analysis, they are based on distinctly different approaches and methodologies (Rose & Casler, 1996; Hoekstra & Bergh, 2003).

The SDA approach is based on the input-output model and requires more detailed data on economy-environment interaction. It is generally based on a 3–10-year time gap since input-output data are, in many cases, not constructed annually. On the other hand, IDA requires only aggregate sector-level data and hence uses annual data, which is widely available. This facilitates a very detailed analysis with respect to change over time periods. Hoekstra and Bergh (2003) specify that SDA is capable of more refined decompositions of economic and technological effects but IDA is capable of more detailed time and country studies. While SDA can only investigate absolute changes in indicators, IDA can address all the indicator forms and is more widely used than SDA due to its broader scope (Xu & Ang, 2013).

2.1 Index Decomposition Analysis (IDA)

There exists a number of methods for Index Decomposition Analysis, broadly based either on the concept of index numbers or Divisia index used widely in economics.

While index numbers measure relative changes based on ratios, the Divisia index is based on the natural logarithms of those ratios. It is the weighted average of relative growth rates, where the weights are the factors' shares in total value, given in the form of a line integral. The most popular IDA decomposition methods are based on Divisia index. The two common Divisia indices are:

(i) *AMDI*: The Arithmetic Mean Divisia Index method (AMDI), proposed by Boyd et al. (1988) uses an arithmetic mean weight function. The formulae for the AMDI method are as follows:

$$\text{Additive:}\quad \Delta V_{v_k} = 0.5\sum_i (V_{i,0} + V_{i,T}) \ln\left(\frac{v_{ik,T}}{v_{ik,0}}\right)$$

$$\text{Multiplicative:}\quad D_{v_k} = \exp\left[\sum_i \left\{0.5\left(\frac{V_{i,0}}{\sum V_{i,0}} + \frac{V_{i,T}}{\sum V_{i,T}}\right)\right\} \ln\left(\frac{v_{ik,T}}{v_{ik,0}}\right)\right]$$

(ii) *LMDI*: The additive method of Logarithmic Mean Divisia Index (LMDI) was proposed by Ang et al. (1998), and the multiplicative version was developed by Ang and Liu (2007). The logarithmic mean is useful when changing from a ratio to a difference and vice versa (Balk, 2003). The formulae for the LMDI method are:

$$\text{Additive:}\quad \Delta V_{v_k} = \sum_i L(V_{i,T}, V_{i,0}) \ln\left(\frac{v_{ik,T}}{v_{ik,0}}\right)$$

$$\text{Multiplicative:}\quad D_{v_k} = \exp\left\{\sum_i \left[\frac{L(V_{i,T}, V_{i,0})}{L(\sum V_{i,T}, \sum V_{i,0})}\right] \ln\left(\frac{v_{ik,T}}{v_{ik,0}}\right)\right\}$$

where the logarithmic mean of two positive numbers is given as

$$\begin{aligned} L(x, y) &= (y - x)/\log(y/x) \quad \text{for } x \neq y \\ &= x \quad\quad\quad\quad\quad\quad\quad\; \text{for } x = y \end{aligned}$$

In order to yield meaningful, complete and effective interpretation, a perfect decomposition technique should be easy to use and interpret, enable perfect decomposition with no residual, consistent in aggregation and satisfy time reversal test and zero value robustness. The AMDI method is simpler than LMDI since it is based on arithmetic mean weight function, while LMDI uses log mean weight function. Though AMDI satisfies the time reversal test, it neither satisfies the factor reversal test nor can it handle zero values.

On the other hand, the LMDI method is time reversal invariant and gives perfect decomposition since it satisfies the factor reversal test. However, due to the logarithmic terms, the method is incapable of dealing with zero and negative values in the data set. This problem can be tackled in two ways. First, Ang and Choi (1997) proposed the replacement of the zero values by a small number δ so that converging results are obtained when δ approaches zero. Secondly, Ang et al. (2000) show that use of analytical limits in the presence of zero values in the data set can address the problem in both additive and multiplicative decomposition. However, Wood and Lenzen (2006) argued that the procedure could produce significant errors if there are large numbers of replacements of zeroes by the constants in the data set.

The variance analysis decomposition model was developed by Pani and Mukhopadhyay (2011) as an alternative to LMDI. The method seems preferable to the LMDI approach owing to several reasons. It is very simple to understand and calculate, not requiring much advanced mathematical or econometric knowledge; unlike other methods, it uses absolute values instead of average values. It possesses all the advantages of LMDI and can also deal with negative values (Pani & Mukhopadhyay, 2011). They used this method to identify the contribution of each of the driving factors in changing the level of emission.

3 Literature Review

There exists a large literature that identifies the main factors affecting CO_2 emissions to be population growth, economic development, energy consumption and technological progress. The Kaya identity is one of the most widely used models that spells out the concurrent impacts of these factors on CO_2 emissions (Duro, 2010). It is the decomposition method that provides the theoretical underpinning to the analysis of emission reduction techniques.

Decomposition studies of CO_2 emissions from electricity generation have been carried out at both regional and country levels. Shrestha et al. (2009) decomposed the emissions from the power sector in select countries in Asia and the Pacific into factors of economic output, electricity intensity of the economy, fuel intensity, and generation structure. Karmellos et al. (2021) studied carbon dioxide emissions from electricity generation in the European Union and found that the main driving factor leading to increased emissions is the economic activity effect counterbalanced mainly by the contribution of the generation structure effect. De Oliveira-De Jesus (2019) analyzed the evolution of carbon accumulation intensity in Latin America and the Caribbean to identify the impact of generating capacity. Kim et al. (2020) in a decomposition analysis of 36 OECD countries found that EU nations fared better in terms of CO_2 reduction compared to that of non-EU nations, with the share of thermal generation as the key driver, and decrease in electricity intensity.

Country-level decomposition analysis of electricity-related CO_2 emissions has been conducted for China (Ding et al., 2020; Yang et al., 2023; Mai et al., 2020; Li et al., 2022), Philippines (Sumabat et al., 2016), Thailand (Muangthai et al., 2014), Nigeria (Emodi & Boo, 2015), Iran (Kouyakhi & Shavvalpour, 2021) and Spain (Alcántara et al., 2022).

4 Research Gap and Thrust of the Chapter

Electricity is the key to development, with countries throughout the world relying more on it for their developmental activities. With the global focus on zero emission (COP22) and about half of the emission being contributed by electricity, curtailing of emission due to electricity generation plays a catalytic role. Since China, the USA and India account for more than 80% of the world emission due to electricity generation, it is imperative to focus on these countries and explore ways how they may cut back their emission. But the extant studies are based on particular countries or regions, without considering their contribution to total emission. Secondly, India is an emerging economy drawing global attraction from investors, accelerating the energy demand, including electricity. However, very little research has been conducted to identify the driving forces of CO_2 emission due to electricity generation in India and related policy prescription to arrest growth in emission.

The contribution of the study in the literature is twofold: First, it analyzes the effects of changes in emission intensity of fuel and fuel mix on the change in carbon dioxide emission in generation of electricity; secondly, it explores the effect of domestic consumption, distribution loss, net export of electricity on emission, which has not been considered earlier.

5 Data Sources and Methodology

The data source for this study is the International Energy Agency (IEA). The emission of CO_2 from generation of electricity is expressed in million metric tons; energy consumption is expressed in terms of quadrillion British thermal unit (1015 Btu), and population is measured in millions. GDP in terms of the 2005 US dollar (Purchasing Power Parity) has been calculated from data on energy intensity. The GDP figures have been derived by dividing total energy consumption by the energy intensity of the GDP of respective countries.

This study is based on variance analysis decomposition model developed by Pani and Mukhopadhyay (2011). The model has been extended to accommodate decomposition of CO_2 emission from electricity generation.

5.1 Decomposition of Emission from Electricity Generation

The starting point of variance analysis method is the IPAT-based (Kaya, 1990) identity that decomposes emission into the major driving factors. We apply the Kaya identity (1990) to CO_2 emission from electricity generation to decompose it into five determining factors as

$$CO_2 = \frac{CO_2}{EF} \times \frac{EF}{ET} \times \frac{ET}{GDP} \times \frac{GDP}{P} \times P \quad (15.1)$$

where CO_2 is the carbon dioxide emitted from electricity generation, EF is the electricity generated from fossil fuel, ET is the total electricity generated, GDP is gross domestic product and P is population.

Equation (15.1) can be rewritten to express CO_2 emission from electricity generation in country 'i' at time period t as

$$I_i(t) = C_i(t).F_i(t).E_i(t).Y_i(t).P_i(t) \quad (15.2)$$

In Eq. (15.2), I_i denotes CO_2 emission from electricity generation; $C_i = \left(\frac{CO_{2i}}{EF_i}\right)$ is the emission intensity, representing per unit emission from electricity generated from fossil fuel; $F_i = \left(\frac{EF_i}{ET_i}\right)$ represents the fossil fuel intensity, which shows the ratio of electricity generated from fossil fuel to total electricity generation; $E_i = \left(\frac{ET_i}{GDP_i}\right)$ is the energy intensity, which represents the electricity generation requirement per unit of GDP; $Y_i = \left(\frac{GDP_i}{P_i}\right)$ is the per capita GDP and P_i is the population.

The factors that are amenable to policy efforts can be broadly grouped into two categories—the supply side and the demand side. Here electricity intensity of income represents the demand side while emission intensity of fossil fuel electricity and fossil fuel intensity of electricity can be considered as comprising of the supply side factors. A change in any of the factors leads to a proportionate change in emission and is termed as variance effect. The deviation in emission due to change in a particular factor can be referred to as the effect of that factor.

Now, the emission from electricity generation at time period '$t + 1$' when there is a change only in population, other factors remaining constant (assuming that per capita GDP, electricity intensity per unit GDP, ratio of fossil fuel and other sources of electricity generation, rate of emission per unit of fossil fuel electricity are unchanged), can be expressed as

$$I_{Pi}(t) = C_i(t).F_{i(t)}.E_i(t).Y_i(t).P_i(t+1) \quad (15.3)$$

The change in emission in power sector due solely to change in population is given by subtracting Eq. (15.3) from (15.2) and is called population effect. It can be calculated as

$$\begin{aligned} P_{\text{eff}} &= I_{Pi}(t) - I_i(t) \\ &= C_i(t).F_i(t).E_i(t).Y_i(t).P_i(t+1) - C_i(t).F_i(t).E_i(t).Y_i(t).P_i(t) \\ &= C_i(t).F_i(t).E_i(t).Y_i(t).\Delta P_i(t) \end{aligned} \quad (15.4)$$

The emission at time period 't + 1' when both population and per capita GDP change, other factors remaining constant, can be given as

$$I_{Yi}(t) = C_i(t).F_i(t).E_i(t).Y_i(t+1).P_i(t+1) \tag{15.5}$$

Income effect (I_{eff}) or the change in emission from electricity generation due solely to change in per capita GDP, with population at period (t + 1) and other factors remaining at the level of period (t) is obtained by subtracting (15.3) from (15.5). Therefore,

$$\begin{aligned} Y_{\text{eff}} &= I_{Yi}(t) - I_{Pi}(t) \\ &= C_i(t).F_i(t).E_i(t).Y_i(t+1).P_i(t+1) - C_i(t).F_i(t).E_i(t).Y_i(t).P_i(t+1) \\ &= C_i(t).F_i(t).E_i(t).\Delta Y_i(t).P_i(t+1) \end{aligned} \tag{15.6}$$

Similarly, electricity intensity effect, fossil fuel intensity effect and emission intensity effect are obtained, respectively, as

$$E_{\text{eff}} = C_i(t).F_i(t).\Delta E_i(t).Y_i(t+1).P_i(t+1) \tag{15.7}$$

$$F_{\text{eff}} = C_i(t).\Delta F_i(t).E_i(t+1).Y_i(t+1).P_i(t+1) \tag{15.8}$$

$$C_{\text{eff}} = \Delta C_i(t).F_i(t+1).E_i(t+1).Y_i(t+1).P_i(t+1) \tag{15.9}$$

Hence, total emission effect (I_{eff}) is determined either by the sum of all the factor effects as

$$P_{\text{eff}} + Y_{\text{eff}} + E_{\text{eff}} + F_{\text{eff}} + C_{\text{eff}} \tag{15.10}$$

or by the difference between the product of the factors at time period (t + 1) and (t) as

$$C_i(t+1).F_i(t+1).E_i(t+1).Y_i(t+1).P_i(t+1) - C_i(t).F_i(t).E_i(t).Y_i(t).P_i(t) \tag{15.11}$$

5.2 *Decomposition of Emission Intensity of Fossil Fuel Electricity*

Fossil fuel primarily consists of coal (C), oil (O) and natural gas (G). Hence, now the emission due to electricity generation from fossil fuel is decomposed on the basis of the source of fossil fuel into emission from coal electricity (CO_{2C}), emission from oil electricity (CO_{2O}) and emission from gas electricity (CO_{2G}). Thus, emission intensity of fossil fuel electricity effect can be expressed as

$$\frac{CO_2}{EF} = \frac{CO_{2C} + CO_{2O} + CO_{2G}}{EF} = \frac{CO_{2C}}{EF_C} \cdot \frac{EF_C}{EF} + \frac{CO_{2O}}{EF_O} \cdot \frac{EF_O}{EF} + \frac{CO_{2G}}{EF_G} \cdot \frac{EF_G}{EF}$$
$$= \sum\nolimits_j \left(\frac{CO_{2j}}{EF_j} \cdot \frac{EF_j}{EF} \right) \tag{15.12}$$

Equation (15.12) depicts that the emission intensity of fossil fuel electricity (C_i) can be decomposed into emission due to electricity generated from use of fossil fuel j or the emission intensity of fuel j (M_{ji}) and the proportion of fossil fuel j in total fossil fuel used in electricity generation or the energy intensity of fuel j (S_{ji}), where $j = C,O,G$. It indicates that emission intensity of electricity generated from fossil fuel depends on the structural composition of fuels and the emission generated per unit of each fuel. Change in emission due to change in structural composition demonstrates mix variances, while change in emission due to change in emission intensity of each fuel shows its yield variance.

Now Eq. (15.2) can be rewritten as

$$I_i(t) = \left\{ \sum\nolimits_{ij} (M_{ij}.S_{ij}) \right\} .F_i(t).E_i(t).Y_i(t).P_i(t) \tag{15.13}$$

Thus, fossil fuel mix variances can be written as

$$C_{\text{Cmvar}} = M_{Ci}(t).\Delta S_{Ci}(t).F_i(t+1).E_i(t+1).Y_i(t+1).P_i(t+1) \tag{15.14}$$

$$C_{\text{Omvar}} = M_{Oi}(t).\Delta S_{Oi}(t).F_i(t+1).E_i(t+1).Y_i(t+1).P_i(t+1) \tag{15.15}$$

$$C_{\text{Gmvar}} = M_{Gi}(t).\Delta S_{Gi}(t).F_i(t+1).E_i(t+1).Y_i(t+1).P_i(t+1) \tag{15.16}$$

And fossil fuel yield variances can be written as

$$C_{\text{Cyvar}} = \Delta M_{Ci}(t).S_{Ci}(t+1).F_i(t+1).E_i(t+1).Y_i(t+1).P_i(t+1) \tag{15.17}$$

$$C_{\text{Oyvar}} = \Delta M_{Oi}(t).S_{Oi}(t+1).F_i(t+1).E_i(t+1).Y_i(t+1).P_i(t+1) \tag{15.18}$$

$$C_{\text{Gyvar}} = \Delta M_{Gi}(t).S_{Gi}(t+1).F_i(t+1).E_i(t+1).Y_i(t+1).P_i(t+1) \tag{15.19}$$

Now total electricity generated (ET) can either be consumed domestically or exported, while some electricity is lost during distribution. Hence,

$$\frac{ET}{GDP} = \frac{DC + DL + X}{GDP} = \sum\nolimits_k \frac{ET_k}{GDP} \tag{15.20}$$

where k = domestic consumption (DC), distribution loss (DL), net export (X).

Now Eq. (15.2) can be rewritten as

$$I_i(t) = C_i(t).F_i(t).\left(\sum\nolimits_k \frac{\mathrm{ET}_k}{\mathrm{GDP}}\right).Y_i(t).P_i(t) \quad (15.21)$$

Thus, domestic consumption, distribution loss and net export variances can be written as

$$E_{\mathrm{Cvar}} = C_i(t).F_i(t).\Delta E_{Ci}(t).Y_i(t+1).P_i(t+1) \quad (15.22)$$

$$E_{\mathrm{Dvar}} = C_i(t).F_i(t).\Delta E_{Di}(t).Y_i(t+1).P_i(t+1) \quad (15.23)$$

$$E_{\mathrm{Xvar}} = C_i(t).F_i(t).\Delta E_{Mi}(t).Y_i(t+1).P_i(t+1) \quad (15.24)$$

6 Result and Discussion

6.1 *Overall Analysis of CO_2 Emission from Electricity Generation*

World CO_2 emission from electricity generation has escalated by nearly 80% (from 7907 million tonnes to 14,182 million tonnes) during 1992 to 2018. Figure 15.1 shows that emission from the sample countries have increased by more than 2.5 times (from 3063 million tonnes to 8114 million tonnes) over the same period. Both the curves in the diagram roughly follow the same path, implying that these three countries drive the world emission and the remaining countries have little role to play (25% rise in emission over the 26 years).

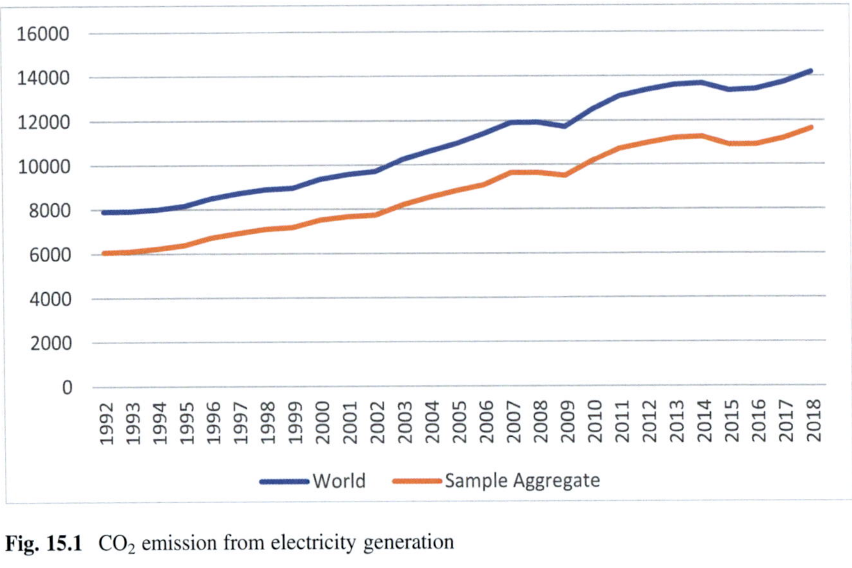

Fig. 15.1 CO_2 emission from electricity generation

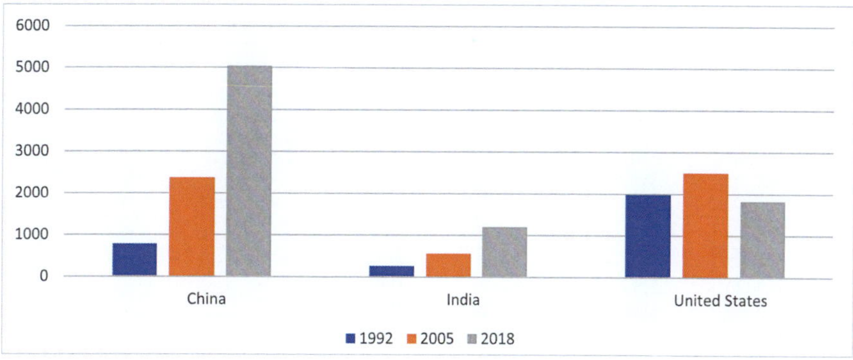

Fig. 15.2 CO_2 emission from electricity generation of sample countries in 1992, 2005, and 2018

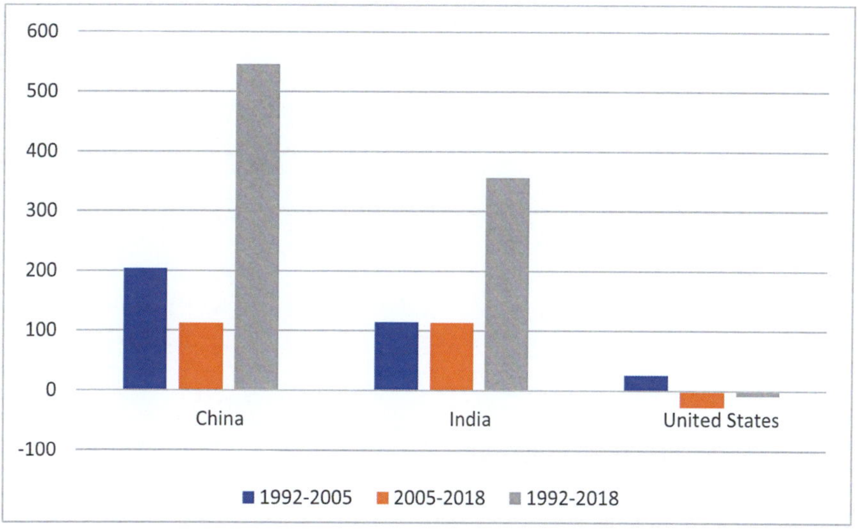

Fig. 15.3 Percentage growth in CO_2 emission from electricity generation

Figure 15.2 shows that there has been a spectacular change in the individual contribution of countries in emission from electricity generation over the last three decades. While emission in China increased about seven times during 1992–2018, that in India escalated by about five times. The USA succeeded in reducing emission from 1986.9 million metric tonnes to 1829.8 million metric tonnes during the period.

The overall emission growth percentage in China during 1992–2018 was remarkably high; nonetheless, the percentage growth reduced considerably in the second phase (2005–2018) than the first phase (1992–2005), although the increase in volume of emission nearly doubled. In India, the percentage growth in emission remained stagnant in both the phases. The USA, on the other hand, demonstrated negative growth in emission in the second phase (Fig. 15.3).

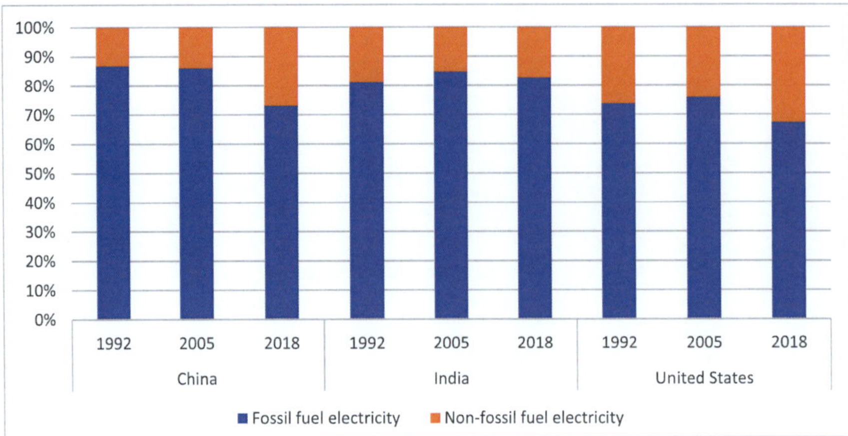

Fig. 15.4 Percentage of electricity generation from fossil fuel and non-fossil fuel sources

All the three countries are highly dependent on fossil fuel energy sources for generation of electricity (Fig. 15.4). China and the USA reduced their dependence on fossil fuel use in 2018.

Thus, the preliminary analysis reveals that both China and the USA have high volume of emission from electricity generation, but they are continuously working on reducing their emission levels. On the other hand, India has been experiencing constant growth in emission. To comprehend the reason and the driving forces behind this, a detailed decomposition analysis is undertaken in the following section.

6.2 Decomposition Analysis

Table 15.1 shows that the population and income effects are positive (emission-increasing) for all the three countries, the latter effect being predominantly higher. The income effect in China is 50 times more than than that of the population, followed by India (nearly 8 times) and the USA (nearly double). The technological effect represented by energy intensity, fossil fuel intensity and emission intensity work favourably to arrest the positive effect of population and income in all the countries; however, in India the fossil fuel effect has been emission-increasing. While technological improvement through reducing electricity intensity is more instrumental in the case of the USA, China has been successful in curtailing fossil fuel intensity and emission intensity.

Although the population and income effects have grown over time in China and India, in the case of the USA the increasing effect has plummeted. China and the USA have successfully reduced the fossil fuel effect over time. India has improved with respect to fossil fuel intensity, but the improvement is not very remarkable.

Table 15.1 Decomposition results

	Countries	P_{eff}	Y_{eff}	E_{eff}	F_{eff}	C_{eff}	ΔCO_2
1992–2018	China	142.82	7277.85	−787.84	−1172.26	−1201.47	4259.10
	India	127.23	977.01	−80.62	24.85	−114.48	934.00
	United States	557.39	1270.54	−1103.83	−235.30	−645.90	−157.10
	Total	1182.54	11101.24	−2485.15	−1331.14	−2984.59	5482.90
1992–2005	China	80.36	1829.52	−103.99	−20.93	−192.07	1592.90
	India	68.45	242.71	−13.02	24.45	−23.19	299.40
	United States	302.56	765.65	−442.07	75.60	−183.24	518.50
	Total	582.97	3491.15	−626.54	169.25	−776.73	2840.10
2005–2018	China	172.22	4689.26	−432.07	−1029.00	−734.21	2666.20
	India	99.87	676.94	−49.69	−30.02	−62.49	634.60
	United States	278.87	344.16	−529.10	−292.35	−477.18	−675.60
	Total	846.60	6498.51	−1420.43	−1490.94	−1790.94	2642.80

During 1992–2018, while the USA achieved reduction in the proportional effect of coal use in fossil fuel mix, China showed a decline in oil effect. To achieve these reductions, they have used more natural gas in the fuel mix. But in India, oil has been substituted with coal in the fuel mix. All the three countries have been successful in reducing emission intensities of coal and gas; however, China has higher rate of reduction with respect to coal, while the emission intensity of gas has plummeted in the USA. There have been moderate reductions in emission intensities of coal and gas in India.

The phase-wise analysis shows that both the USA and China experienced an improvement in reducing emission through the fuel mix effect, while there has been an emission increasing effect in India. This is due to the fact that over time, China has reduced the proportion of coal, the USA shows improvement with regard to both coal and oil, but India has increased its coal use. Again, China has improved its performance in emission intensity of coal in the second phase compared to the first phase, while no remarkable changes can be noticed in the other two countries (Table 15.2).

The USA has been more effective in curtaining emission through consumption efficiency, followed by China (Table 15.3). The performance of China regarding distribution efficiency is better than the USA. India has hardly achieved any remarkable feat in this respect. Over time, China has experienced considerable improvements in consumption as well as distribution efficiencies. Consumption efficiency remained fairly similar in both the phases in the USA, the distribution efficiency nearly doubled in the second phase than in the first one. The performance in India deteriorated with respect to consumption efficiency, but it managed to reduce emission considerably through distribution efficiency.

The net export effect is not very remarkable, except China, which had high effect in the first phase, but it has managed to control it over time. However, in the second

Table 15.2 Mix and yield variances

		Mix variances				Yield variances			
	Countries	Coal	Oil	Gas	Total	Coal	Oil	Gas	Total
1992–2018	China	290.98	−425.50	129.07	−5.45	−1171.81	12.25	−36.46	−1196.02
	India	72.58	−78.04	−9.14	−14.60	−87.46	0.78	−13.20	−99.87
	United States	−855.82	−60.02	590.42	−325.42	−34.46	2.60	−288.62	−320.48
	Total	−125.00	−1493.61	1332.35	−286.26	−1996.43	18.70	−720.60	−2698.33
1992–2005	China	154.05	−127.65	2.78	29.18	−222.45	2.27	−1.07	−221.25
	India	−30.92	−12.33	27.23	−16.01	16.28	−4.61	−18.85	−7.17
	United States	−192.91	−1.90	122.68	−72.13	−21.61	15.35	−104.85	−111.11
	Total	194.61	−671.59	478.53	1.56	−661.36	123.39	−240.32	−778.29
2005–2018	China	−76.43	−119.98	106.71	−89.70	−638.01	11.77	−18.27	−644.51
	India	146.62	−41.65	−44.97	60.01	−128.69	2.35	3.84	−122.50
	United States	−671.05	−67.74	370.52	−368.27	−21.70	−2.06	−85.15	−108.91
	Total	−721.88	−511.37	659.57	−573.68	−869.20	−60.04	−288.02	−1217.27

Table 15.3 Consumption, distribution and net export variances

	Countries	E_{Cvar}	E_{Dvar}	E_{Xvar}
1992–2018	China	−596.37	−260.89	69.42
	India	−46.89	−42.44	8.71
	United States	−985.65	−120.99	2.80
	Total	−2144.40	−440.16	99.41
1992–2005	China	−109.24	−19.22	24.47
	India	−51.67	37.62	1.03
	United States	−414.63	−36.66	9.22
	Total	−693.25	36.04	30.67
2005–2018	China	−241.68	−185.63	−4.76
	India	76.47	−132.43	6.27
	United States	−448.64	−72.11	−8.35
	Total	−960.98	−472.94	13.49

phase, it is noticeable that while both the USA and China have been improving the net export effect, in India, the effect is worsening.

7 Policies Adopted for Emission Cut

According to the report on 'The Long-Term Strategy of the United States: Pathways to Net-Zero Greenhouse Gas Emissions by 2050', published by the United States Department of State in 2020, the zero net emission target by the USA is based largely on two policy pillars: electrification of end uses and decarbonization of electricity. Measures have been taken to electrify most of the economy, from cars to buildings and industrial processes. In recent years, the transition to a clean electricity system has been accelerating due to government policies, declining costs for solar and wind technologies, and conducive consumer demand.

In China too, a power sector dominated by renewables provides the foundation for its clean energy transition. China has been focussing on closing small and outdated units and giving priority to the development of high-efficiency and large-scale coal fire power plant units (Yang & Wang, 2023). New investment in electricity-based transportation has lowered the energy intensity of passengers. China is emerging as a frontrunner in clean energy innovation, with public spending on low-carbon energy research and development rising by 70% since 2015 (IEA, 2021).

India in recent times, has launched several schemes, such as the National Mission for Enhanced Energy Efficiency to accelerate the transition in the power sector. The Jawaharlal Nehru National Solar Mission envisages for tax exemptions, capital subsidies and incentive components of solar energy value chain, Renewable Energy Supply for rural areas whereby viable models for de-centralized energy supply in rural areas from renewable sources are developed, Solar or Green City development to promote the use of renewable energy in urban areas and Akshay Urja programme to encourage establishment of shops to make renewable energy products easily available to people and provide after sales and repair services. Apart from the Central policies, Rajasthan, Gujarat and Karnataka have devised their state solar policies for furtherance of solar mission. The recently enacted Energy Conservation (Amendment) Bill, 2022 seeks to make it obligatory for designated consumers, such as certain polluting industries, transport sector and commercial buildings to meet a minimum share of energy consumption from non-fossil sources.

8 Conclusion and Policy Prescriptions for Electricity Sector

The alarming spiralling of CO_2 emission resulting in global warming and frequent natural disasters have compelled countries throughout the world to desperately reduce CO_2 emission. Data shows that China, followed by the USA and India are the major emitters, with the electricity sector accounting for a major share in total emission. Although electricity-related CO_2 emission is higher in theUSA and China in absolute terms, they have undertaken several measures that have been instrumental in cutting down on emission. The decomposition analysis in the study reflects considerable improvement with respect to the major determining factors in these countries. On the other hand, India is the third largest producer of electricity, with power generation projected to increase nearly threefold from 2017 to 2040 (International Energy Agency, 2019). More than 70% of power generation is sourced from coal causing emission, making it imperative for India to restrict its emission in the electricity sector.

But India has been increasingly contributing to higher growth rate of emission and unable to reduce it. This is due to three reasons: first, its dependence on coal use has been escalating over time; second, there has been no significant improvement in emission intensity of coal; and third, while India's policy thrust has been on solar energy for electricity generation, it is quite expensive and imposes technical barriers in many cases. Due to the topographical diversity of India, there are several opportunities for the policymakers to explore alternate sources of electricity. In addition to solar power, emphasis should be given on wind, hydro, tidal, etc. for generating electricity.

Generation of electricity based on higher efficiency technologies such as modernization of thermal power plants and adoption of smart electricity grid technology are some of the potential ways for emission cut in the power sector. Smart grid generation includes accelerated and strategic deployment of gas and renewable energy sources like wind, solar, or hydropower. Displacement or saving of grid electricity through on-grid solar panels can make substantive reductions in emissions from the power sector.

Electricity requires transmission from large power plants to the consumers through extensive networks. The transmission and distribution (T&D) over long distances creates power losses. The loss can be of two types: technical and commercial. While technical losses are due to old and faulty equipment, undersized conductors and transformers, improper maintenance and operation, faulty meters, etc., the reasons for commercial losses are tempering of meters, power theft, wrong metering and billing, poor vigilance and poor management by utility, etc. In India, the extent of T&D loss is estimated at about 32% (Gaurav et al., 2018). The electricity loss creates a mismatch between demand and supply of power. Minimization of distribution loss can be an effective means to ensure demand-supply balance and prevent unnecessary generation of additional power to meet the losses. Smart transmission and distribution can optimize the existing overhead transmission lines, underground cables, transformers and substations such that minimum generating capacities are required in the future (Abdallah & El-Shennawy, 2013). Instead of large power plants with long-distance coverage, there can be relocation with installation of small power plants in different areas to meet the local demand so that the distance-related power loss can be minimized.

Coal use is still subsidized through different fiscal policies. Gradual shift towards renewables by redirecting subsidies away from fossil fuels to renewable energy sources is inevitable to augment renewable energy in electricity generation. It is necessary to develop an objective action plan for phasing out coal.

However, clean sources of energy have several inherent limitations and constraints on feasibility, requiring huge investment and technological advancement. Proper choices of fossil fuel can also curtail emission to some extent. Pani and Mukhopadhyay (2013) pointed out that on average, coal is 77% more emitting than natural gas and 45% more emitting than petroleum, while petroleum is 22% more emitting than gas. Hence, substitution of coal by gas or petroleum, or of petroleum by gas may be a more viable policy option. Although the government has taken some measures like replacement of petroleum vehicles by CNG/electric vehicles, coal- and diesel-operated trains by electric trains, the extent of such replacement is not noteworthy. For example, electric vehicles are charged from electricity but the main source of electricity still remains to be thermal power. Therefore, more serious steps are desired from the policy makers. The operation of electric vehicles should be sourced from electricity generated from renewable energy.

References

Abdallah, L., & El-Shennawy, T. (2013). Reducing carbon dioxide emissions from electricity sector using smart electric grid applications. *Journal of Engineering*, 845051. https://doi.org/10.1155/2013/845051

Alcántara, V., Padilla, E., & Del Río, P. (2022). The driving factors of CO_2 emissions from electricity generation in Spain: A decomposition analysis. *Energy Sources, Part B: Economics, Planning, and Policy, 17*, 1. https://doi.org/10.1080/15567249.2021.2014604

Ang, B. W., Zhang, F. Q., & Chew, E. P. (2000). A decomposition technique for quantifying real process performance. *Production Planning & Control: The Management of Operations, 11*(4), 314–321.

Ang, B. W., & Liu, N. (2007). Handling zero values in the logarithmic mean Divisia index decomposition approach. *Energy Policy, 35*, 238–246.

Ang, B. W., & Zhang, F. Q. (2000). A survey of index decomposition analysis in energy and environmental studies. *Energy, 25*(12), 1149–1176.
Ang, B. W., & Choi, K. H. (1997). Decomposition of aggregate energy and gas emission intensities for industry: A refined Divisia index method. *Energy Journal, 18*(3), 59–73.
Ang, B. W., Zhang, F. Q., & Choi, K. (1998). Factorizing changes in energy and environmental indicators through decomposition. *Energy, 23*(6), 489–495.
Balk, B. M. (2003). Ideal indices and indicators for two or more factors. *Journal of Economic and Social Measurement, 28*, 203–217.
Boyd, G. A., Hanson, D. A., & Sterner, T. (1988). Decomposition of changes in energy intensity: A comparison of the Divisia index and other methods. *Energy Economics, 10*(4), 309–912.
Central Electricity Authority. (2022). *CO2 baseline database for the Indian power sector*. Ministry of Power, Government of India.
De Oliveira-De Jesus, P. M. (2019). Effect of generation capacity factors on carbon emission intensity of electricity of Latin America & the Caribbean, a temporal IDA-LMDI analysis. *Renewable and Sustainable Energy Reviews, 101*, 516–526.
Ding, Z., Li, K., Zhu, H., Zhou, W., Liu, J., & Bai, Z. (2020). Decomposition analysis of CO_2 emission in China's electric power industry. *Journal of Physics: Conference Series, 1549*. https://doi.org/10.1088/1742-6596/1549/2/022023
Duro, J. A. (2010). Decomposing international polarization of per capita CO2 emissions. *Energy Policy, 38*(11), 6529–6533.
Emodi, N. V., & Boo, K. J. (2015). Decomposition analysis of CO_2 emissions from electricity generation in Nigeria. *International Journal of Energy Economics and Policy, 5*(2), 565–573.
Environmental Protection Agency (2017). *Annual Report and Accounts*. EPA, Ireland.
Gaurav, S., Pal, S., Jain, S., & Katna, R. (2018). Transmission and distribution losses and aggregate technical and commercial losses. *Global Journal of Enterprise Information System*. https://doi.org/10.18311/gjeis/2018/19938
Hoekstra, R., & Bergh, J. J. (2003). Comparing structural decomposition analysis and index. *Energy Economics, 25*, 39–64.
IEA. (2021). *An energy sector roadmap to carbon neutrality in China*. IEA. https://www.iea.org/reports/an-energy-sector-roadmap-to-carbon-neutrality-in-china
International Energy Agency. (2019). *World energy outlook 2019*. https://www.iea.org/reports/world-energy-outlook-2019
Karmellos, M., Kosmadakis, V., Dimas, P., Tsakanikas, A., Fylaktos, N., Taliotis, C., & Zachariadis, T. (2021). Decomposition and decoupling analysis of carbon dioxide emissions from electricity generation: Evidence from the EU-27 and the UK. *Energy, 231*, 120861.
Kaya, Y. (1990). *Impact of carbon dioxide emission control on GNP growth: Interpretation of proposed scenarios*. Paper presented to IPCC Energy and Industry Sub-group Response Strategies Working Group.
Kim, H., Kim, M., Kim, H., & Park, S. (2020). Decomposition analysis of CO_2 emission from electricity generation: Comparison of OECD countries before and after the financial crisis. *Energies, 13*(14), 3522. https://doi.org/10.3390/en13143522
Kouyakhi, N. R., & Shavvalpour, S. (2021). The driving forces of energy consumption and carbon dioxide emissions in Iran's electricity sector: A decomposition analysis based on types of ownership. *Cleaner Environmental Systems, 2*, 100012.
Li, W., Xu, L., & Jin, Y. (2022). The influence factors of interprovincial power transmission on China's CO_2 emissions. *Science Progress, 105*(4). https://doi.org/10.1177/00368504221137466
Mai, L., Ran, Q., & Wu, H. (2020). A LMDI decomposition analysis of carbon dioxide emission from the electric power sector in Northwest China. *Natural Resource Modeling, 33*(4), e12284. https://doi.org/10.1111/nrm.12284
Meyerson, F. B. (1998). Population, development and global warming: Averting the tragedy of the climate commons. *Population Environment, 19*, 443–463.

Muangthai, I., Lewis, C., & Lin, S. J. (2014). Decoupling effects and decomposition analysis of CO2 emissions from Thailand's thermal power sector. *Aerosol and Air Quality Research, 14*, 1929–2938. https://doi.org/10.4209/aaqr.2014.03.0066

Myers, J., & Nakamura, L. (1978). *Saving energy in manufacturing*. Ballinger.

Pachauri, R. K., & Mayer, L. (2014). *Intergovernmental panel on climate change climate change, synthesis report*. IPCC.

Pani, R., & Mukhopadhyay, U. (2011). Variance analysis of global CO_2 emission—A management accounting approach for decomposition study. *Energy, 36*, 486–499.

Pani, R., & Mukhopadhyay, U. (2013). Management accounting approach to analyse energy related CO_2 emission: A variance analysis study of top 10 emitters of the world. *Energy Policy, 52*(2013), 639–655.

Rose, A., & Casler, S. (1996). Input-output structural decomposition analysis: A critical appraisal. *Economic Systems Research, 8*(1), 33–62.

Shrestha, R. M., Anandarajah, G., & Liyanage, M. H. (2009). Factors affecting CO2 emission from the power sector of selected countries in Asia and the Pacific. *Energy Policy, 37*, 2375–2384.

Sumabat, A. K., Lopez, N. S., Yu, K. D., Hao, H., Li, R., Geng, Y., & Chiu, A. S. F. (2016). Decomposition analysis of Philippine CO2 emissions from fuel combustion and electricity generation. *Applied Energy, 164*, 795–804.

UNFCCC. (2016). *Report of the conference of the parties on its twenty-first session, held in Paris from 30 November to 13 December 2015*. UNFCCC.

Wood, R., & Lenzen, M. (2006). Zero-value problems of the logarithmic mean divisia index decomposition method. *Energy Policy, 34*(12), 1326–1331.

Xu, X. Y., & Ang, B. W. (2013). Index decomposition analysis applied to CO2 emission studies. 683. *Ecological Economics, 93*, 313–329.

Yang, F., Yang, X., Tian, X., Wang, X., & Xu, T. (2023). Decomposition analysis of CO_2 emissions of electricity and carbon-reduction policy implication: A study of a province in China based on the logarithmic mean Divisia index method. *Clean Energy, 7*(2), 340–349.

Yang, R., & Wang, W. (2023). Potential of China's national policies on reducing carbon emissions from coal-fired power plants in the period of the 14th Five-Year Plan. *Heliyon, 9*(9), e19868.

Chapter 16
Revisiting Exploratory Research Methodology in the Context of Environment and Development in the Anthropocene

Lakshminarayan Satpati and Anwesha Haldar

Abstract The Anthropocene is an age of dilemma to mankind, as it has been experiencing the dichotomy of fast pace development at the expense of habitable environment which provides biological diversity and food security to all. Understanding this complex situation necessitates incorporation of diverse agenda as well as approaches for research to enhance our knowledge system. This chapter tries to explore the relevance of exploratory research as a pertinent methodology by assessing the different methods in exploratory research that have progressed since the last century in social science in general and analyse their relevance to the present geographical research scenario in India.

1 The Anthropocene Research Paradigm

The opportunities and challenges to man in any geographical region is determined by the risks and rewards associated with the individual as well as community-level psychology of fear and greed, leading to settlements and occupational transformation in it. The quality of land, water, flora and fauna provides the initial base for livelihood support to the inhabitants, to be followed by population growth, intensification of man-environmental relationships, and occupational change synchronized with technological advancements. Human civilizations throughout the ages has witnessed such transformations in various ecosystems, which later on connected among each other to make the people more biospheric and thereby exploitation of the natural resources with scant regard to their sustainability resulting in deterioration of environment. In this context, time-space tele-connections are very important to understand the rootedness as well as hegemony among the communities. Earlier, in the sun-synchronous lifestyle, availability of energy and water controlled the livelihood support systems in an isolated manner over space giving regional

L. Satpati (✉)
UGC-MMTTC, University of Calcutta, Kolkata, India

A. Haldar
Department of Geography, East Calcutta Girls' College, Kolkata, West Bengal, India

S. Hazra et al. (eds.), *Geographies of the Indian Subcontinent*,
https://doi.org/10.1007/978-3-031-74813-4_16

specializations in occupations. However, with the growing mobility and connectivity, the communities became more intensively engaged to exploit resources even at remote places without understanding their environmental processes and consequences resulting in global crises. The recent globalization has accentuated over-exploitation of resources in the context of contemporary climate change and biodiversity loss. These changes, particularly was observed after the period of industrial revolution, and during the Progressive Era during as per the American History in the early 1900s lead to the perception as the greatest social science laboratory in the United States that gave a spurt to research in human-oriented methodologies (Recchiuti, 2007). Even though geologists refer to the Anthropocene times since the 1950s, the influence of mankind in shaping the earth's planetary systems dates back to the 1500s (Ellis et al., 2016). The Anthropocene, a term used to describe the current geological era in which human activities have had a significant impact on the earth's ecosystems, has brought attention to the role of human beings in shaping the environment. It was first proposed by Paul Crutzen and Eugene Stoermer in 2000, as the idea that human activity has had a significant impact on the earth's environment and geology, leading to a new geological epoch. This era is said to be irreversible, necessitating new adaptive and more mitigation measures to be taken by the mankind. This era was instrumental in changing societal interdependence relationships, and sketches foundations of an emerging new paradigm in the social sciences, 'Earth System' governance (Biermann, 2014). Social scientists like Jason W. Moore prefer to interpret this situation with a more pointed terminology as the 'Capitalocene', in which the consumer-oriented capitalistic mode of production has been held responsible for the environment-development dichotomy (Moore, 2016, 2017). Moore in his book *Capitalism in the Web of Life: Ecology and the Accumulation of Capital* (2015) has described the current epoch as a product of capitalism and its adverse impact on the earth's environment and ecology. The Anthropocene and Capitalocene are two different ways of conceptualizing the current epoch of the earth's history. However, traditional research on environment and development has been heavily influenced by anthropocentrism, which prioritizes the needs and perspectives of human beings over those of other species and the natural world (Haraway, 2016). These approaches have led to complex ideas between the environment and development, and human interactions having detrimental consequences for the sustainability of life on earth. The Anthropocene approach typically employs a natural science-based methodology, using data from fields such as geology, palaeontology, and climatology to study the impact of human. This approach has been used to study changes in the earth's climate, the spread of invasive species, and the modification of the earth's surface (Steffen et al., 2011).The Capitalocene approach, employs a social science-based methodology, using data from fields such as political economy, sociology, and history to study the impact of capitalism. This approach emphasizes on the study the relationship between capitalism and environmental degradation, the impact of capitalist production on biodiversity, and the role of capitalist accumulation in driving environmental crisis (Moore, 2015). One of the major consequences of anthropocentrism in research on environment and development is the prioritization of economic growth over ecological sustainability (Latouche, 2009) leading to the exploitation of natural resources and the degradation of the environment in the pursuit of economic development. Additionally, anthropocentrism has led to a lack of attention to the perspectives and needs of other species and the natural world, resulting in a lack of understanding of the ways in which human activities are impacting the environment (Foster, 2000).

In the period of imperialism, the Global North-South politico-economic relationship was extremely hegemonic upon the colonized countries. The decolonization period witnessed rapid transformation of the global economy that strives towards self-reliance in agriculture and food security, industry, trade and commerce, urbanization, and provisioning of the basic amenities. All these activities led to intense pressure on land resulting in degradation of environment. However, since 1970s, global concern for environmental health, without sacrificing social welfare, heralded new paradigm in research, innovation, and planning for sustainable development.

Research being a study towards innovations, principally bases on some underlying philosophical assumptions about what constitutes 'valid' research, and which research method(s) is/are appropriate for the development of knowledge in a given study (Antwi & Hamza, 2015). It is essential in social sciences as it helps in the development of theories and concepts that can be applied to real-world situations. The social sciences are concerned with studying human behaviour and society, and research is critical in gathering empirical evidence to test theories and hypotheses in these fields. According to Babbie (2016), research in the social sciences helps to identify the causal relationships between different social phenomena, which is essential for developing theories that can be applied in practice. Furthermore, research in social sciences is critical in identifying trends and patterns in human behaviour and society. It is an essential tool for understanding the complexity of social phenomena and identifying trends and patterns that would otherwise be difficult to discern (Bryman, 2016). By providing empirical evidence on the effectiveness of different policies and interventions, research can help policymakers make informed decisions and develop effective strategies for addressing social problems. Research is critical in providing policymakers with evidence-based recommendations for addressing social issues and improving the lives of citizens (Peters, 2016).

The paradigm of the methodology governs its conceptual framework, which refers to the different approaches, perspectives, and methods used in conducting research in this field. These paradigms guide researchers in the design, implementation, and interpretation of research studies and reflect different assumptions and worldviews about the natural and human-made environment. Different research paradigms have been developed over time, each with its own set of assumptions, methods, and approaches. Over centuries, scientific studies have evolved from environmentalism approach to Anthropocene approach with methods ranging from quantification-based positivistic method to qualitative humanistic methods and right up to the critical theory methods. In geographical research, 'methods' refer to the specific techniques and procedures that researchers use to collect and analyse data, while 'methodology' refers to the broader theoretical and philosophical framework within which those methods are situated (Kitchin & Tate, 2013). According to Hay

(2016), methods in geographical research include techniques such as surveys, interviews, geographic information systems (GIS) mapping, and statistical analysis, among others. These methods are employed to collect and analyse data on the spatial and temporal patterns of human and physical phenomena. On the other hand, methodology in geographical research encompasses the theoretical and philosophical underpinnings that guide the research process. Methodology includes the researcher's ontological and epistemological assumptions, as well as their theoretical perspective and research design. For example, a researcher who adopts a critical realist methodology may employ a mixed methods approach to collect both quantitative and qualitative data to gain a better understanding of the complex, dynamic, and interrelated social and environmental issues (Mackenzie, 2019; Janowicz et al., 2020).

2 Contemporary Research Methodologies in Geography

In the Anthropocene, geographical research has forayed into various directions, bringing out critical inter-linkages in research. In the present context, one of the most all-encompassing disciplines of social sciences is geography. This is a multidisciplinary subject that involves studying the physical, social, and economic aspects of the world around us. According to Boyle et al. (2015), research in geography helps to identify and explain the physical processes that shape the earth's surface, including climate change, natural hazards, and landform evolution. By using GIS and other spatial analysis tools, researchers can analyse spatial patterns and relationships between different variables such as population density and land use. This knowledge is critical for urban and regional planning, natural resource management, and disaster management (Longley et al., 2015).

This discussion draws upon a range of scholarly sources, including seminal works by prominent social scientists, to provide a comprehensive overview of exploratory research in the social sciences and its role in generating new knowledge and informing subsequent research designs. It also highlights the ongoing need for continued innovation and experimentation in exploring new research methods and techniques to address complex social issues and challenges. A brief chronological study on the use of different research methods and analysis, particularly in geography and its allied subjects, reveals that exploratory research always gained predominance. According to Creswell (2014), exploratory research is characterized by its flexibility and adaptability, and it often involves the use of multiple data sources and methods to gain a comprehensive understanding of a particular phenomenon. This may involve field observations, interviews, surveys, and the collection and analysis of secondary data sources such as maps, photographs, and historical records. It is often used in combination with other types of research, such as descriptive and explanatory research. Descriptive research aims to describe and summarize the characteristics of a particular phenomenon, while explanatory research seeks to explain the relationships between different variables and phenomena (Neuman, 2014).

The evolution of exploratory research has been marked by a gradual shift from descriptive to experimental and analytical approaches, with a growing emphasis on interdisciplinary and collaborative research. Its history can be traced back to the early twentieth century, when social scientists began to use qualitative methods to study human behaviour. Over the years, exploratory research has evolved in response to a range of social, political, and technological changes. In the early 1900s, sociologists such as W. E. B. Du Bois and Jane Addams were among the first to use qualitative methods such as participant observation and ethnography to study social issues such as poverty, race, and gender inequality (Babbie, 2016).

In the 1930s and 1940s, the Chicago School of Sociology led the development of urban ethnography, which aimed to understand the social and cultural dynamics of urban environments as stated by Robert E. Park in 1934 (Shils, 1996). In the early days of social research, exploratory research was often conducted using methods such as case studies, and informal interviewing (Babbie, 2016). These methods allowed researchers to gain insight into the social world and generate hypotheses for further investigation.

In the 1950s and 1960s, sociologists such as Erving Goffman and Howard Becker began to use participant observation and other qualitative methods to study deviant behaviour and the social construction of identity (Becker, 1963; Goffman, 1961). In the mid-twentieth century, survey research became a popular method for exploratory research in the social sciences. Surveys allowed researchers to collect large amounts of data from a representative sample of the population and identify patterns and trends in social attitudes and behaviours (Fowler, 2014).

In the 1960s and early 1970s, exploratory research in geography in India began to shift towards a more critical and analytical approach. Geographers started to explore the social and economic dimensions of different regions of India, using both qualitative and quantitative methods to understand the complex socio-economic processes shaping the country. Critical theory now gained predominance which emphasized the role of power, inequality, and social justice in shaping human-environment interactions. Heynen and Kaika (2017) opined that critical theory assumes social and environmental phenomena to be shaped by social, economic, and political factors, and that research should be oriented towards promoting social justice and environmental sustainability. During this time, a growing focus on issues of poverty, inequality, and regional imbalances gained predominance (Ghosh, 2005).

In the 1970s and 1980s, feminist scholars such as Bell Hooks (1984) and Patricia Hill Collins (1990) introduced new perspectives and methods that emphasized the importance of inter-sectionality and the experiences of marginalized groups, and ethnography became a popular method for exploratory research in the social sciences (Denzin & Lincoln, 2011).

In the 1990s and 2000s, exploratory research in geography in India continued to evolve, with a growing focus on issues of development, environment, and sustainability (Singh, 2018). New technologies such as the internet and social media began to shape the ways in which social scientists conducted exploratory research. Researchers began to use digital methods such as web scraping, data mining, and social network analysis to study online communities and social movements (Boyd & Crawford, 2012; Marres, 2017). This period also saw the emergence of new forms of exploratory research, such as action research, case study research, and narrative research.

In the present scenario of the twenty-first century onwards, exploratory research in geography has continued to grow and diversify, with a growing emphasis on interdisciplinary and collaborative research approaches, and engaging with local communities and stakeholders to explore complex socio-environmental issues facing the country (Dwivedi et al., 2020). During this phase, mixed methods research and machine learning with artificial intelligence have become increasingly popular (Creswell & Plano Clark, 2018).

3 The Methods of Exploratory Research

By selecting appropriate methods and combining them in creative ways, researchers can generate new insights and perspectives that can inform more rigorous research and help to identify potential solutions to the challenges facing communities around the world. Ethnography, which originated in anthropology in 1800s, involves the immersion of the researcher in the culture or group being studied and the collection of data through participant observation and interviews (Hammersley & Atkinson, 2007; Denzin & Lincoln, 2011; Ryan, 2017). Ethnographic research is a qualitative research approach that involves observing and participating in social activities and interactions in their natural settings, as well as interviewing individuals and analysing documents and artefacts (Lofland et al., 2006). Ethnographic research has been widely used in various environmental and developmental contexts, such as exploring the cultural and social dimensions of natural resource management, the perceptions and experiences of environmental risks and hazards, and the social impacts of development projects. Critical ethnography approach later combined the methods of traditional ethnography with critical social theory, and aims to expose power relations and social inequalities within a particular community or social group. In the field of environment and development, critical ethnography can be used to identify the ways in which environmental and development policies and practices impact vulnerable and marginalized communities, and to generate insights that can inform more socially just and equitable solutions (Lather, 1991).

Since the early times, there have been a growing emphasis on the varied forms of participatory research and mixed methods research, which combine qualitative and quantitative methods, aimed to involve community members in the research process and to combine qualitative and quantitative data collection and analysis methods (Creswell & Plano Clark, 2018; Minkler & Wallerstein, 2003). Participatory research, since the early 1940s, is a collaborative approach that involves working with community members and stakeholders to identify research questions and design research projects (Macaulay, 2017; Reason & Bradbury, 2001) and is particularly useful for exploring complex social and ecological issues that are best understood from a community perspective. Participatory research methods, such as participatory rural appraisal and participatory action research, have been used to engage communities in the research process, empower them to identify and address their own environmental and development challenges, and facilitate knowledge sharing and co-learning between researchers and communities (Cornwall & Jewkes, 1995). Social learning is another method that refers to the process by which individuals and groups learn from their own and each other's experiences, and uses this knowledge to inform their decisions and actions (Reed et al., 2010).

Action research, a term coined by Kurt Lewin in 1944, is another dominant exploratory research methodology that involves collaborating with stakeholders to identify and address specific environmental or development challenges (Reason & Bradbury, 2001). Action research can be useful for generating new insights and perspectives, as well as for developing and testing potential solutions to real-world environmental problems. Action research has been used in various environmental and developmental contexts, such as developing sustainable agriculture practices, promoting community-based conservation, and improving access to clean water and sanitation. Action research can provide a powerful exploratory research methodology for engaging with stakeholders and promoting participatory problem-solving and decision-making. By involving stakeholders in the research process, researchers can develop a more nuanced and context-specific understanding of the phenomenon under study and work towards more sustainable and equitable outcomes.

A combination of the above methods gave rise to the critical participatory action research (CPAR). CPAR is a form of action research that aims to challenge power relations and address social inequalities through collective action and critical reflection (Maguire, 1987). CPAR has been used in various environmental and developmental contexts, such as promoting community-based conservation, addressing environmental justice, and promoting sustainable livelihoods. It can aid in challenging power relations and promoting more equitable and sustainable outcomes and by engaging with marginalized and disenfranchised communities, and it can provide a platform for amplifying voices that are often overlooked in conventional research methods.

With the advent of survey research, various methods of data collection evolved to bring out the complexities in human perspectives. Interviews are the most common method used in exploratory research involving one-on-one conversations between a researcher and a participant, and can be structured or unstructured. Structured interviews involve a predetermined set of questions, while unstructured interviews allow for more open-ended and flexible discussions (Cresswell, 2013). A questionnaire is a list of questions given to respondents to gather their views, experiences, or knowledge on a particular topic. These can be particularly useful for exploring personal experiences, attitudes, and beliefs related to the environment and development (Miles & Huberman, 1994).

Case study methods are also widely used in exploratory research for investigating complex and context-specific environmental and developmental issues. They involve a detailed and in-depth analysis of a particular case or a small number of cases such as an individual, group, organization, or community, and can be particularly useful for exploring complex social and ecological issues (Miles & Huberman, 1994). It aims to provide a rich and comprehensive understanding of the complexities, dynamics, and contextual factors that shape the phenomenon under study (Yin, 2017). Focus groups technique was proposed by social scientists like Robert K. Merton and Ernest Dichter, ever since 1940s (Liamputtong, 2011). They involve group discussions, typically with six to ten participants, and are often used to explore shared experiences, attitudes, and beliefs (Cresswell, 2013). Focus groups can be particularly useful for exploring community perspectives on environmental and development issues.

Narrative research emphasizes the importance of individuals' stories or personal accounts and experiences in the development of knowledge, to create a rich and detailed description of individuals' experiences, perceptions, and emotions related to places (Clandinin & Connelly, 2000; Cresswell, 2013). It can be applied to exploring the lived experiences of marginalized communities, exploring the urban green spaces through walking tours, understanding the ways in which people construct meanings and identities through their interactions with places, or examining the ways in which places are represented in popular culture (Rose & Degen, 2014). Observation method, on the other hand, is another widely used qualitative method involving systematically perceiving and recording the behaviour of individuals or groups in a natural setting (Bryman, 2016). Observational research can be used to explore the relationship between people and their environment, as well as to identify patterns and trends in behaviour.

Arts-based research involves the use of creative arts like music, dance, and visual art as a way of exploring environmental and development issues. They being observational research can be used to engage communities in the process, to create more emotional and affective connections to environmental and development issues, and to generate new and innovative solutions that are grounded in local perspectives and creativity (Leavy, 2015).

Another subsidiary of observational research being the visual methods involve using photographs, videos, and other visual media to help participants express their experiences and perspectives (Pink, 2013; Rose, 2016). By incorporating visual aids, researchers can promote more effective communication and collaboration with stakeholders, as well as provide a more nuanced and holistic understanding of environmental and social dynamics. Photovoice contrarily involves giving cameras to participants to take photos that represent their experiences and perspectives on a particular topic, and using these photos as a basis for discussion and analysis (Wang & Burris, 1997). This can be useful for exploring how people perceive and interact with their environment, and for identifying areas of concern or interest.

Another emerging exploratory research methodology that can be used in the field of environment and development is social media analytics. Social media platforms have become a rich source of data that can be used to explore the attitudes, behaviours, and values of communities regarding environmental issues (Delli Carpini, 2019). By analysing social media data, researchers can gain insights into public perceptions and concerns about environmental issues, and identify patterns and trends in public opinion and behaviour.

Content analysis is also a technique that can be used in exploratory research to analyse textual or visual data. Content analysis involves systematically analysing and categorizing data, and can be particularly useful for identifying patterns and themes in large datasets (Krippendorff, 2013). Content analysis can be used to explore media coverage of environmental and development issues, for example, or to identify common themes in interviews or focus group discussions.

At the turn of the 1950s, Norman Dalkey and Olaf Helmer laid the foundations for the Delphi method (Gordon, 1994). Delphi technique involves gathering a group of experts to provide anonymous feedback on a particular topic, with the aim of identifying areas of consensus or disagreement (Linstone & Turoff, 1975). It is useful for exploring complex environmental and development issues that require input from a range of experts with different perspectives and expertise. The 1950s also saw the first idea of Machine learning with the Turing Test, which was described as a 'learning machine' that could acquire information and become artificially intelligent (Lavallin & Downs, 2021).

Experimental research involves the manipulation of one or more variables in a controlled setting in order to observe the effect on a particular outcome. This methodology can be used to explore the impact of different interventions or strategies on environmental and development outcomes, and to identify potential solutions that can be scaled up and implemented in the real world (Campbell & Stanley, 1963).

The goal of grounded theory by Barney Glaser and Anselm Strauss was to discover and generate a new theory that is grounded in the data rather than imposing preconceived theories or concepts on the data (Charmaz, 2014). Grounded theory has been used in understanding the social dimensions of sustainability transitions, environmental behaviour, and climate change adaptation (Glaser & Strauss, 1967; Corbin & Strauss, 2014) and allows for the exploration of new and unexpected phenomena and can lead to the development of innovative theories. One example of grounded theory research in environmental behaviour is a study by Stern et al. (1999) on the factors that influence pro-environmental behaviour among individuals across different domains of everyday life. This methodology involves generating theories and hypotheses based on data collected in the field, rather than testing pre-existing hypotheses (Charmaz, 2014).

4 The Emerging Trends in Methodologies

Recent periods have seen a marked shift from the traditional of analysis to model-based interpretations. System dynamics modelling is a tool for analysing complex systems and exploring the behaviour of these systems over time. This methodology can be useful for exploring the interactions and feedback loops between different environmental and development factors, and for identifying potential leverage points for intervention (Sterman, 2000).

In this regard, network analysis involves mapping the relationships and connections between different individuals, organizations or institutions, or other entities, and can be used to identify key stakeholders, power dynamics, and other factors that shape environmental governance (Borgatti et al., 2009). It can be particularly useful for exploring the ways in which different actors collaborate or compete in the context of environmental decision-making (Wasserman & Faust, 1994; Bodin & Crona, 2009). Social network analysis (SNA) is a method that focusses on the relationships between individuals, organizations, or other entities in a social system. One of the lines of social network analysis could be traced back to Sociometry Method created by social psychologist Jacob Levy Moreno in 1930s, and this method paved the way for quantitative analysis in social network approach. SNA has been used to study a wide range of environmental and developmental issues, such as community-based natural resource management (Cox et al., 2010), climate change adaptation (Bodin & Crona, 2009), and the diffusion of sustainable technologies (Wejnert, 2002).

Scenario planning, introduced by Herman Kahn in 1950s, has been used to explore alternative futures and identify potential solutions to complex environmental and development challenges (Peterson, 2003a). It involves developing and exploring alternative future scenarios for environmental and development issues (Peterson, 2003b). Grove and Baumann (2008) proposed an exploratory modelling and visualization (EMV) approach to environmental impact assessment (EIA), which involves the use of conceptual models, data visualization, and stakeholder participation to develop a more comprehensive understanding of the potential impacts of development projects. By combining EMV and stakeholder participation, researchers can develop a more comprehensive understanding of the potential impacts of development projects and identify more effective and sustainable approaches to managing environmental and social systems.

Today, technology-enabled research has become more prevalent, with the emergence of new technologies, such as online surveys, data mining, and social media analysis, which allow researchers to collect and analyse large amounts of data quickly and efficiently. As exploratory research continues to evolve, there are several promising directions that researchers can take to further advance the field. One such direction is the use of big data and machine learning techniques to analyse large data sets and uncover patterns and relationships that were previously unknown. These techniques can be particularly useful for exploring complex social phenomena and understanding the impact of various factors on human behaviour.

In the field of environment and development, big data analytics can be used to analyse data from remote sensors, social media, and other sources to better understand environmental conditions, predict future trends, and identify potential solutions (Kitchin, 2014). Similarly, citizen science is an emerging methodology that involves engaging members of the public in data collection and analysis often through the use of mobile applications and other digital tools (Bonney et al., 2009; Jordan et al., 2011).

Machine learning and artificial intelligence (AI) methods, particularly in the field of environment and development studies, can help researchers analyse large and complex datasets, identify patterns, and make predictions. The most common application of machine learning in environmental research is remote sensing analysis and predicting environmental outcomes. AI methods such as natural language processing (NLP) and computer vision are increasingly used in geographical research. It provides powerful analytical tools to extract insights and make predictions from vast and diverse datasets. Another emerging method is machine learning, which involves the use of algorithms and statistical models to analyse and interpret large datasets. In the field of environment and development, machine learning can be used to identify patterns and relationships in complex environmental systems, and to make predictions about future environmental conditions (Katurji, 2018).

The geographic information systems (GIS) and remote sensing (RS) technologies can be used to map and analyse environmental conditions, changes, and impacts over time, and to identify the spatial distribution of environmental risks and vulnerabilities (Goodchild & Janelle, 2010). Techniques like participatory mapping can be useful for generating new insights and perspectives on local environmental issues, as well as for engaging communities in the process of environmental monitoring and decision-making (Nayak, 2017; Widgren & Näsholm, 2017).

5 Emergence of Exploratory Research as a Transformative Methodology

A transformative research methodology in social sciences is needed to deconstruct anthropocentrism and better understand the relationships between environment and development (Brosius, 2005). This approach would involve a shift away from traditional research methods, which have been heavily influenced by anthropocentrism, and towards methods that take into account the perspectives and needs of other species and the natural world. One example of a transformative research methodology in social sciences is the use of interdisciplinary approaches (Egoh et al., 2014), which bring together experts from different fields to examine environmental and developmental issues from multiple perspectives. Additionally, this approach can also involve the use of qualitative research methods, such as participatory research (Chambers, 1994), which allow for the inclusion of the perspectives and needs of local communities and other stakeholders.

Researchers have increasingly recognized the need for a multidisciplinary approach to social research to understand complex phenomena (Jasanoff, 2004). For instance, a study of environmental issues may require the involvement of experts from multiple fields such as biology, economics, and sociology. Furthermore, the increasing diversity of research participants has also necessitated the development of new methods that are sensitive to cultural and contextual differences (Israel et al., 2003).

The evolution of methods of exploratory research in social sciences can be attributed to several factors such as advancements in technology, changes in societal structures and norms, and criticisms of existing methods. The following are some examples of how these factors have contributed to the change in methods of exploratory research in social sciences:

- *Advancements in technology*: The advent of new technologies has led to the development of new methods of exploratory research. For example, social media platforms have provided researchers with access to large amounts of data, which can be analysed to gain insights into social phenomena (Boyd & Crawford, 2012). Similarly, advances in data visualization techniques have allowed researchers to identify patterns and trends in data more easily (Wickham, 2016).
- *Changes in societal structures and norms*: Changes in societal structures and norms have led to the development of new methods of exploratory research that are better suited to studying contemporary social phenomena. For example, the rise of subcultures and countercultures has led to the development of new methods such as 'netnography', which involves studying online communities and social networks (Kozinets, 2015).
- *Criticisms of existing methods*: Criticisms of existing methods of exploratory research have led to the development of new methods that are more suited to studying complex social phenomena. For example, criticisms of traditional survey research led to the development of new methods such as cognitive interviewing, which involves using probing techniques to improve the accuracy of survey responses (Willis & Artino, 2013).

6 Limitations of Exploratory Research

It is to be noted that this research method also has some limitations or drawbacks. Here are some elaborations on the drawbacks of exploratory research with in-text citations and references:

1. *Limited generalizability*: Since exploratory research is typically based on a small sample size, the findings may not be representative of the larger population. For example, in a study on consumer behaviour, the findings may only be applicable to a specific group of consumers and not necessarily to the wider population (Kumar & Kansal, 2018).
2. *Lack of reliability*: Exploratory research often relies on qualitative data that is subject to interpretation. This can lead to issues with reliability since different researchers may interpret the data differently. Additionally, the lack of standardization in the data collection process can also contribute to reliability issues (Yadav & Dutta, 2016).

3. *Limited depth*: Exploratory research is often used as a preliminary step to more in-depth research. As a result, it may not provide the level of detail and depth required to fully understand a research problem (Shukla, 2016).
4. *Time-consuming*: It can be time-consuming due to the need to collect and analyse data from multiple sources, especially problematic when researchers are on a tight schedule or have limited resources (Kumar & Kansal, 2018).
5. *Bias*: Exploratory research may be biased due to the researcher's own subjective interpretations or personal biases, particularly when using qualitative data, since the researcher's interpretations can heavily influence the findings (Sharma et al., 2023).
6. *Limited control over variables*: As exploratory research is often conducted in naturalistic settings, researchers may have limited control over the variables that they are studying. This can make it difficult to establish cause-and-effect relationships and limit the researcher's ability to draw meaningful conclusions (Kumar & Kansal, 2018).
7. *Data analysis challenges*: Since exploratory research involves analysing large amounts of data, researchers may face challenges in effectively organizing and interpreting the data which can make it difficult to identify meaningful patterns or trends in it (Kumar & Kansal, 2018).

7 Exploratory Research in the Indian Subcontinent

In India, exploratory research has a rich history in geography, with several milestones marking its evolution over the years. The evolution of exploratory research in geography in the country can be traced through the changing research themes and methods employed by geographers over time. In the early years, geographers focussed on the physical geography of India, conducting descriptive studies of the country's landforms, climate, and vegetation (D'Souza, 2019). However, with the increasing importance of development and regional planning in post-independent India, geographers began to shift their focus towards understanding the social and economic dimensions of different regions of the country. The earliest exploratory research studies in geography in India were conducted in the 1940s and 1950s, when geographers began to explore the physical and cultural landscapes of the country. These studies were largely descriptive in nature, aimed at mapping and classifying the different regions of India based on their physical and cultural features (D'Souza, 2019).

Located in the subtropical region, the Indian subcontinent, comprising of India and its neighbouring countries, has a diverse physical attributes leading to several environment and development challenges in its regional setup. Accordingly human occupation on the basis of physical endowments evolved through ages leading to social-economic and cultural plurality. Historically, the subcontinent was subjected to various invasions, colonization, struggle for freedom, and decolonization. Meanwhile, the administrators tried to map the resources for better exploitation and later on conservation for sustainable development. Therefore, various exploratory researches were conducted at times from different disciplines. Establishment of the institutions like The Anthropological Survey of India, The Survey of India, The Geological Survey of India, India Meteorological Department, The Census of India, the National Atlas and Thematic Mapping Organization, and the Planning Commission was definitely significant milestone in exploratory research in this part of the world.

The colonial period had scant regard to environmental sustainability and socio-economic development of this region. It was a huge challenge for the decolonized nations of the South Asia to provide the basic life-sustaining amenities to its growing population. Planning for agriculture, industry, trade, and commerce were given priority for self-reliance in which designing appropriate curriculum and research agenda along with establishment of new educational institutions was a prominent event. A synergy among university education system with the technical training and research institutions was formulated to ensure, knowledge as an important input to economic activities. In case of social sciences and humanities, exploratory research was strengthened by the Indian Institutions of Management, Central Universities, and many stand-alone academic institutions for human resource developments.

In the context of geographical research, it may be noted that geography as a discipline of study in the higher education of teaching and research started in the undivided India in 1920s which happened to be much later compared to similar disciplines of social sciences. In fact, geography emerged as a scientific study to bridge the gap between the physical sciences and humanities. Maps of various scales and projections were recognized as major tools to capture and depict various thematic information of the earth's surface for taking appropriate decisions on regional planning and development on the basis of resource base along with their opportunities and constraints of their exploitation as well as management. Primarily the maps were produced on the basis of field surveys using conventional tools and techniques but later on, especially in the post-colonial era, remote sensing techniques played significant roles for preparation and upgradation of the maps, in terms of content and visual clarity. Invention of computer technology followed by GIS paved the way for better management of spatial database, and in the recent years, its integration with ICT and AI has given impetus to the exploratory research in geography in the region.

In the post-independent period, necessary steps were taken to promote planning regions for sectoral and multipurpose development to reduce the regional imbalances in the country. The early geographical researches were focussed on regional scale with more emphasis on regional development. Specialized departments and centres came up to conduct researchers in this line. However, since 1970s, as a follow up of the Stockholm Conference of 1972, the environment became an important agenda *vis-a-vis* development, and therefore different branches of geography began incorporating environmental issues as exploratory research. As a consequence, sub-regional and local problems were identified as research problem and accordingly tools and techniques were selected. Examples on use of exploratory research applied in environmental and development research include farmers' perceptions and adaptation strategies related to climate change through interview methods, waste management studies, where researchers organize focus group discussions with residents, waste collectors, and local authorities, and also documenting the traditional ecological knowledge and conservation practices of indigenous communities through the use of ethnography. By the use of action research and participatory mapping, local communities actively participate in mapping their coastal areas, identifying vulnerable zones, and sharing their knowledge of historical changes in the coastline, engage in community management in protecting embankments, etc. Field surveys and experimentations have always gained prominence in environmental and development studies. Across India, the various physically and socially vulnerable locations are conducive for case study methods. There have been numerous research on urban expansions of Indian towns and cities, land use and land cover changes, and/or surface temperature fluctuations analysed through the RS and GIS methods. The impact analysis of degrading environments and disaster management research has been the forerunner for sustainability policies in the last decade. No development policies can be successful without a critical appraisal of the impacts on the physical and social environments. Thus, analysis of both qualitative and quantitative methods in almost all types of research on environment, development, or society within the realm of social sciences is indispensable for humanity to be sustainable.

Exploratory researches in geography in India are now more interdisciplinary, both in their thematic identity and in methodological approaches. This is because of availability of trained manpower to handle sophisticated tools and techniques used in advanced research and publications at international level. The opportunities of the recent globalization should be acknowledged to provide the impetus in accepting the challenges by the contemporary Indian geographers, who are more focussed and connected to the latest technological developments worldwide.

8 Conclusion

Exploratory research on the relationship between environment and development has become increasingly important as the world faces ongoing environmental crises such as climate change, deforestation, and pollution in the Anthropocene era with predominantly the Capitalocene approach. In this type of research, researchers aim to gain a deeper understanding of the complex and multifaceted interactions between environmental and developmental factors, and to identify potential solutions to environmental problems. In the field of environment and development, exploratory research can be a useful tool for understanding complex social and ecological issues. For example, researchers may use exploratory research to better understand the relationship between environmental degradation and poverty, or to identify the cultural, economic, and political factors that influence local attitudes towards conservation. One of the key benefits of exploratory research in the field of environment and development is its ability to generate new insights and perspectives. By engaging with local communities and stakeholders, exploratory research can provide a more nuanced understanding of the social and ecological dynamics at play in a given context. There are several methods that can be used in exploratory research, including interviews, focus groups, observation, and case studies. These methods are often combined to create a more comprehensive picture of the topic under investigation. For example, a researcher conducting exploratory research on the relationship between environmental degradation and poverty might begin by conducting interviews with local residents to better understand their experiences and perspectives. Observations may be conducted on the local environment to identify the ways in which environmental degradation is impacting the community. Finally, taking the advantage of exploratory research, they might use case studies at regional and local scale to explore the potential solutions that have been implemented in other communities facing similar challenges.

References

Antwi, S. K., & Hamza, K. (2015). Qualitative and quantitative research paradigms in business research: A philosophical reflection. *European Journal of Business and Management, 7*(3), 217–225.

Babbie, E. (2016). *The practice of social research*. Cengage Learning.

Becker, H. S. (1963). *Outsiders: Studies in the sociology of deviance*. Free Press.

Biermann, F. (2014). The Anthropocene: A governance perspective. *The Anthropocene Review, 1*(1), 57–61. https://doi.org/10.1177/2053019613516289

Bodin, Ö., & Crona, B. I. (2009). The role of social networks in natural resource governance: What relational patterns make a difference? *Global Environmental Change, 19*(3), 366–374.

Bonney, R., Cooper, C. B., Dickinson, J., Kelling, S., Phillips, T., Rosenberg, K. V., & Shirk, J. (2009). Citizen science: A developing tool for expanding science knowledge and scientific literacy. *Bioscience, 59*(11), 977–984.

Borgatti, S. P., Mehra, A., Brass, D. J., & Labianca, G. (2009). Network analysis in the social sciences. *Science, 323*(5916), 892–895.

Boyd, D., & Crawford, K. (2012). Critical questions for big data: Provocations for a cultural, technological, and scholarly phenomenon. *Information, Communication & Society, 15*(5), 662–679.

Boyle, M., Gamble, M., & Mock, S. (2015). *Physical geography: The global environment*. Oxford University Press.

Brosius, J. P. (2005). Anthropocentrism and the use of traditional ecological knowledge in conservation. *Environmental Conservation, 32*(2), 124–129.

Bryman, A. (2016). *Social research methods*. Oxford University Press.

Campbell, D. T., & Stanley, J. C. (1963). *Experimental and quasi-experimental designs for research*. Houghton Mifflin.

Chambers, R. (1994). Participatory rural appraisal (PRA): Challenges, potentials and paradigm. *World Development, 22*(10), 1437–1454.

Charmaz, K. (2014). *Constructing grounded theory*. Sage publications.

Clandinin, D. J., & Connelly, F. M. (2000). *Narrative inquiry: Experience and story in qualitative research*. Jossey-Bass.

Collins, P. H. (1990). *Black feminist thought: Knowledge, consciousness, and the politics of empowerment*. Routledge.

Corbin, J., & Strauss, A. (2014). *Basics of qualitative research: Techniques and procedures for developing grounded theory*. Sage publications.
Cornwall, A., & Jewkes, R. (1995). What is participatory research? *Social Science & Medicine, 41*(12), 1667–1676.
Cox, M., Arnold, G., Villamor, G. B., & Gavrilis, E. (2010). A review of design principles for community-based natural resource management. *Ecology and Society, 15*(4), 38.
Cresswell, T. (2013). *Geographic narratives: Storytelling and the sense of place*. Routledge.
Creswell, J. W. (2014). *Research design: Qualitative, quantitative, and mixed methods approaches* (p. 185). Sage publications.
Creswell, J. W., & Plano Clark, V. L. (2018). *Designing and conducting mixed methods research*. Sage publications.
Crutzen, P. J., & Stoermer, E. F. (2000). The Anthropocene. *Global Change Newsletter, 41*, 17–18.
Delli Carpini, M. (2019). Social media and environmental communication. In *The Routledge handbook of environment and communication* (pp. 195–206). Routledge.
Denzin, N. K., & Lincoln, Y. S. (2011). *The Sage handbook of qualitative research*. Sage publications.
D'Souza, R. (2019). Indian geography: A historiographical perspective. In *Indian geography* (pp. 1–16). Springer.
Dwivedi, R. S., Mishra, R., & Sinha, R. (2020). Exploratory research in geography in India: A bibliometric analysis. *Journal of the Indian Society of Remote Sensing, 48*(3), 435–449.
Egoh, B., Reyers, B., Richardson, D. M., Rouget, M., Le Maitre, D. C., & Cowling, R. M. (2014). Integrating ecosystem services into conservation assessments: A review of current challenges. *Biological Conservation, 170*, 56–67.
Ellis, E., Maslin, M., Boivin, N., & Bauer, A. (2016). Involve social scientists in defining the Anthropocene. *Nature, 540*(7632), 192–193.
Foster, J. B. (2000). *Marx's ecology: Materialism and nature*. Monthly Review Press.
Fowler, F. J. (2014). *Survey research methods*. Sage Publications.
Ghosh, J. (2005). Geography in India: Historical perspectives and contemporary developments. *Progress in Human Geography, 29*(2), 222–236.
Glaser, B. G., & Strauss, A. L. (1967). *The discovery of grounded theory: Strategies for qualitative research*. Aldine Transaction.
Goffman, E. (1961). *Asylums: Essays on the social situation of mental patients and other inmates*. Doubleday.
Goodchild, M. F., & Janelle, D. G. (2010). Toward critical spatial thinking in the social sciences and humanities. *GeoJournal, 75*(1), 3–13.
Gordon, T. J. (1994). The delphi method. *Futures Research Methodology*, *2*(3), 1–30.
Grove, K. J., & Baumann, M. R. (2008). Environmental impact assessment using an exploratory modeling and visualization approach. *Environmental Modelling & Software, 23*(10–11), 1252–1262.
Hammersley, M., & Atkinson, P. (2007). *Ethnography: Principles in practice*. Routledge.
Haraway, D. (2016). *Staying with the trouble: Making kin in the Chthulucene*. Duke University Press.
Hay, I. (2016). On being ethical in geographical research. *Key Methods in Geography, 3*, 30–43.
Heynen, N., & Kaika, M. (2017). In E. Swyngedouw (Ed.), *In the nature of cities: Urban political ecology and the politics of urban metabolism*. Routledge.
Hooks, B. (1984). *Feminist theory: From margin to center*. South End Press.
Israel, B. A., Schulz, A. J., Parker, E. A., & Becker, A. B. (2003). Review of community-based research: Assessing partnership approaches to improve public health. *Annual Review of Public Health, 24*(1), 5–35.
Janowicz, K., Gao, S., McKenzie, G., Hu, Y., & Bhaduri, B. (2020). GeoAI: spatially explicit artificial intelligence techniques for geographic knowledge discovery and beyond. *International Journal of Geographical Information Science, 34*(4), 625–636.
Jasanoff, S. (2004). *States of knowledge: The co-production of science and social order*. Routledge.

Jordan, R. C., Gray, S. A., Howe, D. V., Brooks, W. R., & Ehrenfeld, J. G. (2011). Knowledge gain and behavioral change in citizen-science programs. *Conservation Biology, 25*(6), 1148–1154.
Katurji, M. (2018). Machine learning for environmental scientists. *Nature, 555*(7695), 445–447.
Kitchin, R. (2014). Big data, new epistemologies and paradigm shifts. *Big Data & Society, 1*(1), 1–12. https://doi.org/10.1177/2053951714528481
Kitchin, R., & Tate, N. (2013). *Conducting research in human geography: Theory, methodology and practice*. Routledge.
Kozinets, R. V. (2015). *Netnography: Redefined*. Sage Publications.
Krippendorff, K. (2013). *Content analysis: An introduction to its methodology*. Sage publications.
Kumar, V., & Kansal, M. (2018). Exploratory research on sustainable business practices in Indian small and medium-sized enterprises. *Journal of Cleaner Production, 171*, 1280–1291.
Lather, P. (1991). *Getting smart: Feminist research and pedagogy with/in the postmodern*. Routledge. https://doi.org/10.4324/9780203451311
Latouche, S. (2009). *Farewell to growth*. Polity Press.
Lavallin, A., & Downs, J. A. (2021). Machine learning in geography–Past, present, and future. *Geography Compass, 15*(5), e12563.
Leavy, P. (2015). Announcements and Plenary Address–Building Research in "New Shapes" and Engaging Diverse Audiences: The Transformative Potential of Arts-Based Research. Published in the Qualitative Report of 6th Annual Conference on Transforming Qualitative Research.
Liamputtong, P. (2011). Focus group methodology: Introduction and history. In *Focus group methodology: Principle and practice* (Vol. 224, pp. 1–14). Sage Publications.
Linstone, H. A., & Turoff, M. (1975). *The Delphi method: Techniques and applications*. Addison-Wesley.
Lofland, J., Snow, D. A., Anderson, L., & Lofland, L. H. (2006). *Analyzing social settings: A guide to qualitative observation and analysis*. Thomson Wadsworth.
Longley, P. A., Goodchild, M. F., Maguire, D. J., & Rhind, D. W. (2015). *Geographic information systems and science*. John Wiley & Sons.
Macaulay, A. C. (2017). Participatory research: What is the history? Has the purpose changed? *Family Practice, 34*(3), 256–258. https://doi.org/10.1093/fampra/cmw117
Mackenzie, H. A. (2019). Understanding performance measurement from a social systems perspective. DoctoralThesis. Department of Design, Manufacturing and Engineering Management. University of Strathclyde. https://doi.org/10.48730/q30y-4f34
Maguire, P. (1987). *Doing participatory research: A feminist approach*. Center for Urban Affairs and Policy Research, Northwestern University.
Marres, N. (2017). *Digital sociology: The reinvention of social research*. John Wiley & Sons.
Miles, M. B., & Huberman, A. M. (1994). *Qualitative data analysis: An expanded sourcebook*. Sage publications.
Minkler, M., & Wallerstein, N. (Eds.). (2003). *Community-based participatory research for health: From process to outcomes*. John Wiley & Sons.
Moore, J. W. (2015). *Capitalism in the web of life: Ecology and the accumulation of capital*. Verso Books.
Moore, J. W. (Ed.). (2016). *Anthropocene or capitalocene?: Nature, history, and the crisis of capitalism*. Pm Press.
Moore, J. W. (2017). The Capitalocene, part I: On the nature and origins of our ecological crisis. *The Journal of Peasant Studies, 44*(3), 594–630.
Nayak, P. K. (2017). Participatory mapping: A method for exploring local knowledge and values in natural resource management. *Journal of Environmental Management, 196*, 35–45.
Neuman, W. L. (2014). *Social research methods: Qualitative and quantitative approaches*. Pearson.
Peters, B. G. (2016). *The politics of problem definition: Putting issues on the agenda*. University of Chicago Press.
Peterson, G. (2003a). Scenario planning: A tool for strategic thinking. *Sloan Management Review, 44*(2), 25–36.

Peterson, G. D. (2003b). Applying resilience thinking to natural resource management through a "planning by doing" framework. *Society & Natural Resources, 16*(6), 523–543.

Pink, S. (2013). *Doing visual ethnography: Images, media and representation in research*. Sage Publications.

Reason, P., & Bradbury, H. (2001). *Handbook of action research: Participative inquiry and practice*. Sage publications.

Recchiuti, J. L. (2007). *Civic engagement: Social science and progressive-era reform in New York City*. University of Pennsylvania Press.

Reed, M. S., Evely, A. C., Cundill, G., Fazey, I., Glass, J., Laing, A., et al. (2010). What is social learning? *Ecology and Society, 15*(4), r1.

Rose, G. (2016). *Visual methodologies: An introduction to researching with visual materials*. Sage publications.

Rose, G., & Degen, M. (2014). Going beyond the visual in the study of urban experience: A multisensory approach. *Environment and Planning A, 46*(4), 970–988.

Ryan, G. S. (2017). An introduction to the origins, history and principles of ethnography. *Nurse Researcher, 24*(4), 15–21.

Sharma, L. R., Bidari, S., Bidari, D., Neupane, S., & Sapkota, R. (2023). Exploring the mixed methods research design: Types, purposes, strengths, challenges, and criticisms. *Global Academic Journal of Linguistics and Literature, 5*.

Shils, E. (1996). The sociology of Robert E. Park. *The American Sociologist, 27*(4), 88–106.

Shukla, P. (2016). Exploratory research on factors influencing online shopping behavior of consumers in India. *International Journal of Management Research and Review, 6*(7), 845–853.

Singh, R. B. (2018). Geography in India: Retrospect and prospects. *Journal of Human Ecology, 62*(3–4), 177–182.

Steffen, W., Crutzen, P. J., & McNeill, J. R. (2011). The Anthropocene: Are humans now overwhelming the great forces of nature? *Ambio, 40*(7), 739–761.

Sterman, J. D. (2000). *Business dynamics: Systems thinking and modeling for a complex world*. McGraw-Hill Education.

Stern, P. C., Dietz, T., Abel, T., Guagnano, G. A., & Kalof, L. (1999). A value-belief-norm theory of support for social movements: The case of environmentalism. *Human Ecology Review, 6*(2), 81–97.

Wang, C. C., & Burris, M. A. (1997). Photovoice: Concept, methodology, and use for participatory needs assessment. *Health Education & Behavior, 24*(3), 369–387.

Wasserman, S., & Faust, K. (1994). *Social network analysis: Methods and applications*. Cambridge University Press.

Wejnert, B. (2002). Integrating models of diffusion of innovations: A conceptual framework. *Annual Review of Sociology, 28*(1), 297–326.

Wickham, H. (2016). *ggplot2: Elegant graphics for data analysis*. Springer.

Widgren, M., & Näsholm, T. (2017). Participatory mapping in the age of big data: Mediating controversies of the Anthropocene. *Area, 49*(1), 104–110.

Willis, G. B., & Artino, A. R. (2013). What do our respondents think we're asking? Using cognitive interviewing to improve medical education surveys. *Journal of Graduate Medical Education, 5*(3), 353–356.

Yadav, R., & Dutta, S. (2016). Exploratory research: An overview. *International Journal of Applied Research, 2*(10), 496–499.

Yin, R. K. (2017). *Case study research and applications: Design and methods*. Sage publications.

Index

S. Hazra et al. (eds.), *Geographies of the Indian Subcontinent*,
https://doi.org/10.1007/978-3-031-74813-4

The manufacturer's authorised representative in the EU is Springer Nature Customer Service Centre GmbH, Europaplatz 3, 69115 Heidelberg, Germany. If you have any concerns regarding our products, please contact ProductSafety@springernature.com

Printed and bound by CPI Group (UK) Ltd, Croydon, CR0 4YY

12/08/2026

02192150-0001